Coherent Optics

T0202972

Advanced Texts in Physics

This program of advanced texts covers a broad spectrum of topics which are of current and emerging interest in physics. Each book provides a comprehensive and yet current and emerging interest in physics. Each book provides a comprehensive and yet accessible introduction to a field at the forefront of modern research. As such, these texts are intended for senior undergraduate and graduate students at the MS and PhD level; however, research scientists seeking an introduction to particular areas of physics will also benefit from the titles in this collection.

Springer

Berlin
Heidelberg
New York
Hong Kong
London
Milan
Paris
Tokyo

Physics and Astronomy

ONLINE LIBRARY

http://www.springer.de/phys/

W. Lauterborn   T. Kurz

Coherent Optics

Fundamentals and Applications

Second Edition
With 183 Figures,
73 Problems and Complete Solutions

 Springer

Professor Dr. Werner Lauterborn
Dr. rer. nat. Thomas Kurz
Universität Göttingen
Drittes Physikalisches Institut
Bürgerstrasse 42–44
Germany

Title of the original German edition:
W. Lauterborn, T. Kurz, and M. Wiesenfeldt: Kohärente Optik

© Springer-Verlag Berlin Heidelberg 1993

Library of Congress Cataloging-in-Publication Data applied for.
Die Deutsche Bibliothek - CIP-Einheitsaufnahme
Lauterborn, Werner:
Coherent optics: fundamentals and applications/W. Lauterborn; T. Kurz. –
2. ed. – Berlin; Heidelberg; New York; Hong Kong; London; Milan; Paris; Tokyo:
Springer, 2003 (Advanced texts in physics)
(Physics and astronomy online library)
Einheitssacht.: Kohärente Optik <engl.>

ISSN 1439-2674

ISBN 978-3-642-07877-4

Springer-Verlag Berlin Heidelberg New York
a member of BertelsmannSpringer Science+Business Media GmbH

http://www.springer.de

© Springer-Verlag Berlin Heidelberg 2010
Printed in Germany

Cover design: *design & production* GmbH, Heidelberg

Preface to the First Edition

Since the advent of the laser, coherent optics has developed at an ever increasing pace. There is no doubt about the reason. Coherent light, with its properties so different from the light we are surrounded by, lends itself to numerous applications in science, technology, and life. The bandwidth of coherent optics reaches from holography and interferometry, with its gravitational wave detectors, to the CD player for music, movies, and computers; from the laser scalpel, which allows surgical cutting in the interior of the eye without destruction of the layers penetrated in front of it, to optical information and data processing with its great impact on society. According to its importance, the foundations of coherent optics should be conveyed to students of natural sciences as early as possible to better prepare them for their future careers as physicists or engineers.

The present book tries to serve this need: to promote the foundations of coherent optics. Special attention is paid to a thorough presentation of the fundamentals. This should enable the reader to follow the contemporary literature from a firm basis. The wealth of material, of course, makes necessary a restriction of the topics included. Therefore, from the main areas of optics, wave optics and the classical description of light is given most of the space available. The book starts with a quick trip through the history of physics from the viewpoint of optics. Thereby, the contributions of optics to virtually all fundamental issues of physics are addressed, and a short overview of the four main areas of optics is given.

The notion of coherence gets its own chapter. The different forms of coherence are presented, starting from the most basic description possible, including the applications they give rise to. The phenomenon of speckle formation, often disturbing but also of potential usefulness, is treated, as is multiple-beam interference. Special attention is paid to holography, with its applications, and to Fourier optics, an elegant method for treating diffraction phenomena[1]. The laser, of course, cannot be missed in a book on coherent optics. But it has become an instrument so widely used that it has its own detailed literature. Therefore, just one modern aspect of the laser is put forward here: its nonlinear dynamics. The fundamentals

---

[1] The hologram included demonstrates the state of the art in producing bright holograms that can be viewed in white light.

of nonlinear optics are presented in a separate chapter. There, photons are split and combined! Fiber optics, including solitons, finds its home in the chapter on optical information and data processing.

A book is not born in a day and not without the interaction of the authors with their environment. Thus, the selection of topics is strongly biased by the experience of one of the authors (W. L.) with 29 two–week courses on laser physics and holography given at the University of Göttingen, and repeatedly given lectures on optics at the Technical University of Darmstadt. He takes this opportunity to thank the numerous coworkers making the courses a success, in particular K. Hinsch, K.-J.Ebeling, W. Hentschel and A. Vogel, and also the participants who, with their questions, add to a clear presentation. The illustrations mainly stem from these courses. Input also came from the research work of the authors on nonlinear oscillators, for example.

This text was set by the authors using LaTeX and Postscript[2].

May this book be of value to the reader, and may he or she also read it with pleasure.

Göttingen,   *Werner Lauterborn*
January 1995   *Thomas Kurz*
                                                            Martin Wiesenfeldt

---

[2] Only the white-light hologram on the cover page could not be translated into electronic form although some means exist in the form of digital holograms that are described in depth.

Preface to the Second Edition

As the first edition of Coherent Optics was well received and soon sold out, both the publisher and the authors were confronted with the situation of either reprinting the text or updating it. After a first reprint the authors felt the need to serve the community better by leaving their research for a while and sitting down for an update.

A first approach is seldom perfect. Thus, omissions of knowledge ranked basic were filled in, one example being evanescent waves today strongly used in fluorescence microscopy. As the future is becoming past in the course of time, the history chapter got some modifications. The main changes, however, were initiated by the progress in the field. Thus holograms go more and more digital and are recorded on CCD chips for special purposes. A new development in fiber optics to measure strong shock waves in liquids has been included. The laser chapter has been considerably augmented by including a section on transient dynamics and on synchronisation, a topic heavily discussed for secure communication. A totally new chapter on ultrafast optics has been included where ultrashort pulse properties, their generation, measurement and applications in optical gating and optical coherence tomography are presented.

Finally it is a great pleasure for the authors to thank R. Geisler, O. Lindau, U. Parlitz, I. Wedekind and D. Zeidler for supply of material included in the new edition.

Göttingen,
September 2002

Werner Lauterborn
Thomas Kurz

Contents

1. History of Optics

Optics has developed from human sight. Sight is one of our most important faculties, whereby we interact with nature and gain knowledge of the physical world surrounding us. Therefore optical phenomena have attracted attention since ancient times.

1.1 Past

The first optical instrument could have been a mirror. In any case, mirrors were known to the ancient Egyptians and Chinese and have been found in excavations, often in good condition.

Besides mirrors, the Greeks were in possession of burning glasses. We also know that they contemplated about the nature of light and found a few laws by observing the propagation of light. They knew about the rectilinear propagation and the law of reflection. *Claudius Ptolemaeus* (about 100–170 A.D.) had found a law of refraction for small angles (angle of refraction proportional to the angle of incidence) and knew that the propagation of light should proceed incredibly fast. *Empedocles* (about 495–435 B.C.) was of the opinion that the speed of light is finite.

The Romans mainly conserved the knowledge of the Greeks. It is believed that they used the magnifying properties of lenses in arts and crafts for producing delicate parts. After the decline of the Western Roman Empire (475 A.D.) the knowledge of optics was preserved and augmented in the Arabic–Moslemic world, in particular by *Ibn al Haitham* (965–1039 A.D.), named *Alhazen* in the Middle Ages. He found that the law of refraction of *Ptolemaeus* is not valid for large angles, but was unable to formulate the correct law.

Alhazen was an experimental physicist, as we would call him today, because he conducted experiments for all optical questions he posed, long before *Sir Francis Bacon* (1215–1294) and *Galileo Galilei* (1564–1642), whom we consider the fathers of experimental science and who both have furthered optics. Progress in the Middle Ages was slow. It was not before *Leonardo da Vinci* (1452–1519) that the set of optical instruments was enlarged by the camera obscura (pinhole camera).

In the 17th century the telescope and the microscope were invented in the Netherlands. *Snellius* (1591–1626), a professor in Leyden, discovered the law of refraction in 1621. It was cast into its present form by *René Descartes* (1596–1650). This was a big step forward as the engineering of optical instruments now became feasible.

The diffraction of light was mentioned first by *Francesco Maria Grimaldi* (1618–1663) in Bologna. He observed, among other things, diffraction seams (fringes) in the shadow of a rod illuminated by a small light source. *Robert Hooke* (1635–1703) also observed diffraction effects and investigated diffraction phenomena at thin sheets. As an explanation he suggested the interference of light reflected from the front with the light reflected from the back. He further suggested that we should imagine light as being made up of fast oscillations; that is, he suggested a wave theory of light. He was unable, however, to explain the colors of thin sheets. Fundamental insight into colored light was obtained by *Isaac Newton* (1642–1727) in 1666. He discovered that white light can be decomposed into a spectrum of colored light and put forward a corpuscular theory of light [1.1].

Christiaan Huygens (1629–1695) furthered the development of the wave theory and discovered the polarization of light. In 1676 the speed of light was measured by *Olaf Römer* (1644–1710).

Optics did not advance again until 1801 with *Thomas Young* (1773–1829). He introduced the principle of interference, which can be considered as a principle of linear superposition of waves, and roughly estimated the wavelength light should have. Combining this principle with Huygens' principle, *Auguste Jean Fresnel* (1788–1827) calculated the diffraction of light at various objects. At that time light was considered as being composed of longitudinal waves. It was the phenomenon of polarization which led *Young* to the assumption that light is a transverse wave. In 1850 *Léon Bernard Foucault* (1819–1868) discovered that the speed of light in water is lower than in air. This was considered as the final triumph of the wave theory of light, as the corpuscular theory had postulated a larger velocity to explain refraction towards the normal for light passing from air to water.

Further progress in optics came from a new direction: the investigation of electricity and magnetism. In 1845 *Michael Faraday* (1791–1867) discovered a connection between electromagnetism and light. He found that the direction of polarization of light in a crystal can be altered by a strong magnet. *James Clerk Maxwell* (1831–1879) cast the then known phenomena into a theory: Maxwell's equations. From this theory he postulated the existence of electromagnetic waves. They were experimentally detected by *Heinrich Hertz* (1857–1894) in 1888. Around this time it was strongly believed that light, too, was an electromagnetic disturbance in the form of a wave propagating in a supporting medium, called the ether, with the finite speed $c = \nu\lambda$ (ν = frequency, λ = wavelength).

The next task was obvious: to determine the properties of the medium supporting electromagnetic waves, that means, also of light. Soon difficulties were encountered, since the postulated ether had to have strange properties. It had to be very transmissive because celestial bodies obviously move through it undisturbed, but simultaneously had to have strong restoring forces to produce the extremely high frequencies ($\approx 10^{15}$ Hz) and the high speed of light. Experimental efforts to measure the motion of the earth with respect to the ether culminated in the experiment of *Albert Abraham Michelson* (1852–1931). The result, published in 1881, was negative. No influence of the motion of the earth on the propagation of light in the ether could be found. But since *James Bradley* (1693–1762), the stellar aberration was known. To explain this phenomenon with the help of the wave theory a relative motion between earth and the ether had to be postulated. Additional difficulties arose with the phenomenon of light carried in moving media (experiments of *Armand Hippolyte Louis Fizault* and *Sir George Briddell Airy*). The solution to these difficulties was given by *Albert Einstein* (1879–1955) in 1905 in a surprising way. In his theory of special relativity he declares the ether to be superfluous:

> Die Einführung eines "Lichtäthers" wird sich insofern als überflüssig erweisen, als nach der zu entwickelnden Auffassung weder ein mit besonderen Eigenschaften ausgestatteter "absoluter Raum" eingeführt, noch einem Punkte des leeren Raumes, in welchem elektromagnetische Prozesse stattfinden, ein Geschwindigkeitsvektor zugeordnet wird [1.2].[1]

The new state of knowledge was: light is an electromagnetic wave propagating in empty space, the vacuum.

In some sense, mankind had come full circle with respect to the ancient thoughts of the Greeks handed down to us by *Lucretius* (about 55 B.C.) in his fundamental work *De rerum natura* (On the nature of things) [1.3]. Therein, the postulate is attributed to *Epicurus* (341–270 B.C.): there are only objects and empty space.

But the theory of light was not at all finished. Many open questions remained, for instance, concerning the spectrum of blackbody radiation or the generation of light (absorption and emission, existence of sharp spectral lines). They led to a further great revolution in our knowledge about nature: quantum theory. Quantum theory was initiated in 1900 by *Max Planck* (1858–1947). He stated that for an explanation of the

---

[1] The introduction of a light ether will turn out to be superfluous insofar as, according to the theory to be developed, neither an "absolute space" with special features will be introduced, nor to a point in empty space, in which electromagnetic processes take place, will a vector of velocity be associated.

measured spectrum of blackbody radiation it must be supposed that the electromagnetic field can exchange energy only in discrete portions of size

$$E = h\nu,\tag{1.1}$$

h $= 6.626176 \cdot 10^{-34}$ Js being a constant, now Planck's constant, and ν being the frequency of the light wave.

Planck first found the formula for the spectrum of blackbody radiation by fitting a curve to experimental data. Written for the spectral energy density, that is, the energy per volume and frequency interval, it reads:

$$\rho(\nu)d\nu = \frac{8\pi}{c^3}\nu^2\frac{h\nu}{\exp\left(\frac{h\nu}{kT}\right)-1}d\nu,\tag{1.2}$$

with T being the temperature in K and k $= 1.380662 \cdot 10^{-23}$ JK^{-1} being the Boltzmann constant.

In the attempt to derive theoretically the radiation formula, *Planck* put forward his famous quantum hypothesis:

> Wir betrachten aber – und dies ist der wesentlichste Punkt der ganzen Berechnung – E als zusammengesetzt aus einer ganz bestimmten Anzahl endlicher gleicher Teile und bedienen uns dazu der Naturkonstanten h $= 6,55 \cdot 10^{-27}$ (erg·s) [1.4].[2]

In 1905, *Einstein* went a step further and postulated that the electromagnetic field itself is composed of quanta of energy $E = h\nu$. Then the photoelectric effect could be explained easily [1.5]. These energy portions were later given the name "photon".

Quantum theory rapidly developed in the years 1925 – 1930 with the wave mechanics of *Erwin Schrödinger* (1887–1961) and the matrix mechanics of *Werner Heisenberg* (1901–1976). The equivalence of both theories was shown by *John von Neumann* (1903–1957) in Göttingen. Beautifully simple relations were found. Particles now also acquired a wave character: a particle of momentum p has a wavelength $\lambda = h/p$ (*Louis de Broglie*, 1892–1987). This discovery gave rise to the construction of the electron microscope (*Ernst Ruska*, 1906–1988). *Einstein* had shown earlier that mass is just a different form of energy: $E = mc^2$. The particles of light have energy, therefore a mass can be attributed to them:

$$E = h\nu = mc^2 \rightarrow m = \frac{h\nu}{c^2}.\tag{1.3}$$

Similarly, they have the momentum

$$p = \frac{h}{\lambda} = \frac{h\nu}{c} = mc.\tag{1.4}$$

---

[2] But we consider – and this is the most important point in the whole calculation – E as being composed of a certain number of finite equal parts, and for that use the constant of nature h $= 6.55 \cdot 10^{-27}$ (erg · s)

The old controversy between the corpuscular and the wave theory of light is overcome in the quantum theory of light. Light has particle and wave aspects depending on the phenomenon considered. They find a unified description only in quantum mechanics. This wave–particle dualism is not only a speciality of light but pervades all phenomena. Optics initiated the discovery of this fact.

1.2 Present

Optics made a big step forward through the invention of the laser by *Theodore Harold Maiman* (born 1927) in 1960. Coherent light, that is, light of extremely high spectral purity, is available today with high intensity in the visible and far into the deep ultraviolet and infrared spectral regions. This has enriched optics with new methods and devices. They have influence on our daily life, for instance in telecommunications, and also allow us to gain insight into deep and subtle physical questions, for instance with respect to the foundations of quantum mechanics. The deepest impression on the general public has surely been left by holography [1.6] with its possibility of recording and reconstructing three-dimensional images as they directly appeal to our visual abilities.

But of deeper importance is the fact that the speed of light in vacuum has gained a place as a constant of nature:

$$c_0 = 299\,792\,458\,\text{ms}^{-1} \tag{1.5}$$

in the present system of units. The unit of length, the meter, thereby has become a derived unit. It is defined, with the help of c_0, as the distance light travels in vacuum in the time interval of $1/299\,792\,458$ second.

Optical techniques and devices for measuring virtually all types of variables have been invented, employed, and new techniques are constantly being developed. Just a few are listed here:

- holographic interferometry and speckle photography to measure small deformations of even rough surfaces;
- dual-recycling interferometry proposed for the detection of gravitational waves;
- laser-doppler (LDA) and phase-doppler anemometry (PDA) for measuring the velocities of particles in fluid flows;
- photon correlation techniques for determining the diameter of stars, for instance;
- laser vibrometer for measuring the vibration of object surfaces;
- scanning-laser and near-field-scanning optical microscopy that allows unprecedented resolution up to the visualization of individual molecules;

- fiber-optic sensors such as optical microphones and hydrophones for measuring sound waves in gases and liquids;
- all-optical filtering of images for picture processing and pattern recognition;
- the generation of harmonic waves (more than 135 times the fundamental frequency) and wave mixing for coherent light generation at high frequencies;
- phase conjugation in real time for obtaining phase conjugating mirrors;
- optical chaos and synchronization for secure communication;
- nonlinear spectroscopy;
- femtosecond technology for ultrafast optics and metrology;
- optical coherence tomography for looking into diffusing media;
- photonic crystals.

This variety has a reason: light, in particular in the form of coherent light, has proven to be an ideal measuring and scanning instrument, as it normally does not disturb the object investigated. Optics is combined with electronics, acoustics, and mechanics. There are acoustooptic deflectors and electrooptic devices such as Pockels cells. Mechanically deformable mirrors are used in adaptive optics to get high-resolution images of the sky. Optics gets "integrated", that is, miniaturized on chips, and more and more takes over tasks in telecommunications. Information is transmitted via glass fibers and stored in optically readable format on compact and magnetooptic disks. Electronics meets with optics as "photonics" [1.7].

Besides refractive optics, making use of the different refractive indices of materials, diffractive optics is developed which is based on the diffraction of light at fine structures. The number of diffracting structures is practically unlimited, thereby opening up novel ways for using light. A hologram, for instance, that has stored a three-dimensional image, is a diffracting structure and thus belongs to diffractive optics in a wider sense. Hummingbirds make use of diffractive optics with their brilliant iridescent plumage. In technical applications, diffractive optics is used in the X-ray region of the spectrum, for instance in X-ray microscopy, as materials with large refractive indices are not known in this wavelength region. Diffractive optical elements are calculated and manufactured for many special applications in the form of digital holograms, then called holographic optical elements. Examples are beam splitters, deflectors, and holographic lenses with several focal points.

The digital electronic computer has become a device indispensable for the lens designer and can be considered itself as a universal optical element since it is capable of modifying images and of performing filter operations or image transforming tasks. It is capable of calculating the Fourier transform of images, a task that can also be done by a lens. Powerful parallel computers, equipped with many processors, and optics mix in a peculiar way to create a new type of reality, called "virtual reality". Wearing special glasses or a helmet display, and a data glove or even suit, you can move interactively in a computer-generated world to explore complex structures such as a DNA molecule or the interior of the human body, a totally new way of optical experience.

1.3 Future

What does the future carry in its cornucopia? Several challenging areas can be identified where optics will have a great impact.

In technical applications, optics will expand deeply into the areas of communications and information processing. Fast communication channels on the basis of fiber networks will be developed further and installed worldwide. Electronic data processing will increasingly be supported by optical data processing. Thereby diffractive and integrated optics will play an important role. As photons do not interact when propagating in a linear medium, much more complex, in particular parallel, circuits are realizable optically than electronically. Neural nets with optical elements are presently implemented and will remain a challenge for some time to come. First arrangements of a digital, purely optical computer demonstrate the possibility of computing with photons. The peculiar laws of quantum mechanics will be explored further with quantum computation, quantum cryptography and quantum cloning [1.8, 1.9].

In physics, nonlinear optics, with its main areas of laser physics and quantum optics, promises new insight into the laws of nature. The X-ray laser and laser fusion wait to become feasible. Femtosecond lasers open up new intensity regions in the laboratory for experimentation. Unprecedented time resolution is now becoming available with laser-generated attosecond pulses to probe the internal dynamics of atoms and molecules. Laser instabilities and chaotic laser-light dynamics are expected to add to our understanding of nature, of deterministic chaos in particular. Both the question of how determinism and predictability are related and the not quite intuitive world of quantum physics presumably can best be demonstrated with experiments in laser physics and nonlinear optics.

Light thus may help us in answering the question: what will we be able to know? But do we already know the answer to the question: what is light?

Problems

1.1 Calculate the frequency, the energy, the momentum, and the (dynamical) mass of a photon of wavelength $\lambda = 550$ nm (yellow light).

1.2 Consider Planck's radiation law given by (1.2).
(a) Derive an expression for the spectral energy density $\rho(\lambda)$ as a function of wavelength.
(b) Locate the maxima ν_{max} and λ_{max} of the spectral energy densities $\rho(\nu)$ and $\rho(\lambda)$, respectively. Explain why $\nu_{max} \neq c/\lambda_{max}$.
(c) How does the total energy density ρ, that is, $\rho(\nu)$ integrated over the frequency ν, depend on the temperature T?
Hints: the algebraic equation $e^z(3-z) = 3$ has roots $z = 0$ and $z \approx 2.821$. Likewise, the equation $e^z(5-z) = 5$ has roots $z = 0$ and $z \approx 4.965$. The following identity holds: $\int_0^\infty z^3 [\exp(z) - 1]^{-1} dz = \pi^4/15$.

2. The Main Areas of Optics

Optical phenomena may be divided into four main areas: geometrical optics, wave optics, quantum optics, and statistical optics. In this book we will mainly discuss wave optics.

2.1 Geometrical Optics

Geometrical optics applies in those cases where diffraction and interference phenomena can be neglected, that is, when the dimensions of mirrors, lenses, and other objects in the optical path as well as the cross section of the light beam itself are large compared to the wavelength of the light. In this simplest theory of light one works with "light rays" propagating in a transparent medium. The main laws used are the law of reflection and the law of refraction.

Geometrical optics has found a new interesting application in the method of "ray tracing" widely used in computer graphics. When properly used it also allows us to calculate diffraction phenomena. Thereby, light rays are used for superimposing partial waves. The elongations of the partial waves are calculated via the propagation distance which is taken modulo the wavelength. These elongations are used when superimposing several rays. In this way digital holograms (holographic optical elements) are calculated (Fig. 2.1). They are explained in Sect. 7.5.

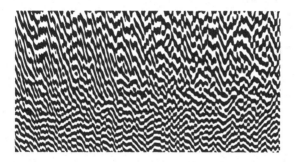

Fig. 2.1. Section from a digital hologram.

2.2 Wave Optics

With the help of wave optics, diffraction and interference phenomena can be described. It is founded on Maxwell's equations that were derived from electromagnetic phenomena independently of optics. According to this theory, light is composed of electromagnetic waves of different frequencies. The optical region only forms a minor part of the frequency spectrum(Fig. 2.2).

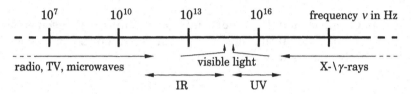

Fig. 2.2. The spectrum of electromagnetic radiation.

Wave optics covers interesting areas such as holography and Fourier optics, with their possibilities of two-dimensional or even three-dimensional signal processing, fiber optics, with their physically interesting solitons, and large parts of the optical computer. Interferometric devices, with their high sensitivity, essentially rely on wave optics. The state of the art in length measurement by interferometry has reached a precision of $\Delta L/L = 10^{-19}$, that is, shifts of a really tiny fraction of the diameter of an atom are observable. It is coherent light that expounds its wave properties by covering everything with speckles. In connection with nonlinearities, wave optics may exhibit surprising phenomena. An example is given by optical phase conjugation that can easily be realized holographically (see Sect. 7.1.4). Phase conjugating mirrors can be built from nonlinear crystals. They have the ability to reflect light back into itself (Fig. 2.3). It is not possible to view oneself in such a mirror. Just the aperture of the eye may be seen. In this case the law of reflection is no longer valid.

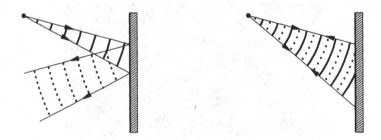

Fig. 2.3. Action of a normal mirror (*left*) and a phase conjugating mirror (*right*).

2.3 Quantum Optics

Absorption and emission processes cannot be described with just Maxwell's equations. This is the realm of quantum optics or, to a greater extent, quantum electrodynamics and quantum field theory. Quantum optics uses the fact that the electromagnetic field can interact with its surroundings only in energy humps, $E = h\nu$, or multiples thereof and that energy differences in atoms are emitted in these discrete amounts as light or electromagnetic radiation. A simple description of a large number of phenomena can be gained this way, for instance, of absorption and of stimulated and spontaneous emission. Spontaneous emission, however, needs a quantization of the light field for a thorough description.

The most important quantum optical instrument is the laser. The laser makes use of the three basic interaction mechanisms just mentioned (Fig. 2.4). It is the main tool for coherent optics. With the help of the laser it

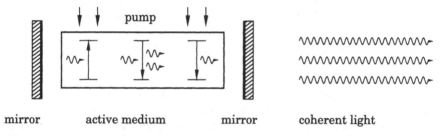

mirror   active medium   mirror   coherent light

Fig. 2.4. Scheme of a laser with its basic physical processes of absorption, stimulated emission, and spontaneous emission.

is even possible to detect quantum states and quantum jumps of single atoms. The atom (ion) is kept in a Paul trap, and a suitable transition is resonantly excited [2.1]. Depending on whether the atom is in the ground state or a third, excited state fluorescent light is observed or not (Fig. 2.5). In this way, the question of whether quantum jumps are observable or not could be answered by an optical experiment.

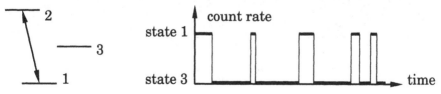

Fig. 2.5. Energy levels and the result of an experiment on laser-induced resonance fluorescence.

2.4 Statistical Optics

Statistical optics is a relatively new branch of optics. It gained a new impetus after the invention of the laser, which provided a new quality of light. In many measurements the question of photon statistics in fluctuating light fields arises. The fluctuations are brought about by the superposition of many light waves. The light waves may have different frequencies or may be generated by scattering at a rough surface. When using monofrequency (monochromatic) light the phenomenon of speckle pattern formation arises (see the chapter on speckles).

A typical measurement in statistical optics looks as follows (Fig. 2.6). A photodiode with a shutter in front of it is placed in a light field. First

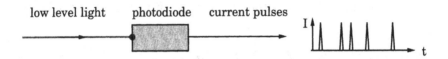

Fig. 2.6. Basic arrangement for measuring the statistical properties of light.

the shutter is closed, so that no light can reach the detector. Then the shutter is opened for some time to count the incoming photons. This simple measurement on the statistics of incoming photons already allows for discerning different types of light (Fig. 2.7).

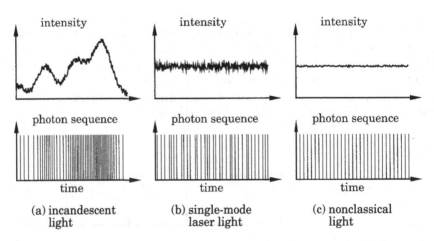

Fig. 2.7. Photon counts and intensity fluctuations for incandescent light, single-mode laser light, and nonclassical light, respectively. Intensity here means number of photons (current pulse counts) per fixed small time interval.

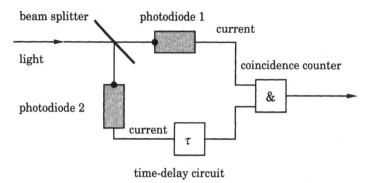

Fig. 2.8. The classical experiment of Hanbury Brown and Twiss for measuring temporal correlations of photons.

With incandescent light, the photons appear "bunched", whereas for single-mode laser light the fluctuations in photon counts are smaller. Light with fluctuations even smaller than those of laser light is possible according to quantum mechanics. This nonclassical light, as it is called, has also been realized.

With two (or even more) photodetectors more complex properties of a light field can be probed (Fig. 2.8), for instance temporal correlations, that is, correlation of two photon counts with a time shift in between. One may also look for spatial correlations when photons are counted at two different locations in a light field [2.2].

3. Fundamentals of Wave Optics

Light has, as we know today, a wave and a quantum aspect. The wave aspect dominates in the propagation of light, the quantum aspect dominates in absorption and emission. The theory covering the wave aspect is the classical theory of electromagnetic fields as elaborated by *Maxwell*. Its fundamentals are presented in this chapter.

3.1 Maxwell's Equations

Wave optics is based on Maxwell's equations (see, for instance, [3.1]). When written in SI units with only a single unit for the electric field and a single one for the magnetic field they read:

$$\operatorname{div} \boldsymbol{E}_\mathrm{m} = \frac{\rho}{\varepsilon_0}, \tag{3.1}$$

$$\operatorname{div} \boldsymbol{B}_\mathrm{m} = 0, \tag{3.2}$$

$$\operatorname{curl} \boldsymbol{E} = -\frac{\partial \boldsymbol{B}_\mathrm{m}}{\partial t}, \tag{3.3}$$

$$\operatorname{curl} \boldsymbol{B} = \varepsilon_0 \mu_0 \frac{\partial \boldsymbol{E}_\mathrm{m}}{\partial t} + \mu_0 \boldsymbol{J}. \tag{3.4}$$

The individual equations represent Coulomb's law, the nonexistence of magnetic monopoles, Faraday's law, and Ampère's law with the displacement current introduced by Maxwell, respectively. The variables and constants appearing in the equations have the following meaning:

$\boldsymbol{E}$ electric field in Vm^{-1},

$\boldsymbol{E}_\mathrm{m}$ electric field corresponding to the dielectric displacement in Vm^{-1},

$\boldsymbol{B}$ magnetic field in Vsm^{-2},

$\boldsymbol{B}_\mathrm{m}$ magnetic field in the presence of a medium in Vsm^{-2},

ρ charge density in Asm^{-3},

ε_0 electric field constant $= 1/\mu_0 c_0^2$ in $\mathrm{AsV}^{-1}\mathrm{m}^{-1}$ (exact),

μ_0 magnetic field constant $= 4\pi\, 10^{-7}\ \mathrm{VsA}^{-1}\mathrm{m}^{-1}$ (exact),

c_0 velocity of light in a vacuum $= 299\,792\,458\ \mathrm{ms}^{-1}$ (exact),

$\boldsymbol{J}$ current density in Am^{-2}.

In matter, additional fields are present, giving rise to the notation

$$E_m = E + \frac{1}{\varepsilon_0}P,\tag{3.5}$$

$$B_m = B + \mu_0 M,\tag{3.6}$$

$$J = \sigma E.\tag{3.7}$$

The properties of the medium are given by P, M, and σ, which depend on the fields E and B:

P polarization in Asm^{-2},
M magnetization in Am^{-1},
σ specific conductivity in $AV^{-1}m^{-1}$.

For small fields the specific conductivity σ is constant (Ohm's law) and the dielectric polarization P and magnetization M are proportional to the applied field:

$$P = \chi \varepsilon_0 E,\tag{3.8}$$

$$M = \frac{\kappa}{\mu_0}B.\tag{3.9}$$

The factors χ and κ are called the dielectricand magnetic susceptibility,respectively. In nonlinear media they depend on the field strength in the material (see the chapter on nonlinear optics).

For light in a vacuum, Maxwell's equations are substantially simpler since most source terms are absent:

$$\rho = 0, \text{ no free charges exist,}$$
$$P = 0, \text{ the vacuum is not polarizable,}$$
$$M = 0, \text{ the vacuum is not magnetizable,}$$
$$\sigma = 0, \text{ the vacuum is nonconducting.}$$

Then $E_m = E$, $B_m = B$, and $J = 0$, and the field equations (3.1)–(3.4) take the following simple form:

$$\text{div } E = 0,\tag{3.10}$$

$$\text{div } B = 0,\tag{3.11}$$

$$\text{curl } E = -\frac{\partial B}{\partial t},\tag{3.12}$$

$$\text{curl } B = \varepsilon_0 \mu_0 \frac{\partial E}{\partial t}.\tag{3.13}$$

In homogeneous, nonconducting, and linear media $\rho = 0$ and $\sigma = 0$ are also valid, whereas the polarization and magnetization act as source terms. This leads to different effective field constants $\varepsilon \varepsilon_0$ and $\mu \mu_0$ and thus to a different, smaller propagation velocity $c_m = c_0/\sqrt{\varepsilon \mu} = c_0/n$; $\varepsilon = 1 + \chi$ being the dielectric constant, $\mu = 1 + \kappa$ being the (relative) permeability, and n being the refractive index of the medium.

3.2 The Wave Equation

We consider the propagation of light in a vacuum as being described by Maxwell's equations (3.10)–(3.13). They can be cast into the form of the wave equation. To begin, we apply the operation 'curl' to (3.12) and obtain

$$\text{curl curl } \boldsymbol{E} = -\text{curl } \frac{\partial \boldsymbol{B}}{\partial t} = -\frac{\partial}{\partial t} \text{curl } \boldsymbol{B}. \tag{3.14}$$

Thereby we have assumed that differentiation with respect to time and space can be interchanged. Next, we introduce curl $\boldsymbol{B}$ from (3.13). This yields

$$\text{curl curl } \boldsymbol{E} = -\mu_0 \varepsilon_0 \frac{\partial^2 \boldsymbol{E}}{\partial t^2}. \tag{3.15}$$

Now, the identity

$$\text{curl curl } = \text{grad div } - \Delta \tag{3.16}$$

is valid, where Δ designates the Laplacian. In Cartesian coordinates the Laplacian reads:

$$\Delta = \frac{\partial^2}{\partial x^2} + \frac{\partial^2}{\partial y^2} + \frac{\partial^2}{\partial z^2}. \tag{3.17}$$

We apply (3.16) to the electric field $\boldsymbol{E}$ and use (3.10) to get

$$\text{curl curl } \boldsymbol{E} = \text{grad div } \boldsymbol{E} - \Delta \boldsymbol{E} = -\Delta \boldsymbol{E}. \tag{3.18}$$

Then (3.15) takes the form

$$\Delta \boldsymbol{E} - \varepsilon_0 \mu_0 \frac{\partial^2 \boldsymbol{E}}{\partial t^2} = 0. \tag{3.19}$$

This equation is called the wave equation. It has been derived here for the propagation of electromagnetic waves in a vacuum. The wave equation appears in many different areas of physics; for instance, in acoustics where it describes the propagation of low amplitude sound waves. From dimensional arguments we easily see that

$$\varepsilon_0 \mu_0 = \frac{1}{v^2}, \tag{3.20}$$

with v being the propagation velocity of a wave or perturbation obeying the wave equation. In this notation the wave equation reads:

$$\Delta \boldsymbol{E} - \frac{1}{v^2} \frac{\partial^2 \boldsymbol{E}}{\partial t^2} = 0. \tag{3.21}$$

For a scalar wave E propagating in the z-direction, the equation simplifies to

$$\frac{\partial^2 E}{\partial z^2} - \frac{1}{v^2} \frac{\partial^2 E}{\partial t^2} = 0. \tag{3.22}$$

It is easy to verify that this equation possesses solutions of the form

$$E(z,t) = f(z - vt) \quad \text{or} \quad E(z,t) = g(z + vt) \tag{3.23}$$

with largely arbitrary functions f and g. A perturbation or wave of this kind keeps its form during propagation and, at a later time, is found just at another location.

This reasoning led Maxwell to postulate the existence of electromagnetic waves and to predict their propagation velocity v from the values of ε_0 and μ_0. He used the values *Rudolph Kohlrausch* (1809–1858) and *Wilhelm Weber* (1804–1891) had measured in 1856 in Leipzig and obtained

$$v = \frac{1}{\sqrt{\varepsilon_0 \mu_0}} \approx 3 \cdot 10^8 \, \text{ms}^{-1}. \tag{3.24}$$

This value coincided so perfectly with the then known velocity of light in a vacuum that he dared to predict that light is nothing else than an electromagnetic wave.

Then one task of wave optics is obvious: the wave equation has to be solved with different boundary conditions given by apertures and obstacles. When the results agree with the corresponding experiments with light, this will confirm that light is an electromagnetic wave. Diffraction experiments brilliantly confirmed Maxwell's theory.

For simplicity, we derived the wave equation for light in a vacuum. When the conductivity does not vanish, $\sigma \neq 0$, the derivation given above yields the telegraph equation

$$\Delta \boldsymbol{E} = \varepsilon_0 \mu_0 \frac{\partial^2 \boldsymbol{E}}{\partial t^2} + \mu_0 \sigma \frac{\partial \boldsymbol{E}}{\partial t} . \tag{3.25}$$

In nonlinear optics, the relations become much more complex as the term curl curl $\boldsymbol{E}$ cannot be substituted by $-\Delta \boldsymbol{E}$.

3.3 Waves

To understand the properties of light we must study the theory of wave propagation. That we will do now.

3.3.1 One-Dimensional Waves

To keep the initial considerations simple, we start with one-dimensional, scalar waves in a vacuum. We restrict our consideration to the propagation of a linearly polarized light wave in the z-direction with a velocity c:

$$\frac{\partial^2 E}{\partial z^2} - \frac{1}{c^2} \frac{\partial^2 E}{\partial t^2} = 0 , \tag{3.26}$$

or

$$\left(\frac{\partial}{\partial z} + \frac{1}{c}\frac{\partial}{\partial t}\right)\left(\frac{\partial}{\partial z} - \frac{1}{c}\frac{\partial}{\partial t}\right)E = 0. \tag{3.27}$$

The second way of writing the wave equation immediately yields that both

$$E(z,t) = f(z - ct) \tag{3.28}$$

and

$$E(z,t) = g(z + ct) \tag{3.29}$$

are solutions of the one-dimensional wave equation (3.26). Because of the linearity of the wave equation we also have

$$E(z,t) = af(z - ct) + bg(z + ct) \tag{3.30}$$

as a solution with two arbitrary functions f and g and arbitrary constants a and b. This is the superposition principle for waves. It is also valid for the three-dimensional wave equation (3.19) or (3.21), and, more generally, for linear differential equations.

The interference of light is described by the superposition of light waves; that is, the addition of light-wave amplitudes. Accordingly, interference phenomena should be of a very simple nature. But some remarks seem appropriate. At this level of description, interference is indeed given by one of the simplest mathematical operations. However, we are not able to follow directly the oscillations of a light wave, let alone recognize them with the eye. We can only perceive the intensity of a light wave (this is different with acoustic waves). Thus optical interference phenomena are more involved as the transition from amplitudes to intensities requires a nonlinear operation. Moreover, interference phenomena in nonlinear media are even more complex since the superposition principle at the amplitude level is no longer valid. However, in some cases (for instance, for solitons in fibers that obey a nonlinear Schrödinger equation),special nonlinear superposition principles have been found [3.2].

The validity of the superposition principle for light waves in a vacuum makes possible the composition of arbitrary wave forms from simple fundamental waves. The most important type is the harmonic wave. Assuming propagation in the positive z-direction and an appropriate definition of time zero, it is given by

$$E(z,t) = A\sin(kz - \omega t). \tag{3.31}$$

Here A is the amplitude of the wave and $(kz - \omega t)$ is its phase. The wave number k is related to the wavelength λ by

$$k = \frac{2\pi}{\lambda}. \tag{3.32}$$

The circular frequency ω is connected with the usual frequency ν, the number of oscillations per second, by

$$\omega = 2\pi\nu \qquad (3.33)$$

or, with the period T, the time one oscillation needs to complete its cycle, by

$$\omega = \frac{2\pi}{T}. \qquad (3.34)$$

Therefore, the harmonic wave (3.31) can be written with the more intuitive attributes of wavelength and period:

$$E(z,t) = A \sin\left(\frac{2\pi}{\lambda}z - \frac{2\pi}{T}t\right). \qquad (3.35)$$

Figure 3.1 gives two different views of the wave.

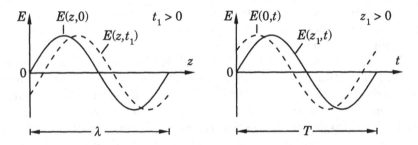

Fig. 3.1. Harmonic wave E depending on space and time.

A point of constant phase moves with the velocity of the wave,

$$c = \frac{\lambda}{T} = \nu\lambda = \frac{\omega}{k}, \qquad (3.36)$$

called phase velocity. Thus, at a fixed propagation velocity c, ω, and k are not independent. The relation between ω and k is called the dispersion relation. In this case it is very simple:

$$\omega = ck. \qquad (3.37)$$

The use of trigonometric functions leads to cumbersome calculations when waves have to be superimposed. Therefore a complex notation is used. To this end, the sine function is replaced by exponential functions according to Euler's formula

$$\sin\alpha = \frac{1}{2i}\left[\exp(i\alpha) - \exp(-i\alpha)\right]. \qquad (3.38)$$

Then the harmonic wave (3.31) reads:

$$E(z,t) = \tfrac{1}{2}E_0 \exp\left[i(kz - \omega t)\right] + \tfrac{1}{2}E_0^* \exp\left[i(-kz + \omega t)\right], \qquad (3.39)$$

with the complex amplitude

$$E_0 = \frac{A}{i}.\tag{3.40}$$

The quantity E_0^* is the complex conjugate amplitude of E_0. For convenience, (3.39) is usually abbreviated as

$$E(z,t) = \tfrac{1}{2}E_0 \exp\left[i(kz - \omega t)\right] + \text{c.c.},\tag{3.41}$$

whereby c.c. means the complex conjugate of the term appearing in front of it.

When only linear operations are performed, we can go one step further and work with the purely complex notation

$$E(z,t) = E_0 \exp\left[i(kz - \omega t)\right].\tag{3.42}$$

At the end of the calculations, the result has to be converted into a real form, by taking the real part or adding the complex conjugate and dividing by two, for instance. In this notation one has to be cautious when calculating the intensity because of the nonlinearity involved.

3.3.2 Plane Waves

A plane wave is a wave whose phase at a fixed time t is constant in every plane perpendicular to the direction of propagation. To get an expression for a wave with this property, one only needs to know that

$$\boldsymbol{k} \cdot \boldsymbol{r} = \text{const}\tag{3.43}$$

is the equation for a plane in space, where $\boldsymbol{k} = (k_x, k_y, k_z)$ is the wave vector and $\boldsymbol{r} = (x, y, z)$ is the displacement vector.

A plane, harmonic wave at time t is given by

$$E(\boldsymbol{r}) = A\,\sin(\boldsymbol{k} \cdot \boldsymbol{r}),\tag{3.44}$$

$$\text{or}\qquad E(\boldsymbol{r}) = \tfrac{1}{2}E_0\,\exp\left(i\boldsymbol{k} \cdot \boldsymbol{r}\right) + \text{c.c.},\tag{3.45}$$

$$\text{or}\qquad E(\boldsymbol{r}) = E_0\,\exp\left(i\boldsymbol{k} \cdot \boldsymbol{r}\right).\tag{3.46}$$

The wave repeats itself in the direction of $\boldsymbol{k}$ after a distance λ, the wavelength. This can easily be verified. With $|\boldsymbol{k}| = k$, $\boldsymbol{k} \cdot \boldsymbol{k} = k^2$, and $k = 2\pi/\lambda$ we have

$$
\begin{aligned}
E\left(\boldsymbol{r} + \lambda\frac{\boldsymbol{k}}{k}\right) &= A\sin\left[\boldsymbol{k} \cdot \left(\boldsymbol{r} + \lambda\frac{\boldsymbol{k}}{k}\right)\right]\\
&= A\sin(\boldsymbol{k} \cdot \boldsymbol{r} + \lambda k)\\
&= A\sin(\boldsymbol{k} \cdot \boldsymbol{r} + 2\pi)\\
&= A\sin(\boldsymbol{k} \cdot \boldsymbol{r})\\
&= E(\boldsymbol{r}).
\end{aligned}\tag{3.47}
$$

Up to now, we have looked at the wave at a fixed time in the way of a snapshot. Introducing a time dependence as in the case of one-dimensional waves we get

$$E(r,t) = A \sin(k \cdot r \mp \omega t), \tag{3.48}$$

written in complex notation as

$$E(r,t) = \tfrac{1}{2} E_0 \exp\left[i(k \cdot r \mp \omega t)\right] + \text{c.c.}, \tag{3.49}$$

$$\text{or} \quad E(r,t) = E_0 \exp\left[i(k \cdot r \mp \omega t)\right]. \tag{3.50}$$

The '$-$' sign applies for a wave propagating in the direction of the wave vector k, the '$+$' sign for a wave propagating in the opposite direction.

As the trigonometric and the complex exponential functions constitute a complete function system each, more complicated waveforms can be represented as a superposition of plane waves.

3.3.3 Spherical Waves

A further, often needed waveform is the spherical wave. Its importance stems from Huygens' principle. It states that each point in space excited by a wave becomes the origin of a spherical wave.

To arrive at a representation of a spherical wave with its constant phase over a spherical surface, it is convenient to introduce spherical coordinates (r, Θ, φ):

$$x = r \sin\Theta \cos\varphi,$$
$$y = r \sin\Theta \sin\varphi, \tag{3.51}$$
$$z = r \cos\Theta.$$

In spherical coordinates the Laplacian Δ reads:

$$\Delta = \frac{1}{r^2} \frac{\partial}{\partial r} \left(r^2 \frac{\partial}{\partial r} \right) + \frac{1}{r^2 \sin\Theta} \frac{\partial}{\partial \Theta} \left(\sin\Theta \frac{\partial}{\partial \Theta} \right) + \frac{1}{r^2 \sin^2\Theta} \frac{\partial^2}{\partial \varphi^2}. \tag{3.52}$$

This is quite a complicated expression. Fortunately, a spherical wave is spherically symmetric; that is, it shows no dependence on Θ and φ. Therefore, the expression for the Laplacian simplifies to

$$\Delta = \frac{1}{r^2} \frac{\partial}{\partial r} \left(r^2 \frac{\partial}{\partial r} \right) = \frac{\partial^2}{\partial r^2} + \frac{2}{r} \frac{\partial}{\partial r} \tag{3.53}$$

or, written with a scalar field amplitude E,

$$\Delta E = \frac{1}{r} \frac{\partial^2}{\partial r^2} (rE). \tag{3.54}$$

The wave equation then attains the form

$$\frac{1}{r} \frac{\partial^2}{\partial r^2} (rE) - \frac{1}{c^2} \frac{\partial^2 E}{\partial t^2} = 0. \tag{3.55}$$

Multiplying the equation by r yields

$$\frac{\partial^2}{\partial r^2}(rE) - \frac{1}{c^2}\frac{\partial^2}{\partial t^2}(rE) = 0. \tag{3.56}$$

This is the one-dimensional wave equation for the quantity rE. We therefore can take the general solutions (3.28) and (3.29) for rE:

$$rE(r,t) = f(r - ct) \quad \text{and} \quad rE(r,t) = g(r + ct). \tag{3.57}$$

The wave

$$E(r,t) = \frac{1}{r}f(r - ct) \tag{3.58}$$

is called an outgoing spherical wave, as it propagates from the origin $r = 0$ radially outwards, and the wave

$$E(r,t) = \frac{1}{r}g(r + ct) \tag{3.59}$$

is called an incoming spherical wave, as it converges towards the origin.

Again, the harmonic spherical waves are of special importance:

$$E(r,t) = \frac{A}{r}\sin(kr \mp \omega t) \tag{3.60}$$

or $\quad E(r,t) = \frac{\frac{1}{2}E_0}{r}\exp\left[i(kr \mp \omega t)\right] + \text{c.c.} \tag{3.61}$

or $\quad E(r,t) = \frac{E_0}{r}\exp\left[i(kr \mp \omega t)\right]. \tag{3.62}$

The amplitude E_0/r of a spherical wave decreases proportionally to $1/r$. Furthermore, at a large distance from the origin a spherical wave locally approaches a plane wave.

3.3.4 Bessel Waves

The wave equation

$$\Delta E(r,t) - \frac{1}{c^2}\frac{\partial^2 E(r,t)}{\partial t^2} = 0 \tag{3.63}$$

has a class of interesting solutions that were detected just recently: the class of diffraction-free waves [3.3]. "Diffraction-free" means that the wave keeps its intensity distribution in the (x,y)-plane independently from z, while propagating in the z-direction (for a definition of intensity see the next section):

$$I(x,y,z) = I(x,y,0), \quad z \geq 0. \tag{3.64}$$

An example is the plane wave. But it is only a limiting case where $I(x,y)$ is constant across the beam. Having knowledge of scalar diffraction theory one would expect that a strongly localized light wave cannot keep its form

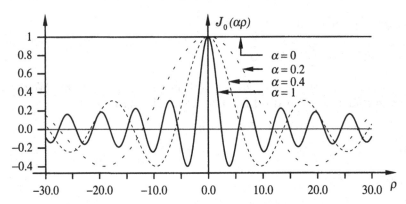

Fig. 3.2. Radial distribution of the field amplitude $J_0(\alpha\rho)$ for Bessel waves.

upon propagation. But solutions with exactly this property exist and can approximately be realized [3.4].

The simplest, nontrivial example of a diffraction-free wave is the fundamental Bessel wave

$$E(\mathbf{r},t) = E_0 J_0(\alpha\rho) \exp\left[-i(\omega t - \beta z)\right], \tag{3.65}$$

with $\beta^2 = k^2 - \alpha^2$, $\beta \geq 0$, $0 \leq \alpha \leq k = 2\pi/\lambda = \omega/c$, $\rho^2 = x^2 + y^2$, $\mathbf{r} = (x,y,z)$, as can be verified by substitution into the wave equation. The Bessel function J_0 is given by [3.5]

$$J_0(s) = \frac{1}{2\pi} \int\limits_0^{2\pi} \exp(is\sin\xi)\,d\xi. \tag{3.66}$$

Figure 3.2 shows the graph of the function $J_0(\alpha\rho)$ for different values of α. The wave is rotationally symmetric in the plane perpendicular to the propagation direction. The parameter α is connected with the width of the central peak, the width being smaller the larger α is, $\alpha \leq k = 2\pi/\lambda$. In the limit $\alpha = k$, it reaches the size of a wavelength; that is, it can become quite small. Nevertheless, no broadening occurs upon propagation through free space. Hence the name "diffraction-free".

The plane wave is contained in (3.65) for $\alpha = 0$. The Bessel wave, like the plane wave, has no finite energy, but has a finite energy density. Therefore it can be realized only approximately. The simplest way for generating a Bessel wave consists in making use of an axicon, that is, a glass cone that is illuminated with a plane wave in the direction of the cone tip (Fig. 3.3).

The length z_B, over which the Bessel wave is obtained, is given by the superposition area of the waves refracted by the axicon. It is proportional to the radius R of the axicon and, for small angles γ, inversely proportional to the axicon angle γ. Only very large axicons yield approximately diffraction-free Bessel waves over a longer distance, and only in the limit

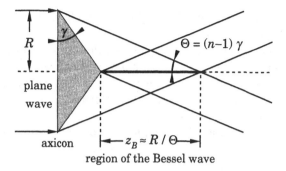

Fig. 3.3. Experimental realization of Bessel waves with an axicon of radius R, refractive index n, and angle γ.

of infinitely large axicons does the diffraction-free distance become infinitely long.

Bessel waves have been proposed for material processing, for instance for drilling deep holes or for inscriptions on curved surfaces. The axicon is also used for focusing light along a line to drive X-ray lasers.

3.3.5 Evanescent Waves

Evanescent waves are very special in that they are nonpropagating. They appear in connection with total internal reflection at a dielectric interface. When a light beam is directed from a transparent medium of higher refractive index n_1 (glass) towards an interface with a medium of lower refractive index n_2 (air) total internal reflection occurs beyond a critical angle Θ_c, the angle traditionally given with respect to the normal to the interface (Fig. 3.4). At Θ_c the refracted beam propagates along the inter-

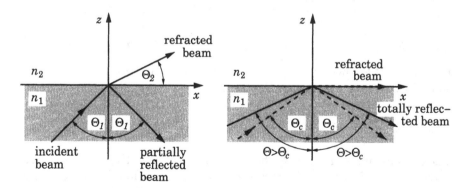

Fig. 3.4. Refraction (*left*) and reflection (*right*) at a dielectric interface for light beams propagating towards the interface from the optically denser medium (refractive index n_1). Beyond the critical angle Θ_c total internal reflection occurs.

face (x-direction), and for an angle $\Theta_1 > \Theta_c$ all light is reflected back into the optically denser medium.

It seems that no wave is entering the less dense medium n_2 in the case of total reflection, but this is not true. No wave field can stop suddenly at an interface (if it is not an — unphysical — infinite barrier): the wave field amplitude has a finite value beyond the interface. A simple description of this phenomenon runs as follows.

The angles between light beams and the normal to the interface at refraction and reflection at an interface are given by Snell's law (*Willebrord Snell*, 1591–1626)

$$\frac{\sin \Theta_1}{\sin \Theta_2} = \frac{n_2}{n_1}, \tag{3.67}$$

where Θ_1 is the angle of incidence from medium 1, Θ_2 is the angle of refraction and n_1, n_2 are the indices of refraction given by

$$n = \frac{c_0}{c_m}, \tag{3.68}$$

where c_0 is the velocity of light in vacuum and c_m is the velocity of light in the medium where the light actually propagates.

The critical angle $\Theta_1 = \Theta_c$, where the refracted beam attains the angle of $\Theta_2 = 90°$, is then given via ($\sin \Theta_2 = 1$)

$$\sin \Theta_c = \frac{n_2}{n_1}. \tag{3.69}$$

A plane wave propagating in the (x,z)-plane in the direction of the refracted beam is given by (amplitude and $\exp(i\omega t)$ omitted)

$$
\begin{aligned}
\exp(i\mathbf{k} \cdot \mathbf{r}) &= \exp(i k_x x + i k_z z) \\
&= \exp(i kx \sin \Theta_2 + i kz \cos \Theta_2). \tag{3.70}
\end{aligned}
$$

To calculate the z-component we look at

$$\cos \Theta_2 = \sqrt{1 - \sin^2 \Theta_2} = \sqrt{1 - \frac{n_1^2}{n_2^2} \sin^2 \Theta_1} = \sqrt{1 - \left(\frac{\sin \Theta_1}{\sin \Theta_c}\right)^2}. \tag{3.71}$$

For $\Theta_1 > \Theta_c$ it follows $\sin \Theta_1 > \sin \Theta_c$ and thus the root becomes imaginary, whereby $\cos \Theta_2$ is replaced by the term

$$i\sqrt{\left(\frac{\sin \Theta_1}{\sin \Theta_c}\right)^2 - 1}.$$

The complex phase $i kz \cos \Theta_2$ then becomes real, i. e., the wave becomes exponentially decaying in the z-direction and is no longer propagating into this direction. With Eq. (3.70) we obtain for $z > 0$ and $\Theta_1 > \Theta_c$:

$$\exp(i\mathbf{k} \cdot \mathbf{r}) = \exp\left(-kz\sqrt{\left(\frac{\sin \Theta_1}{\sin \Theta_c}\right)^2 - 1}\right), \tag{3.72}$$

where $k = 2\pi/(n_2\lambda_0)$ is the wave number in the optically thinner medium (refractive index n_2) and λ_0 is the wavelength of the light in vacuum. We thus have an exponentially decaying wave field.

The intensity (see Sect. 3.4) can be written as

$$I(z) = I_0 \exp\left(-\frac{z}{d}\right) \tag{3.73}$$

with

$$d = \frac{1}{2k\sqrt{\left(\dfrac{\sin\Theta_1}{\sin\Theta_c}\right)^2 - 1}} = \frac{\lambda_0}{4\pi\sqrt{n_1^2\sin^2\Theta_1 - n_2^2}}. \tag{3.74}$$

The decay constant d of the evanescent field intensity varies between about 30 nm and 300 nm depending on the wavelength λ_0, the refractive indices n_1 and n_2 and the angle of incidence Θ_1.

The presence of this wave can be probed by placing objects into the field. When they pick up some energy this is replenished from the incident wave and thus energy flows through the interface. Figure 3.5 shows an experimental arrangement where two prisms are separated by a small gap. When prism 2 is several decay constants d of the field away from prism 1 then all light is reflected by prism 1 into the direction 1. When prism 2 is approached near to prism 1 more and more light is leaking through the gap into direction 2. This leaking of the wave through a small gap is called frustrated total internal reflection. In the photon picture this is the tunneling of particles. The effect is used in ring lasers for adjusting the quality of the cavity and the tapping of light out of the cavity. A scientific application is total internal reflection fluorescence microscopy. When molecules that absorb and reemit light (fluorophores) are brought into an evanescent wave field their fluorescence is excited. The advantage

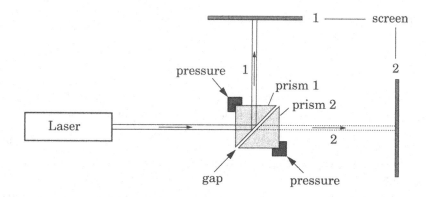

Fig. 3.5. Demonstration experiment for frustrated total internal reflection. By pressing the two prisms firmly together light from beam 1 can be coupled to the direction 2.

of the method is twofold. First, only a very thin layer of material near the interface is excited giving rise to easy single molecule detection. Second, the very weak fluorescence signal is not masked by the usually strong excitation signal because excitation is via a nonpropagating exponentially decaying wave. The axial resolution of this type of microscope can thus be enhanced down to the decay constant of the evanescent field.

3.3.6 Polarized Waves

So far we have considered scalar waves, that is waves with one component of the electric field vector $E = (E_x, E_y, E_z)$ only. Dropping this restriction leads to a larger variety of waves. One new property of these waves is that the direction of the electric field vector E in space is of importance and how this direction changes with time. This property is captured by the notion of the state of polarization of a wave.

We start from the three-dimensional wave equation (3.19), valid in vacuum,

$$\Delta E - \varepsilon_0 \mu_0 \frac{\partial^2 E}{\partial t^2} = 0$$

with $\varepsilon_0 \mu_0 = 1/c_0^2$, c_0 being the velocity of light in vacuum. We consider plane waves propagating in the z-direction. Because plane waves have constant phases and amplitudes in any plane perpendicular to the propagation direction, the components of the electric field vector $E(x,y,z,t) = \big(E_x(x,y,z,t), E_y(x,y,z,t), E_z(x,y,z,t)\big)$ can only vary in the z-direction, and the derivatives $\partial E_x/\partial x$, $\partial E_y/\partial x$, $\partial E_z/\partial x$, $\partial E_x/\partial y$, $\partial E_y/\partial y$, $\partial E_z/\partial y$ are all vanishing. The electric field vector E of a three-dimensional wave propagating in the z-direction thus can be written in the form

$$E(x,y,z,t) = E(z,t) = E_x(z,t)\hat{x} + E_y(z,t)\hat{y} + E_z(z,t)\hat{z},$$

where $\hat{x}$, $\hat{y}$, $\hat{z}$ are unit vectors in the x-, y- and z-directions, respectively. Moreover, from (3.10) we have

$$\text{div}\, E = \frac{\partial E_x}{\partial x} + \frac{\partial E_y}{\partial y} + \frac{\partial E_z}{\partial z} = \frac{\partial E_z}{\partial z} = 0. \tag{3.75}$$

Thus there is no z-dependence of E_z: E_z must be a constant field. When we dispense with this constant field to focus on a plane wave we get

$$E(z,t) = E_x(z,t)\hat{x} + E_y(z,t)\hat{y}. \tag{3.76}$$

The corresponding relation can be derived for the wave's magnetic field vector B.

The interpretation of this result is that Maxwell's equations in vacuum, additionally to giving rise to the three-dimensional wave equation,

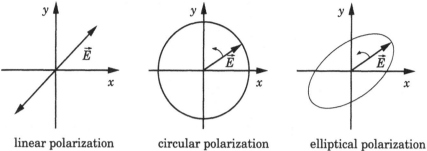

linear polarization   circular polarization   elliptical polarization

Fig. 3.6. Traces of the electric field vector E in the (x, y)-plane for the case of linear polarization (*left*), circular polarization (*center*) and elliptical polarization (*right*).

predict that electromagnetic plane waves are transverse waves. For a harmonic plane wave we have

$$E(z,t) = E_0 \exp[i(kz - \omega t)] = E_{0x} \exp[i(kz - \omega t)]\hat{x} + E_{0y} \exp[i(kz - \omega t)]\hat{y}$$
(3.77)

with the complex amplitude vector $E_0 = (E_{0x}, E_{0y}, 0)$.

The complex amplitudes E_{0x} and E_{0y} can be chosen independently and contain a phase of which only one can be made zero by appropriate choice of time zero. We are thus left with three real numbers determining the state of polarization of the wave. The general case is called elliptical polarization because E is scanning an ellipse in the (x, y)-plane with time. Two important special cases arise when the ellipse degenerates to a line giving linear polarization and when it has the form of a circle giving circular polarization (Fig. 3.6).

The plane defined by the electric field vector E and the propagation direction k (here $\hat{z}$) is called plane of polarization. This plane is fixed in space for linear polarization but may have any direction of E in the (x, y)-plane, and turning around k in the other cases.

Maxwell's equations in vacuum are linear equations so that, as stated above, the superposition principle holds. Each component of (3.77) progresses independently of the other component. Media in which this statement is true are called linear media. In these media light waves do not interact, they just sum up by vector addition of the fields present.

The state of polarization enters many physical phenomena, starting with the basic phenomenon of reflection at an interface. The amount of reflected and transmitted light strongly depends on the state of polarization. We consider the case of a light wave incident from an optically thinner medium 2 onto a plane interface with an optically thicker medium 1. We consider a linearly polarized wave with the E-vector lying in the plane of the propagation vector k and the normal of the interface $\hat{n} = \hat{z}$. This type of linear polarization is called parallel polarization in this type of configuration: $E = E_\parallel$. In this case it can be shown by recourse to Maxwell's

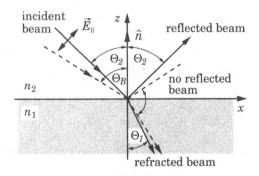

Fig. 3.7. The Brewster angle θ_B of no reflected beam for plane parallel polarized light.

equations together with boundary conditions that there exists an angle of incidence $\Theta_2 = \theta_B$ where no light is reflected (Brewster angle, see Fig. 3.7). It is given by

$$\tan \Theta_2 = \tan \theta_B = \frac{n_1}{n_2}. \tag{3.78}$$

A Brewster angle window is used in laser physics as end plate of the tube of a He-Ne laser, for instance, to get a defined linear polarization of the output beam. Parallel polarized light leads to less losses, thus has more gain and wins the competition of taking photons from inversion (see Chap. 10). Similarly, stacks of parallel glass plates inclined at the Brewster angle (about 56.5° for an air-glass interface) are used in high-power lasers to get a polarized light beam.

When speaking of a polarized light beam it is in general assumed that the state of polarization does not change over the cross-section of the beam. Note, however, that this need not be the case. The state of polarization can vary from place to place in a light field, and its spatial structure can be tailored by suitable placement of polarizing optics to yield, e. g., purely radially or azimuthally polarized beams [3.6].

3.4 Intensity of a Light Wave

Unfortunately, we cannot yet follow the instantaneous amplitude of a light wave, only its intensity. The intensity therefore is an important quantity. It belongs to the more difficult concepts since the field amplitude enters nonlinearly. For instance, we may not add the intensities of harmonic waves to arrive at a proper description of an optical phenomenon, except for special cases.

The intensity is a measure of the energy flowing through an area in a given time interval:

$$\text{intensity} = \frac{\text{energy}}{\text{area} \cdot \text{time interval}} .$$

It is connected with the field amplitude E. To obtain the relation, we consider the energy density u, that is, the energy in a volume of the electromagnetic field. The electric field alone yields (remember the plate capacitor [3.7])

$$u_E = \frac{\varepsilon_0}{2} E^2, \qquad E \text{ real}, \tag{3.79}$$

and the magnetic field alone (remember the coil [3.7]) gives

$$u_B = \frac{1}{2\mu_0} B^2, \qquad B \text{ real}. \tag{3.80}$$

We consider scalar, that is, linearly polarized, fields. It can be shown that for a plane wave

$$E = cB. \tag{3.81}$$

It follows that the electric and magnetic energy densities are equal,

$$u_E = u_B, \tag{3.82}$$

a relation which is generally valid for propagating electromagnetic fields. Then, the total energy density can be written as

$$u = u_E + u_B = 2u_E = \varepsilon_0 E^2. \tag{3.83}$$

To obtain the intensity, we look at the energy that flows through an area ΔA perpendicular to the propagation direction in a time interval Δt.

To this end, we multiply the energy density u with the volume $c\Delta t\Delta A$, to give an energy, and divide by $\Delta t\Delta A$. This yields

$$I = \frac{uc\Delta t\Delta A}{\Delta t\Delta A} = uc \tag{3.84}$$

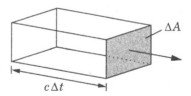

and thus

$$I = \varepsilon_0 c E^2. \tag{3.85}$$

The most important fact in this relation is that I is proportional to E^2:

$$I \propto E^2 \quad (E \text{ real}). \tag{3.86}$$

In its present form (3.86), the intensity, which we will call the instantaneous intensity, cannot be used for measuring purposes because no detector exists that could follow the light frequencies. Therefore, any intensity measurement is an average taken over a measuring time T_m which is large compared to the period $T = 2\pi/\omega$ of the light wave.

For a one-dimensional harmonic plane wave, when we use the complex notation (3.39)

$$E(z,t) = \tfrac{1}{2}E_0 \exp\left[i(kz - \omega t)\right] + \tfrac{1}{2}E_0^* \exp\left[-i(kz - \omega t)\right]$$
$$= \tfrac{1}{2}E_R \exp(-i\omega t) + \tfrac{1}{2}E_R^* \exp(i\omega t), \tag{3.87}$$

we get

$$I(z,t) \propto E^2(z,t) = \tfrac{1}{4}\left[E_R^2 \exp(-2i\omega t) + E_R^{*2} \exp(2i\omega t) + 2E_R E_R^*\right]. \tag{3.88}$$

Herein we used the abbreviation

$$E_R = E_R(z) = E_0 \exp\left(ikz\right) \tag{3.89}$$

for the spatial dependence of the wave. When we form the average over the measuring time T_m, we get

$$\bar{I}(z) = \frac{1}{T_m} \int\limits_{-T_m/2}^{+T_m/2} I(z,t)\,\mathrm{d}t, \tag{3.90}$$

and, omitting a factor $\tfrac{1}{4}$:

$$\bar{I}(z) \propto \frac{1}{T_m} \int\limits_{-T_m/2}^{+T_m/2} \left[E_R^2 \exp(-2i\omega t) + E_R^{*2} \exp(2i\omega t) + 2E_R E_R^*\right]\,\mathrm{d}t$$

$$= \frac{E_R^2}{2i\omega T_m}\left[\exp(+i\omega T_m) - \exp(-i\omega T_m)\right]$$

$$+ \frac{E_R^{*2}}{2i\omega T_m}\left[\exp(+i\omega T_m) - \exp(-i\omega T_m)\right] + 2E_R E_R^* \,. \tag{3.91}$$

When the measuring time T_m is large compared to the period T of the light oscillation,

$$T_m \gg T = \frac{2\pi}{\omega} \quad \text{or} \quad \omega T_m \gg 1, \tag{3.92}$$

the first two terms in (3.91) can be neglected compared to $2E_R E_R^*$, and, with (3.89), we obtain

$$\bar{I}(z) \propto 2E_R E_R^* = 2E_0 \exp\left(ikz\right) E_0^* \exp\left(-ikz\right) = 2E_0 E_0^* \,. \tag{3.93}$$

A harmonic plane wave has a constant intensity with respect to space and time. Therefore, omitting factors of proportionality, intensity is defined as

$$I = E_0 E_0^* = |E_0|^2 \,. \tag{3.94}$$

This definition includes the temporal averaging procedure.

More general wave fields can also be represented by a complex wave amplitude $E(r,t)$, the analytic signal (see Section 4.4). Then, assuming a stationary wave field, the definition of the intensity is formulated as follows:

$$I(r) = \lim_{T_m \to \infty} \frac{1}{T_m} \int_{-T_m/2}^{+T_m/2} E(r,t') E^*(r,t') \, dt' \overset{\text{def}}{=} \langle EE^* \rangle_\infty \,. \tag{3.95}$$

This theoretically satisfactory definition proves to be too narrow for many purposes. Fluctuations and other pulse-like phenomena cannot be described adequately. Therefore, a short-term intensity is introduced:

$$I(r,t;T_m) = \frac{1}{T_m} \int_{t-T_m/2}^{t+T_m/2} E(r,t') E^*(r,t') \, dt' . \tag{3.96}$$

It corresponds to the moving average with a window of width T_m centered at t. The averaging time interval T_m is usually large compared to the period of the light wave and has to be smaller than the time scale of the phenomenon to be investigated.

Problems

3.1 Two one-dimensional harmonic waves having equal frequencies and amplitudes are superimposed. Write this superposition using real notation,

$$E(z,t) = E_1 \sin(kz - \omega t + \varphi_1) + E_2 \sin(kz - \omega t + \varphi_2),$$

and also using complex notation,

$$E(z,t) = \tilde{E}_1 \exp\left[i\left(kz - \omega t\right)\right] + \tilde{E}_2 \exp\left[i\left(kz - \omega t\right)\right],$$

with $\tilde{E}_{1,2} = E_{1,2} \exp(i\varphi_{1,2}')$. In each case, state the amplitude and phase of the resulting wave.

3.2 A wave is given by

$$E(z,t) = 4 \sin\left[2\pi(3z - 5t)\right] + 3\cos\left[\pi(6z - 10t)\right].$$

What are the frequency, the oscillation period, the wave number, the wavelength, the amplitude, the phase velocity, and the direction of propagation? Specify the complex amplitude of this wave using the notation of (3.42).

3.3 A point source emits a spherical monofrequency wave with $\lambda = 550$ nm. At what distance d from the source does the wave front deviate from that of a plane wave by less than $\lambda/10$ over a circular area of diameter 1 cm?

3.4 In a dispersive medium the following wave equation holds:

$$\frac{\partial^2 E}{\partial z^2} + \eta \frac{\partial^4 E}{\partial z^4} - \frac{1}{c^2} \frac{\partial^2 E}{\partial t^2} = 0.$$

Specify the dispersion relation of the medium.

3.5 Give a mathematical expression for a scalar monofrequency wave being cylindrically symmetric about the z-axis. Using energy conservation, find out how the amplitude $E(\rho)$ of the cylindrical wave should depend on the radial distance ρ from the symmetry axis for large ρ. In cylindrical coordinates (ρ, Θ, z) the Laplacian Δ reads:

$$\Delta = \frac{1}{\rho}\frac{\partial}{\partial\rho}\left(\rho\frac{\partial}{\partial\rho}\right) + \frac{1}{\rho^2}\frac{\partial^2}{\partial\Theta^2} + \frac{\partial^2}{\partial z^2} .$$

Derive a differential equation for $E(\rho)$ and show that it is solved by the function $E(\rho) = J_0(k\rho)$. Use the hints given in Problem 3.7.

3.6 We consider an axicon with $n = 1.5$ and $\gamma = 15°$ (see Fig. 3.3). Which diameter should the axicon have to produce a Bessel wave $z_B = 1$ m long?

3.7 Verify that the Bessel wave (3.65) solves the three-dimensional, scalar wave equation. Describe and outline the solution for $\alpha = \pm k$. The Bessel wave (3.65) can be written as a superposition of plane waves $\exp[i(\boldsymbol{kr} - \omega t)]$. Which wave vectors $\boldsymbol{k}$ do appear in the superposition? Using this result, explain the operation of an axicon and derive the expressions for Θ and z_B given in Fig. 3.3.

Hints: $\mathrm{d}J_0(z)/\mathrm{d}z = -J_1(z)$, $\mathrm{d}^2J_0/\mathrm{d}z^2 = J_1(z)/z - J_0(z)$, $J_0(s) = (2\pi)^{-1}\int_0^{2\pi}\exp(is\cos\xi)\mathrm{d}\xi$.

3.8 A 100 W light bulb converts 2% of its electric power into light. The radiation is distributed evenly over a solid angle $\Omega = 1$ sterad by a reflector. Calculate the light intensity I and the amplitude E of the light field at a distance of $d = 50$ cm from the bulb. Assume harmonic light waves.

3.9 Consider the superposition of two harmonic oscillations with equal (real) amplitudes E_0 and frequencies ω_1, ω_2:

$$E(t) = E_0\exp(-i\omega_1 t) + E_0\exp(-i\omega_2 t) .$$

Rewrite the superposition as a product of two oscillations. Assume that the two frequencies are close, that is, $|\omega_1 - \omega_2| \ll (\omega_1 + \omega_2)/2$. Calculate the instantaneous intensity, the short-term intensity (measuring time T_m) and the intensity of the beat signal $E(z,t)$. Outline the time dependence of the short-term intensity for $T_m \ll T_b$, for $T_m \approx T_b$ and for $T_m \gg T_b$, where $T_b = 2\pi/|\omega_1 - \omega_2|$ denotes the beat period.

4. Coherence

In optics, the original sense of the word coherence was attributed to the ability of radiation to produce interference phenomena. Today, the notion of coherence is defined more generally by the correlation properties between quantities of an optical field. Usual interference is the simplest phenomenon revealing correlations between light waves.

Two limiting cases of a more general description exist: temporal and spatial coherence. This is partly due to historical reasons, partly, because these limiting cases can be well realized experimentally.

Michelson introduced a technique for measuring the temporal coherence. The instrument nowadays is called the Michelson interferometer (Sect. 4.1). Spatial coherence is best illustrated by the double-slit experiment of *Young* (Sect. 4.2).

4.1 Temporal Coherence

We consider the path of rays in a Michelson interferometer (Fig. 4.1). The light to be investigated is divided into two beams by a beam splitter. One beam is reflected back onto itself by a fixed mirror, the other one is also reflected back by a mirror, but one that can be shifted along the beam. Both reflected beams are divided again into two by the beam splitter, whereby one beam from each mirror propagates to a screen. The idea of this arrangement is to superimpose a light wave with a time-shifted copy of itself.

We now set out for a mathematical description. On the screen, we have a superposition of two wave fields, E_1 and E_2. Let E_1 be the light wave that reaches the screen via the first mirror and E_2 the one that reaches the screen via the movable mirror. Then we have at a point on the screen, when the incoming wave has been split evenly at the beam splitter,

$$E_2(t) = E_1(t+\tau), \quad \text{or} \quad E_1(t) = E_2(t-\tau). \tag{4.1}$$

The wave E_2 thus has to start earlier to reach the screen at time t because of the additional path of length $2d$. The quantity τ depends on the mirror

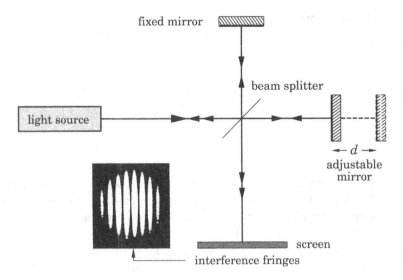

Fig. 4.1. The Michelson interferometer.

displacement d according to

$$\tau = \frac{2d}{c}. \tag{4.2}$$

On the screen, we observe the interference of both waves given by the superposition of the wave amplitudes:

$$E(t) = E_1(t) + E_2(t) = E_1(t) + E_1(t+\tau). \tag{4.3}$$

This superposition is not directly visible, but only the intensity:

$$
\begin{aligned}
I = \langle EE^* \rangle &= \langle (E_1 + E_2)(E_1 + E_2)^* \rangle \\
&= \langle E_1 E_1^* \rangle + \langle E_2 E_2^* \rangle + \langle E_2 E_1^* \rangle + \langle E_1 E_2^* \rangle \\
&= I_1 + I_2 + 2\,\mathrm{Re}\{\langle E_1^* E_2 \rangle\} \\
&= 2I_1 + 2\,\mathrm{Re}\{\langle E_1^* E_2 \rangle\}.
\end{aligned} \tag{4.4}
$$

It can be seen that the total intensity on the screen is given by the sum of the intensity I_1 of the first wave and I_2 of the second wave and an additional term, the interference term. The important information is contained in the expression $\langle E_1^* E_2 \rangle$. With $E_2(t) = E_1(t+\tau)$ this gives rise to the definition

$$
\begin{aligned}
\Gamma(\tau) &= \langle E_1^*(t) E_1(t+\tau) \rangle \\
&= \lim_{T_{\mathrm{m}} \to \infty} \frac{1}{T_{\mathrm{m}}} \int\limits_{-T_{\mathrm{m}}/2}^{+T_{\mathrm{m}}/2} E_1^*(t) E_1(t+\tau)\,\mathrm{d}t.
\end{aligned} \tag{4.5}
$$

$\Gamma(\tau)$ is called the complex self coherence function. It is the autocorrelation function of the complex light wave $E_1(t)$. For the intensity $I(\tau)$ we then get

$$I(\tau) = I_1 + I_2 + 2\,\mathrm{Re}\{\Gamma(\tau)\} = 2I_1 + 2\,\mathrm{Re}\{\Gamma(\tau)\}. \tag{4.6}$$

As an example we take the harmonic wave

$$E_1(t) = E_0 \exp(-i\omega t). \tag{4.7}$$

Then

$$\Gamma(\tau) = \lim_{T_m \to \infty} \frac{1}{T_m} \int_{-T_m/2}^{+T_m/2} E_1^*(t) E_1(t+\tau)\,dt$$

$$= \lim_{T_m \to \infty} \frac{1}{T_m} \int_{-T_m/2}^{+T_m/2} |E_0|^2 \exp(i\omega t)\exp\left[-i\omega(t+\tau)\right] dt$$

$$= |E_0|^2 \exp(-i\omega\tau) = I_1 \exp(-i\omega\tau), \tag{4.8}$$

that is, the self coherence function harmonically depends on the time delay τ. The intensity is given, with (4.6), as

$$\begin{aligned}
I(\tau) &= 2I_1 + 2\,\mathrm{Re}\{\Gamma(\tau)\} \\
&= 2I_1 + 2I_1\,\mathrm{Re}\{\exp(-i\omega\tau)\} \\
&= 2I_1 + 2I_1 \cos\omega\tau \\
&= 2I_1(1 + \cos\omega\tau). \tag{4.9}
\end{aligned}$$

The graph of $I(\tau)$ is plotted in Fig. 4.2. In a Michelson interferometer with a slightly tilted mirror it can be observed as a fringe pattern, see Fig. 4.1.

It is easy to envisage that it is possible to superimpose on the screen not two time-shifted light waves from the same source but two light waves from different sources whose coherence is to be tested. Interference experiments with lasers may lead to nontrivial results [4.1]. For such cases,

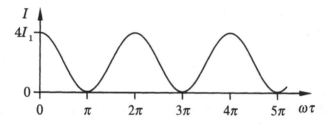

Fig. 4.2. Graph of the intensity $I(\tau)$ in a Michelson interferometer for a harmonic wave in dependence on the phase shift $\omega\tau$, where $\tau = 2d/c$, d being the mirror displacement and c being the velocity of light.

the above definition (4.5) for describing temporal coherence must be extended, leading to the definition of the cross coherence function

$$\Gamma(\tau) = \langle E_1^*(t)E_2(t+\tau)\rangle. \tag{4.10}$$

It is the crosscorrelation function of the two light waves. It is taken at one fixed location in space as is the self coherence function.

The complex self coherence function $\Gamma(\tau)$ may be normalized:

$$\gamma(\tau) = \frac{\Gamma(\tau)}{\Gamma(0)}. \tag{4.11}$$

The magnitude $\gamma(\tau)$ is called the complex degree of self coherence. Because $\Gamma(0) = I_1$ is always real and is the largest value that occurs when we take the modulus of the autocorrelation function $\Gamma(\tau)$, we have

$$|\gamma(\tau)| \leq 1. \tag{4.12}$$

The intensity $I(\tau)$ then reads

$$\begin{aligned}
I(\tau) &= 2I_1 + 2I_1 \operatorname{Re}\{\gamma(\tau)\} \\
&= 2I_1(1 + \operatorname{Re}\{\gamma(\tau)\}). \tag{4.13}
\end{aligned}$$

The functions $\Gamma(\tau)$ and $\gamma(\tau)$ are contained in the interference term coming into existence only when we take the intensity. They are not directly obtainable. It is, however, easy to determine the contrast K between interference fringes. This quantity has already been used by *Michelson* who called it visibility and defined it via the maximum and minimum intensity $I_{\max}$ and $I_{\min}$ as

$$K = \frac{I_{\max} - I_{\min}}{I_{\max} + I_{\min}}. \tag{4.14}$$

The contrast K obviously depends on the time shift τ between the light waves; that is, K is a function of τ. A precise definition of the contrast has to take into account the fact that the maximum and the minimum intensity of the interference fringes do not occur at the same time shift of the light waves (see Fig. 4.2). Let τ_1 and τ_2, $\tau_2 > \tau_1$, be the time shifts belonging to adjacent interference fringes of maximum and minimum intensity, $I_{\max}(\tau_1)$ and $I_{\min}(\tau_2)$. Then the contrast $K(\tau)$ is defined on the interval $[\tau_1, \tau_2)$ by

$$K(\tau) = \frac{I_{\max}(\tau_1) - I_{\min}(\tau_2)}{I_{\max}(\tau_1) + I_{\min}(\tau_2)}. \tag{4.15}$$

Usually $\tau_2 - \tau_1$, corresponding to half a mean wavelength, is small compared to the duration of the wave train to be investigated. Only in this case does the definition make sense. Then the contrast function $K(\tau)$ can be expressed in terms of the self coherence function $\Gamma(\tau)$.

We demonstrate the connection between $K(\tau)$ and $\Gamma(\tau)$ by way of example and use quasimonochromatic light, that is, light of relatively small bandwidth ($\Delta\omega/\omega \ll 1$). The typical dependence of the self coherence function $\Gamma(\tau)$ on the time shift τ for this case is given in Fig. 4.3. We observe that according to (4.6) the maximum intensity is attained at maximum $\text{Re}\{\Gamma(\tau)\}$, occurring at τ_1, and the minimum intensity at minimum $\text{Re}\{\Gamma(\tau)\}$, occurring at τ_2. Moreover, we see that the modulus of $\Gamma(\tau)$ practically stays constant in the interval $[\tau_1, \tau_2[$. It follows, for τ taken from this interval, that

$$\text{Re}\{\Gamma(\tau_1)\} = |\Gamma(\tau)| \qquad \text{and} \qquad \text{Re}\{\Gamma(\tau_2)\} = -|\Gamma(\tau)|. \qquad (4.16)$$

This leads to the intensities

$$I_{\max}(\tau_1) = 2I_1 + 2\,\text{Re}\{\Gamma(\tau_1)\} = 2I_1 + 2|\Gamma(\tau)|, \qquad (4.17)$$

$$I_{\min}(\tau_2) = 2I_1 + 2\,\text{Re}\{\Gamma(\tau_2)\} = 2I_1 - 2|\Gamma(\tau)|, \qquad (4.18)$$

and to the contrast function

$$\begin{aligned} K(\tau) &= \frac{2I_1 + 2|\Gamma(\tau)| - 2I_1 + 2|\Gamma(\tau)|}{2I_1 + 2|\Gamma(\tau)| + 2I_1 - 2|\Gamma(\tau)|} \\ &= \frac{4|\Gamma(\tau)|}{4I_1} = \frac{|\Gamma(\tau)|}{I_1} = \frac{|\Gamma(\tau)|}{\Gamma(0)} \\ &= |\gamma(\tau)|. \end{aligned} \qquad (4.19)$$

The contrast function thus equals the modulus of the complex degree of coherence. This is valid for two waves of equal intensity, otherwise some prefactors arise.

For quasimonochromatic light whose self coherence function slowly spirals into the origin (see Fig. 4.3) it is easily seen that a monotonously decreasing contrast function is obtained, as the modulus of $\Gamma(\tau)$ continuously decreases.

For a harmonic wave we found

$$\gamma(\tau) = \exp(-i\omega\tau). \qquad (4.20)$$

Therefore the contrast function is just

$$K(\tau) = |\gamma(\tau)| = 1. \qquad (4.21)$$

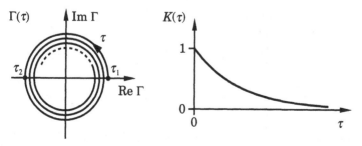

Fig. 4.3. Self coherence function $\Gamma(\tau)$ in the complex plane (*left*) and contrast function $K(\tau)$ (*right*) for quasimonochromatic light.

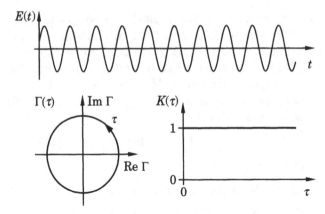

Fig. 4.4. Graph of the field amplitude E, the self coherence function Γ, and the contrast function K for completely coherent light.

A harmonic wave thus can be shifted arbitrarily in time and superimposed with itself without altering its interference properties. Light with this property is called completely coherent. This, of course, is a limiting case. It can be realized approximately, for instance, with a stabilized single-mode laser.

The graph of the contrast function can attain very different shapes. A further limiting case is completely incoherent light, characterized by $|\gamma(\tau)| = 0$ for $\tau \neq 0$ ($\gamma(0) = 1$ in all cases). The corresponding light field is made up of a mixture of light waves of all wavelengths with a statistical distribution of phases. This case, too, is realized only approximately. Good examples are daylight and incandescent light.

The two limiting cases of completely coherent and completely incoherent light are depicted in Fig. 4.4 and Fig. 4.5, respectively, with the graphs

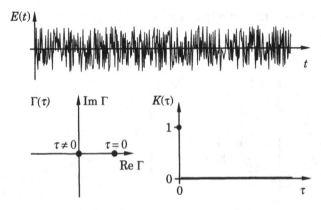

Fig. 4.5. Graph of the field amplitude E, the self coherence function Γ, and the contrast function K for completely incoherent light.

of the field amplitude versus time, the self coherence function, and the contrast function.

Light in the large range in between these two limiting cases is called partially coherent. Therefore, the following cases are distinguished ($\tau \neq 0$, $|\gamma(0)| = 1$):

$$|\gamma(\tau)| \equiv 1 \text{ completely coherent,}$$
$$0 \leq |\gamma(\tau)| \leq 1 \text{ partially coherent,}$$
$$|\gamma(\tau)| \equiv 0 \text{ completely incoherent.}$$

Many natural and artificial light sources have a monotonously decreasing contrast function; for instance, the light from a spectral lamp. Figure 4.6 shows typical graphs of the field amplitude, the self coherence function, and the contrast function for light from a mercury lamp. To characterize the decay of the contrast function, the coherence time τ_c is introduced. It is defined as the time shift when the contrast function has decayed to the value $1/e$. In optical arrangements, such as the Michelson interferometer, the time shift between the waves to be superimposed is effected by different optical path lengths. Thus, equivalently to the coherence time, the coherence length,

$$l_c = c\tau_c , \tag{4.22}$$

is used for characterizing the interference properties of light. Typical values of the coherence length are some micrometers for incandescent light and some kilometers for single-mode laser light.

The notions of coherence time and coherence length can be introduced without difficulties for all those light sources that show a monotonously decreasing contrast function (see Fig. 4.6).

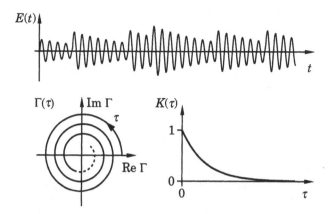

Fig. 4.6. Graph of the field amplitude E, the self coherence function Γ, and the contrast function K for light from a mercury-vapor lamp.

The decay, however, need not proceed monotonously. For instance, when we consider the superposition of two harmonic waves of different frequency, the field amplitude varies in the form of beats (Fig. 4.7). This case is approximately realized in a two-mode laser. What does the contrast function look like for this type of light? For simplicity, we consider two harmonic waves of equal amplitude:

$$E(t) = E_0 \exp(-i\omega_1 t) + E_0 \exp(-i\omega_2 t) . \tag{4.23}$$

Then, with (4.5), the self coherence function is given by

$$\Gamma(\tau) = \lim_{T_m \to \infty} \frac{1}{T_m} \int\limits_{-T_m/2}^{+T_m/2} \left[E_0^* \exp(i\omega_1 t) + E_0^* \exp(i\omega_2 t) \right] \cdot$$

$$\cdot \left(E_0 \exp\left[-i\omega_1(t+\tau)\right] + E_0 \exp\left[-i\omega_2(t+\tau)\right] \right) dt$$

$$= \lim_{T_m \to \infty} \frac{|E_0|^2}{T_m} \int\limits_{-T_m/2}^{+T_m/2} \left(\exp\left[-i\omega_1\tau\right] + \exp\left[-i\omega_2\tau\right] + \right.$$

$$\underbrace{\exp(-i\omega_1\tau)\exp\left[-i(\omega_1-\omega_2)t\right] + \exp(-i\omega_2\tau)\exp\left[-i(\omega_2-\omega_1)t\right]}_{\text{no contribution, since zero mean}} \left. \right) dt$$

$$= |E_0|^2 \left[\exp(-i\omega_1\tau) + \exp(-i\omega_2\tau) \right] . \tag{4.24}$$

With (4.11) and $\Gamma(0) = 2|E_0|^2$ we get:

$$\gamma(\tau) = \frac{1}{2} \left[\exp(-i\omega_1\tau) + \exp(-i\omega_2\tau) \right] . \tag{4.25}$$

Finally, with (4.19) we obtain:

$$K(\tau) = |\gamma(\tau)| = \frac{1}{2} \left| \exp(-i\omega_1\tau) + \exp(-i\omega_2\tau) \right|$$

$$= \frac{1}{2} \sqrt{\left[\exp(-i\omega_1\tau) + \exp(-i\omega_2\tau) \right] \left[\exp(i\omega_1\tau) + \exp(i\omega_2\tau) \right]}$$

$$= \frac{1}{2} \sqrt{2 + 2\cos(\omega_1 - \omega_2)\tau} = \frac{1}{2} \sqrt{4\cos^2 \frac{(\omega_1 - \omega_2)}{2}\tau}$$

$$= \left| \cos\left(\frac{\omega_1 - \omega_2}{2}\tau \right) \right| . \tag{4.26}$$

In this case, the contrast function $K(\tau)$ takes a periodic dependence on the time shift τ (Fig. 4.7). A coherence time or coherence length in the sense defined above does not seem meaningful as the contrast again and again attains the maximum value of unity. Here, the location of the first root or the first minimum may be taken as a measure of the coherence time or length.

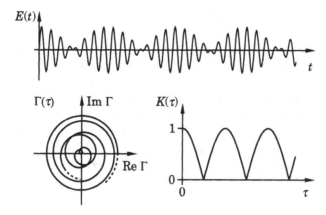

Fig. 4.7. Graph of the field amplitude E, the self coherence function Γ, and the contrast function K for light from a two-mode laser.

The result obtained for the self coherence function of two harmonic waves of different frequency can easily be extended to a sum of many harmonic waves of different frequency. Let

$$E(t) = \sum_{m=1}^{M} E_{0m} \exp(-i\omega_m t), \tag{4.27}$$

then immediately

$$\Gamma(\tau) = \sum_{m=1}^{M} |E_{0m}|^2 \exp(-i\omega_m \tau) \tag{4.28}$$

is obtained. In the limit of arbitrarily densely spaced harmonic waves we have

$$E(t) = \int_0^\infty E_0(\nu) \exp(-i2\pi\nu t)\,d\nu. \tag{4.29}$$

Then we get for the self coherence function

$$\Gamma(\tau) = \int_0^\infty |E_0(\nu)|^2 \exp(-i2\pi\nu\tau)\,d\nu$$

$$= \int_0^\infty W(\nu) \exp(-i2\pi\nu\tau)\,d\nu. \tag{4.30}$$

The function $W(\nu) = |E_0(\nu)|^2$ is the power spectrum of the complex light field [4.2].

4.2 Spatial Coherence

For many light sources the interference property of the light emitted gets lost when they are spatially extended, as with incandescent light, spectral lamps, and some lasers (ruby laser, copper-vapor laser). Extended light sources therefore are subject to limitations in optical arrangements, when interference experiments are planned. For these sources the interference of light waves coming from different points in space has to be considered. This property is kept in the notion of spatial coherence. It is best explained with the help of Young's interference experiment (see Fig. 4.8).

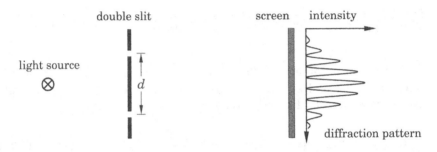

Fig. 4.8. Young's double-slit experiment.

To prepare for the discussion of the experiment we consider the interference of two spherical waves in the geometry of Fig. 4.9. Two point sources L_1 and L_2 separated by a distance d emit spherical waves that are superimposed on a screen at a distance z_L. We want to know about the fringe pattern on the screen. To this end, we consider the point P on the screen with coordinates $(x, y, 0)$. There, the spherical wave from L_1 generates the field amplitude

$$E_1(s_1, t) = A(s_1) \exp\left[i(ks_1 - \omega t + \varphi_1)\right]. \tag{4.31}$$

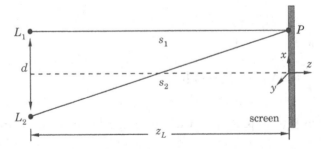

Fig. 4.9. Geometry for discussing the interference of two spherical waves.

In this equation we have expressed the complex amplitude E_0 used previously by its modulus $A(s_1)$ (real) and its phase φ_1. Similarly, the source L_2 generates the field amplitude

$$E_2(s_2,t) = A(s_2)\exp\left[i(ks_2 - \omega t + \varphi_2)\right]. \tag{4.32}$$

Then the total field at the point P is given by

$$E(x,y,0,t) = E_1(s_1,t) + E_2(s_2,t). \tag{4.33}$$

Again, we only observe the intensity $I(x,y,0)$:

$$\begin{aligned} I &= \langle EE^* \rangle = \langle (E_1+E_2)(E_1+E_2)^* \rangle \\ &= A^2(s_1) + A^2(s_2) + 2A(s_1)A(s_2)\cos(k(s_2-s_1)+\varphi_2-\varphi_1). \end{aligned} \tag{4.34}$$

This expression does not immediately reveal what the interference phenomenon on the screen looks like, since the dependence of the path difference $s_2 - s_1$ on the position of the point $P = (x,y,0)$ is not obvious. We therefore express the quantity $s_2 - s_1$ by x and y. To this end, we make a few assumptions on the geometry of the arrangement and its size in relation to the wavelength of light (paraxial approximation). First, let the distance of the two point sources be small compared to the distance z_L to the screen:

$$\frac{d}{z_L} \ll 1. \tag{4.35}$$

Second, let the observation area on the screen be limited by

$$\frac{x}{z_L} \ll 1, \quad \frac{y}{z_L} \ll 1. \tag{4.36}$$

Then the expressions for s_1 and s_2 can be approximated, and the interference pattern can be described in simple terms. With $L_1 = (d/2,0,-z_L)$ and $L_2 = (-d/2,0,-z_L)$ as coordinates of the two point sources we get for the distances s_1 and s_2:

$$s_1 = \sqrt{(x-d/2)^2 + y^2 + z_L^2}, \qquad s_2 = \sqrt{(x+d/2)^2 + y^2 + z_L^2}. \tag{4.37}$$

With the assumptions (4.35) and (4.36) the roots can be approximated:

$$\begin{aligned} s_{1,2} &= z_L\sqrt{1 + \frac{y^2}{z_L^2} + \left(\frac{x\mp d/2}{z_L}\right)^2} \\ &= z_L\left(1 + \frac{1}{2}\frac{y^2}{z_L^2} + \frac{1}{2}\frac{(x\mp d/2)^2}{z_L^2}\right). \end{aligned} \tag{4.38}$$

This yields

$$s_2 - s_1 = \frac{1}{2}\frac{xd}{z_L} + \frac{1}{2}\frac{xd}{z_L} = \frac{d}{z_L}x. \tag{4.39}$$

The difference is linear in x, the factor of proportionality being d/z_L. To further simplify the analysis we make a third assumption. Let the spher-

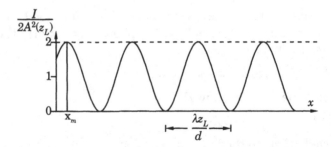

Fig. 4.10. Interference pattern obtained upon superposition of two spherical waves along the x-direction (paraxial approximation).

ical waves leaving L_1 and L_2 be of the same intensity. Then, the wave amplitudes read:

$$A(s_1) = \frac{A_0(L_1)}{s_1} = \frac{A_0}{s_1} \approx \frac{A_0}{z_L} = A(z_L), \tag{4.40}$$

$$A(s_2) = \frac{A_0(L_2)}{s_2} = \frac{A_0}{s_2} \approx \frac{A_0}{z_L} = A(z_L). \tag{4.41}$$

That is, the amplitudes of the two spherical waves are constant and equal across the screen. Introducing these amplitudes into (4.34) we get for the intensity distribution on the screen:

$$I(x,y,0) = 2A^2(z_L)\left[1 + \cos\left(\frac{2\pi d}{\lambda z_L}x + \varphi_2 - \varphi_1\right)\right]. \tag{4.42}$$

This is a fringe pattern, modulated in the x-direction, extended in the y-direction (Fig. 4.10). The fringe separation is given by

$$a = \frac{\lambda z_L}{d}. \tag{4.43}$$

The first maximum near the optical axis is shifted proportionally to the phase difference $\varphi_1 - \varphi_2$:

$$x_m = \frac{\lambda z_L}{d}\frac{\varphi_1 - \varphi_2}{2\pi} = a\frac{\varphi_1 - \varphi_2}{2\pi}. \tag{4.44}$$

Thus, the phase shift between the two spherical waves is recorded in the shift of the first maximum of the intensity distribution (Fig. 4.11).

Having treated the interference pattern of two spherical waves, we now proceed with the discussion of spatial coherence. The two point sources L_1 and L_2 of Fig. 4.11 are replaced by a stop with two small holes as illustrated in Fig. 4.12. This configuration is illuminated by an extended light source of diameter L. The light source can be thought of as being composed of single (independent) point sources each being the origin of harmonic spherical waves. We are interested in the interference pattern on the screen produced by the superposition of the waves leaving the holes L_1 and L_2.

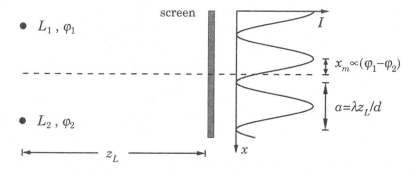

Fig. 4.11. Location of the interference fringes on the screen.

To begin with, we consider the point Q_0 of the light source on the optical axis. From this point a spherical wave is emitted that reaches the two holes L_1 and L_2 with the same phase. According to our previous analysis we get a fringe system with a maximum intensity on the optical axis. An off-axis point Q_1 generates a fringe system shifted laterally. This is immediately seen from the fact that the distances $r_1 = \overline{Q_1 L_1}$ and $r_2 = \overline{Q_1 L_2}$ are not of equal length. Therefore, a phase difference exists between the two waves originating at L_1 and L_2:

$$\varphi_1 - \varphi_2 = \frac{2\pi}{\lambda}(r_1 - r_2). \tag{4.45}$$

According to our previous analysis we get a shift of the fringes of size

$$x_{\mathrm{m}} = \frac{a}{\lambda}(r_1 - r_2). \tag{4.46}$$

Now, we let Q_0 and Q_1 emit simultaneously. When there exists a fixed phase difference between Q_0 and Q_1, we will get an interference pattern similar to that from Q_0 or Q_1, just a shifted fringe system. But when the source is "incoherent" with statistically fluctuating phase between Q_0 and Q_1, then the two wave fields superimpose in such a way that only the sum of the intensities of the two wave fields is measured. The interference terms, whose values are continuously fluctuating, have zero mean. The final result is that we have to add the intensities on the screen.

Thus, the condition for interference fringes to be visible will be that the two fringe systems generated by Q_0 and Q_1 are not shifted too far. The maximum of one system must not coincide with the minimum of the other. This requires, when only Q_0 and Q_1 radiate, that

$$|x_{\mathrm{m}}| < \frac{a}{2}. \tag{4.47}$$

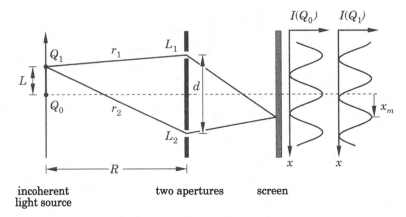

Fig. 4.12. Young's interference experiment.

Along with (4.46) this condition reads:

$$|r_1 - r_2| < \frac{\lambda}{2}. \tag{4.48}$$

This relation can be expressed in terms of the geometrical quantities of the arrangement. These are:

R = distance between the light source and the two-hole aperture,
L = lateral size of the light source,
d = distance between the two apertures.

Now we calculate, as we did before with the two spherical waves, the distance between the point Q_1 of the light source and the two apertures. For simplicity the y-coordinate is not considered. We then get (see Fig. 4.12)

$$r_{1,2} = \sqrt{(L \mp d/2)^2 + R^2} \tag{4.49}$$

and approximately

$$r_{1,2} = R \left(1 + \frac{1}{2} \frac{(L \mp d/2)^2}{R^2} \right). \tag{4.50}$$

This yields

$$|r_1 - r_2| = \frac{dL}{R}, \tag{4.51}$$

and thus with (4.48):

$$\frac{dL}{R} < \frac{\lambda}{2}. \tag{4.52}$$

Now, if all points Q_i of the extended light source emit light waves, we have to add the shifted interference patterns of all points. As the outermost points of the light source yield the largest shift of the fringes, (4.52) is valid to a good approximation also in this case. One last reasoning has to be done with respect to the lateral location of the light source. We have considered a light source extending along $\overline{Q_0 Q_1}$, that is, a source not symmetrical to the optical axis. The precise location, however, is not important for the visibility of the interference pattern since the total interference pattern is just shifted with a shift of the light source. Thus

$$\frac{dL}{R} \lesssim \frac{\lambda}{2} \tag{4.53}$$

is the condition of spatial coherence when working with extended, incoherent light sources.

Often this condition is formulated differently, not as symmetrically as in (4.53). When the angles Θ and φ are introduced according to Fig. 4.13, we can write

$$\frac{d}{R} = 2\tan\frac{\varphi}{2} \approx \varphi \approx \sin\varphi,$$

$$\frac{L}{R} = 2\tan\frac{\Theta}{2} \approx \Theta \approx \sin\Theta,$$

and get

$$L\sin\varphi \lesssim \frac{\lambda}{2} \tag{4.54}$$

$$\text{or } d\sin\Theta \lesssim \frac{\lambda}{2}. \tag{4.55}$$

These are two equivalent formulations of the spatial coherence condition, as can be seen by recourse to (4.53).

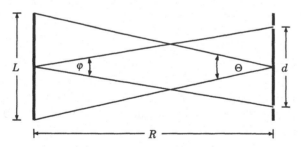

Fig. 4.13. Geometry for the spatial coherence condition.

We observe that the coherence properties of the light field are solely determined by the field at the two apertures. The remaining components are ingredients. The screen, for instance, serves for measuring the intensity of the superposition of the partial waves originating from the apertures. The superposition is done on the screen, but the wave fields of the partial waves there are the same as those at the apertures except for a time shift.

When we consider the interference pattern near the optical axis, the paths from the two apertures to the screen are of nearly equal lengths. Therefore, the pattern contains information about the similarity of the wave forms $E(r_1,t)$ and $E(r_2,t)$ at the locations r_1 and r_2 of the apertures without any time shift. This similarity is caught in the crosscorrelation function

$$\Gamma(r_1,r_2,0) = \Gamma_{12}(0) = \langle E(r_1,t)E^*(r_2,t)\rangle. \tag{4.56}$$

In this context, it is called the spatial coherence function. The two points r_1 and r_2 therein correspond to the two apertures in Young's interference experiment.

4.3 Spatiotemporal Coherence

When considering temporal coherence we compare a light wave with itself at different times. When considering spatial coherence we compare two light waves at different points in space. To describe these two cases the self coherence function and the spatial coherence function have been introduced. Both concepts, however, do not cover the interference properties of a light field in their entirety. In Young's experiment, for instance, the light paths of the waves from the two apertures reaching the screen farther away from the axis are different, that is, when the two waves are superimposed they are time shifted. To deal with this more general case, the notion of spatiotemporal coherence is introduced as an extension both to temporal and to spatial coherence. For that, the (complex) mutual coherence function is defined:

$$\Gamma(r_1,r_2,t_1,t_2) = \Gamma_{12}(t_2 - t_1) = \Gamma_{12}(\tau) = \langle E(r_1,t+\tau)E^*(r_2,t)\rangle. \tag{4.57}$$

Thereby, a stationary wave field is supposed. Then, besides on r_1 and r_2, the coherence function depends on the time shift $\tau = t_2 - t_1$ only. For $r_1 = r_2$ purely temporal coherence is obtained, for $\tau = 0$ it is purely spatial. The mutual coherence function includes the previously introduced notions as special cases:

$\Gamma_{11}(\tau)$ = self coherence function at r_1,

$\Gamma_{22}(\tau)$ = self coherence function at r_2,

$\Gamma_{12}(0)$ = spatial coherence function,

$\Gamma_{11}(0)$ = intensity at r_1,

$\Gamma_{22}(0)$ = intensity at r_2.

As with the self coherence function the mutual coherence function can

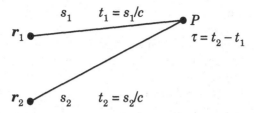

Fig. 4.14. Notation used in the derivation of Eq. (4.59).

be normalized. The quantity

$$\gamma_{12}(\tau) = \frac{\Gamma_{12}(\tau)}{\sqrt{\Gamma_{11}(0)\Gamma_{22}(0)}} \tag{4.58}$$

is called the (complex) degree of mutual coherence. Also, as with the degree of self coherence, the intensity I_P at a point P (see Fig. 4.14) can be written as

$$I_P = I_P^{(1)} + I_P^{(2)} + 2\sqrt{I_P^{(1)}I_P^{(2)}}\,\mathrm{Re}\{\gamma_{12}(\tau)\}. \tag{4.59}$$

As with the degree of self coherence, the contrast of interference fringes can be determined from the degree of mutual coherence

$$K = \frac{I_{\max} - I_{\min}}{I_{\max} + I_{\min}} = 2\frac{\sqrt{\Gamma_{11}(0)\Gamma_{22}(0)}}{\Gamma_{11}(0) + \Gamma_{22}(0)}|\gamma_{12}(\tau)|. \tag{4.60}$$

When the two interfering waves are of equal intensity, we again have

$$K = |\gamma_{12}(\tau)|. \tag{4.61}$$

The mutual coherence function and the degree of mutual coherence contain the influence of both temporal and spatial coherence.

As in the case of temporal coherence the following cases may be distinguished:

$|\gamma_{12}| \equiv 1$ (complete) coherence,
$0 \leq |\gamma_{12}| \leq 1$ partial coherence,
$|\gamma_{12}| \equiv 0$ (complete) incoherence.

When speaking of complete incoherence we have to omit the entries $\tau = 0$ and $r_1 = r_2$, as then, for a nonvanishing field, γ_{12} always equals unity.

Without proof we state some properties of the degree of mutual coherence. More details can be found in the literature (see, for instance, [4.3]).

When $|\gamma_{12}(\tau)| \equiv 1$ for all τ and for all r_1 and r_2, then the light field contains one frequency only.

When $|\gamma_{12}(\tau)| \equiv 0$, then the light field must vanish; that is, a completely incoherent light field cannot exist. This can be understood intuitively as follows. Suppose, we have a light source radiating completely incoherent light. When we choose the two points r_1 and r_2 far away from the source and at a small enough separation, then, according to the coherence condition (4.53), we can generate interference phenomena; that is, the light shows properties of coherence. This contradicts the assumption that the light field should be completely incoherent. Thus, $|\gamma_{12}(\tau)| \equiv 0$ can only be valid for a vanishing light field.

Light can become more coherent upon propagation. Coherence or incoherence therefore are not properties that can be ascribed to a light source itself without precaution. An experiment that shows no signs of interference phenomena near a given light source, will do so at a sufficiently large distance. A star for instance, a large thermal light source, yields light on the Earth that nevertheless shows interference (see Sect. 4.5 on stellar interferometry).

The degree of mutual coherence, $\gamma_{12}(\tau)$, obeys two wave equations. In Cartesian coordinates $r_j = (x_j, y_j, z_j), j = 1, 2$, they read:

$$\Delta_j \gamma_{12}(\tau) - \frac{1}{c^2} \frac{\partial^2 \gamma_{12}(\tau)}{\partial \tau^2} = 0, \tag{4.62}$$

where

$$\Delta_j = \frac{\partial^2}{\partial x_j^2} + \frac{\partial^2}{\partial y_j^2} + \frac{\partial^2}{\partial z_j^2}. \tag{4.63}$$

In this chapter we have treated the notion of coherence in the framework of classical wave theory. In the quantum theory of light we also find states of light that cannot be described classically [4.4]. The coherence properties of this "nonclassical light" are not covered by our present description.

4.4 Complex Representation of the Light Field

We consider a harmonic plane wave

$$E(r,t) = A \sin(k \cdot r - \omega t + \varphi). \tag{4.64}$$

Because of its sinusoidal space and time dependence, we could use the Euler relation (3.38) to find a natural complex representation

$$E(r,t) = E_0 \exp\left[i(k \cdot r - \omega t)\right]. \tag{4.65}$$

What does a more general wave form look like in a complex notation? We start with the Fourier representation of a given real signal $E^{(r)}(t)$:

$$E^{(r)}(t) = \int_{-\infty}^{+\infty} H(\nu) \exp(-i2\pi\nu t) d\nu. \tag{4.66}$$

As $E^{(r)}(t)$ is real, the following relation holds:

$$H(-\nu) = H^*(\nu), \tag{4.67}$$

that is, the Fourier coefficients at negative frequencies are the complex conjugate of those at positive frequencies. Thus the complete information about the signal is already contained in the positive or negative half of the spectrum.

With the harmonic wave in mind, we give the real signal $E^{(r)}(t)$ a complex representation of the form

$$\tilde{E}(t) = 2 \int\limits_0^\infty H(\nu) \exp(-i2\pi\nu t)\mathrm{d}\nu, \tag{4.68}$$

whereby the integration ranges over the positive half of the spectrum $H(\nu)$ only. The function $\tilde{E}(t)$ is called the complex analytic signal.

A quasimonochromatic wave can be written as

$$E^{(r)}(t) = E_0(t) \cos[2\pi\nu t + \alpha(t)] \tag{4.69}$$

with slowly varying amplitude $E_0(t)$ and phase $\alpha(t)$. It can be shown that the definition (4.68) is the only possible one when the envelope representation (4.69) should fluctuate as slowly as possible in the sense of least squares. Independently of this property the power of the definition shows up in the elegance of the notations and in the correspondence with the quantum mechanical description of light (see [4.5]).

The analytic signal has been defined in a way such that its real part just corresponds to the real signal given:

$$\tilde{E}(t) = E^{(r)}(t) + iE^{(i)}(t). \tag{4.70}$$

The imaginary part $E^{(i)}(t)$ contains no new information about the light field. Therefore it must be calculable from the real part $E^{(r)}$. The following relationship holds between the real and imaginary part:

$$E^{(i)}(t) = \frac{1}{\pi} \int\limits_{-\infty}^{+\infty} \frac{E^{(r)}(t')}{t'-t} \mathrm{d}t'. \tag{4.71}$$

The integral has the form of a convolution integral and is known as the Hilbert transform of the real signal. Conversely, we have

$$E^{(r)}(t) = -\frac{1}{\pi} \int\limits_{-\infty}^{+\infty} \frac{E^{(i)}(t')}{t'-t} \mathrm{d}t'. \tag{4.72}$$

The self coherence function Γ is an analytic signal (see (4.30)) as are the power spectrum and the mutual coherence function Γ_{12}.

4.5 Stellar Interferometry

The properties of the spatial coherence of light fields, in particular the fact that the coherence changes with the propagation of light, can be applied

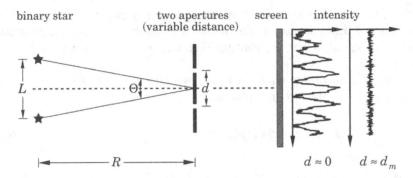

Fig. 4.15. Sketch of the arrangement for stellar interferometry.

for measurement purposes. That way, Michelson was able to determine the diameter of stars and the separation of binary stars. The principle is illustrated for the simpler example of a binary star. The binary star as a light source is analysed with a double aperture as an interference device, the same one we got acquainted with in Young's experiment with "incoherent" sources (see Fig. 4.15). Each star forms a fringe system in a sufficiently small spectral range. The two fringe systems are shifted with respect to each other. Because of the incoherence of the radiation from both sources the final result of the superposition of the light waves on the screen is as if both fringe systems had been superimposed.

The displacement depends on the aperture separation d. The total path length difference (see (4.51) and Fig. 4.12) is given by

$$\Delta r = \frac{dL}{R}.$$

This corresponds to a phase shift of

$$\Delta\varphi = \bar{k} \cdot \frac{L}{R}d, \tag{4.73}$$

where $\bar{k} = 2\pi/\bar{\lambda}$ is the mean wave number of the light, $\bar{\lambda}$ being the mean wavelength. This leads to a shift of the two fringe systems with respect to each other, which is proportional to the aperture separation d. For $d \to 0$, both stars form the same fringe system with no shift. For increasing d, finally a size d_m is reached so that the maxima of one fringe system fall onto the minima of the other fringe system. Then the contrast disappears. That is the case at $\Delta\varphi = \pi$:

$$\Delta\varphi = \pi = \frac{2\pi}{\bar{\lambda}}\frac{L}{R}d_m, \tag{4.74}$$

or, resolved for d_m, at

$$d_m = \bar{\lambda}\frac{R}{2L} = \frac{\bar{\lambda}}{2\Theta}. \tag{4.75}$$

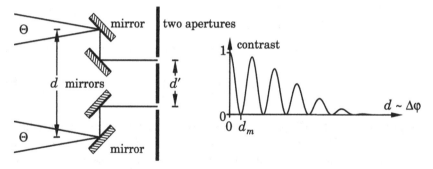

Fig. 4.16. Michelson stellar interferometer.

The quantity $\Theta = L/R$ is the angular separation of the two stars as seen from the earth.

The real arrangement looks somewhat different (Fig. 4.16). As Θ usually is very small, d_m gets very large. To arrive at comfortably observable interference fringes the aperture separation is transformed down to a (fixed) distance d'. Then, the distance d' determines the fringe separation a on the screen; the distance d determines the phase shift $\Delta\varphi$.

4.6 Fourier Spectroscopy

When discussing temporal coherence in the context of the Michelson interferometer, we considered a monofrequency wave

$$E_1(t) = E_0 \exp(-i2\pi\nu t) \tag{4.76}$$

and obtained for the output intensity in its dependence on the time shift τ (see (4.9)):

$$I(\tau) = 2I_1(1 + \cos 2\pi\nu\tau). \tag{4.77}$$

Here, $I_1 = |E_0|^2$ and $\tau = 2d/c$, where d is the mirror displacement of the movable mirror and c the velocity of light. The intensity $I(\tau)$ oscillates more or less rapidly depending on the frequency of the harmonic wave. Now we take a light field being composed of a mixture of waves of different frequencies:

$$E(t) = \int_0^\infty E_0(\nu) \exp(-i2\pi\nu t)\,d\nu. \tag{4.78}$$

At the exit of the interferometer we get the intensity function, also called the interference function,

$$I(\tau) = 2 \int\limits_0^\infty |E_0(\nu)|^2 \, (1 + \cos 2\pi\nu\tau) \, d\nu$$

$$= 2 \int\limits_0^\infty W(\nu) \, d\nu + 2 \int\limits_0^\infty W(\nu) \cos(2\pi\nu\tau) \, d\nu. \tag{4.79}$$

The quantity $W(\nu) = |E_0(\nu)|^2$ is the power spectrum of the complex light wave, as in (4.30). The first expression on the right-hand side is a constant giving the total brightness of the light source. The second, variable, expression (for brevity denoted as $\Delta I(\tau)$) is called the interferogram and is given by a Fourier transform.

This can directly be seen as follows. According to (4.78) the power spectrum $W(\nu)$ is only given for positive frequencies. By setting $W(-\nu) = W(\nu)$, it is extended to negative frequencies ($W(0)$ being counted twice for correctness). We get the result stated:

$$\Delta I(\tau) = 2 \int\limits_0^\infty W(\nu) \cos(2\pi\nu\tau) \, d\nu$$

$$= \int\limits_{-\infty}^\infty W(\nu) \exp(-i2\pi\nu\tau) \, d\nu. \tag{4.80}$$

By transforming back a measured interferogram $\Delta I(\tau)$, we obtain the power spectrum of the light source investigated:

$$W(\nu) = \int\limits_{-\infty}^\infty \Delta I(\tau) \exp(+i2\pi\nu\tau) \, d\tau$$

$$= 2 \int\limits_0^\infty \Delta I(\tau) \cos(2\pi\nu\tau) \, d\tau. \tag{4.81}$$

Thus, by using a Michelson interferometer and measuring the output intensity as a function of the mirror displacement, we can determine the power spectrum of a light wave. This principle underlies Fourier spectroscopy. As an illustration, Fig. 4.17 shows an interferogram $\Delta I(\tau)$ and its corresponding power spectrum $W(\nu)$.

Ideally, the measurement should be extended to an infinite time shift. For practical reasons, the interferometer arm length is limited, leading to a maximum shift τ_{max}. This value determines the frequency resolution $\Delta\nu_{min} \propto 1/\tau_{max}$.

Fourier spectroscopy is preferably used in the infrared spectral region. Despite the low photon energies there, fast measurements at high resolution, and at higher signal-to-noise ratio compared to wavelength-

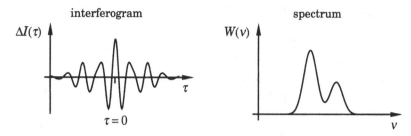

Fig. 4.17. Interferogram and the corresponding power spectrum.

selective spectrometers, can be done. The increased signal-to-noise ratio is obtained because at each sample point τ the total intensity of the incoming light is used for the measurement and not just that of a small frequency band according to the frequency resolution.

4.7 Intensity Correlation

So far we have used the correlation of the field amplitudes at two points in space with an additional time shift to describe the interference of light. One can now ask whether a light field may possess nontrivial higher correlations; that is, whether expressions relating more than two field amplitudes may give additional insight into the nature of the wave field. Indeed, higher correlation functions can be defined. From these, the intensity correlation function $K_{12}(\tau)$ has proven as important and useful. It is defined as the crosscorrelation function of the intensities $I_1(t)$ and $I_2(t)$ present at the locations r_1 and r_2, respectively:

$$K(r_1, r_2, \tau) = K_{12}(\tau) = \langle I_1(t+\tau)I_2(t) \rangle. \tag{4.82}$$

It has the advantage that it can be measured easily. Just two photon detectors are placed into the wave field at r_1 and r_2 to measure the intensities. The correlation is done electronically (Fig. 4.7). This is a modified version of the experiment of Hanbury Brown and Twiss, mentioned in chapter 2.

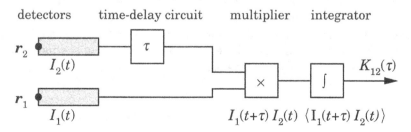

Fig. 4.18. Intensity correlation experiment of Hanbury Brown and Twiss.

Now the correlation function $\langle I_1(t+\tau)I_2(t)\rangle$ can be expressed as

$$\langle I_1(t+\tau)I_2(t)\rangle = \langle E_1(t+\tau)E_1^*(t+\tau)E_2(t)E_2^*(t)\rangle, \tag{4.83}$$

that is, we have a fourfold correlation of the wave amplitudes in the special form in which only two locations are considered. It can be shown that for statistically emitting light sources (more exactly, those with Gaussian statistics) the following relation holds:

$$K_{12}(\tau) = \langle I_1\rangle\langle I_2\rangle + |\Gamma_{12}(\tau)|^2, \tag{4.84}$$

$$\text{or}\quad K_{12}(\tau) = \langle I_1\rangle\langle I_2\rangle(1+|\gamma_{12}(\tau)|^2). \tag{4.85}$$

In this case, the quantity $K_{12}(\tau)$ is completely determined by $\Gamma_{12}(\tau)$. Conversely, measuring $K_{12}(\tau)$ opens up the possibility to determine $|\Gamma_{12}(\tau)|$. In the experiment, often only the fluctuations of the intensity $\Delta I(t)$ are considered:

$$\Delta I(t) = I(t) - \langle I\rangle. \tag{4.86}$$

Correlating only these fluctuations, we get the correlation function of the intensity fluctuations, R_{12}:

$$\begin{aligned}
R_{12}(\tau) &= \langle \Delta I_1(t+\tau)\Delta I_2(t)\rangle \\
&= \langle I_1(t+\tau)I_2(t)\rangle - \langle I_1\rangle\langle I_2\rangle \\
&= |\Gamma_{12}(\tau)|^2 \tag{4.87}
\end{aligned}$$

$$\text{or}\quad R_{12}(\tau) = \langle I_1\rangle\langle I_2\rangle|\gamma_{12}|^2. \tag{4.88}$$

Therefore, the intensity correlation function $K_{12}(\tau)$ and the correlation function of the intensity fluctuations, $R_{12}(\tau)$, may serve in determining $|\Gamma_{12}(\tau)|$ and the contrast function. This measuring technique is called correlation interferometry.

In this way, stellar diameters and the distances between binary stars can be measured, in a manner similar to that for the Michelson stellar interferometer, just much simpler and more precisely. The advantage is that no phase-sensitive adjustments are necessary and that the shift τ need not be realized by different optical path lengths but can be done by electronic means. An arrangement is depicted in Fig. 4.19.

Correlation interferometry is also used in other wavelength regions, in particular in radio astronomy [4.6]. With very long baseline interferometry (VLBI), radio images with a resolution down to 0.001 arc seconds can be obtained from radio telescopes located far away from each other on different continents. The technique has reached a level of sophistication that, conversely, even the continental drift can be measured.

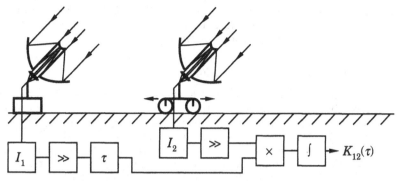

Fig. 4.19. Stellar correlation interferometer.

Problems

4.1 Two harmonic signals with frequencies ω_1 and ω_2 ($\neq \omega_1$), and amplitudes E_{01} and E_{02}, respectively, are superimposed. For the resulting signal, compute and outline the field amplitude $E(t)$ as a function of time, the self coherence function $\Gamma(\tau)$, the complex degree of coherence $\gamma(\tau)$, and the contrast function $K(\tau)$.

4.2 How has the definition of the self coherence function $\Gamma(\tau)$ to be modified to apply to signals of finite total energy? Using this altered definition, compute the complex degree of coherence $\gamma(\tau)$ and the contrast function $K(\tau)$ for a Gaussian wave packet of width σ and mean frequency ω,

$$E(t) = \frac{E_0}{\sqrt{2\pi}\sigma} \exp(-i\omega t) \exp\left(-\frac{t^2}{2\sigma^2}\right).$$

Determine the coherence time τ_c of the signal.
Hint: $\int_{-\infty}^{+\infty} \exp(-t^2/\sigma^2)\,dt = \sigma\sqrt{\pi}$.

4.3 Verify expression (4.30) for the self coherence function of a superposition of harmonic oscillations (4.29) by inserting the definition of the self coherence function. Use (A.23).

4.4 A screen with two pinholes a distance d apart is illuminated with yellow–filtered sunlight ($\lambda = 550$ nm). What is the maximum separation of the pinholes to yield interference fringes on a screen behind the apertures? The angular diameter of the sun is $\varphi \approx 30$ arc minutes.

4.5 In the arrangement of Young's experiment (Fig. 4.12) a circular light source of radius R and of constant surface brightness is used. Its distance from the aperture plane is denoted by L_1, the separation of the apertures by d. The source may be imagined as a collection of independently emitting, mutually incoherent monofrequency point sources (wavelength λ). Show that on a screen a distance L_2 behind the apertures the following diffraction pattern is observed:

$$I(\xi,\eta) \propto \pi R^2 \left[1 + 2\cos(kd\xi/L_2)\frac{J_1(kdR/L_1)}{kdR/L_1}\right]. \tag{4.89}$$

The variables (ξ,η) denote coordinates on the observation screen, with the ξ-axis running parallel to the line connecting the apertures.
Hints: $J_0(z) = (2\pi)^{-1}\int_0^{2\pi} \cos(z\cos(t))\,dt$ and $\int_a^b zJ_0(z)\,dz = zJ_1(z)|_a^b$.

4.6 For quasimonochromatic light, derive relation (4.60) from the definition of contrast. Calculate the degree of mutual coherence $\gamma_{12}(\tau)$ and the contrast function $K(\tau)$ for a monofrequency spherical wave.

4.7 Calculate the Fourier spectrum and the analytic signal of a square pulse $f(t) = \text{rect}\,(t)$ (see the Appendix on the Fourier transform).

4.8 When observing a single star with Michelson's stellar interferometer the intensity distribution (4.89) given in Problem 4.5 approximately applies. Which term of this formula contains the mirror separation d, which one the separation d' of the apertures? Upon measuring the diameter of the red giant star Betelgeuse (α Orionis), Michelson observed that the interference pattern disappeared at a mirror separation of $d = 306.5$ cm (at $\lambda = 570$ nm). Give the diameter of Betelgeuse, which is at a distance of approximately 400 light years from the earth. *Hint:* the first root of $J_1(z)$ is $z = 3.83$.

4.9 An optical correlation interferometer with two mirrors ($30\,\text{m}^2$ collecting area) and with a variable base line of 10 to 188 m is situated at Narrabri, Australia. What is the smallest angular diameter of a star that can be measured with this instrument at a wavelength of $\lambda = 550$ nm? Use (4.89).

4.10 Explain why the frequency resolution of a Fourier spectrometer is given by $\Delta \nu = 1/\tau_{\max}$. A Fourier spectrometer for the near infrared spectral range ($\lambda = 0.78$ to $3.0\,\mu$m) is required to have a resolution of $A = \lambda/\Delta\lambda = 1000$. What is the maximum time shift $\tau_{\max}$ that must be realized to achieve this resolution? How many (equally spaced) samples of the intensity must be taken in the interval $[0, \tau_{\max}]$ to compute the power spectrum correctly?

4.11 Consider the following decaying wave pulse,

$$E(t) = \begin{cases} 0 & \text{for } t \le 0, \\ \sin(\omega_0 t)\exp(-\sigma t) & \text{for } t \ge 0, \end{cases}$$

and assume $\sigma \ll \omega$. Calculate its interferogram and its power spectrum. Determine the coherence time τ_c and the band width $\Delta \nu$ (full width at half maximum of the power spectrum). What relation exists between τ_c and $\Delta \nu$?

5. Multiple-Beam Interference

When discussing the coherence properties of light, we mainly compared two light waves with each other. In the course of deriving the condition for spatial coherence, we in fact considered light sources comprising many independent point sources. Their waves, however, were incoherently summed by adding the intensities. We now consider phenomena occurring when many mutually coherent waves are superimposed: multiple-beam interference. To them belong Newton's rings, the color of thin films, Bragg reflection, and also the principle underlying interference filters. An important instrument, whose operation is based on multiple superposition of light waves, is the Fabry–Perot interferometer.

5.1 Fabry–Perot Interferometer

The operation of a Fabry–Perot interferometer has been known since more than a hundred years. But it is by no way an obsolete device. It makes use of multiple-beam interference and has many applications in modern optics. It serves as a device of extremely high resolution in spectroscopy, and, as a resonator, forms part of the laser.

A Fabry–Perot interferometer basically consists of just two plane parallel, highly reflecting surfaces together with some means to alter the distance between them (Fig. 5.1). In the form of a plane parallel glass plate made highly reflecting on both sides, it is called an etalon (Fig. 5.2).

The effect a Fabry–Perot interferometer has on a light wave comes about through the interference of the light waves that are reflected back and forth between the surfaces. To arrive at a mathematical description, we consider a plane wave propagating perpendicularly to the plane parallel mirrors, which are seperated by the distance L and have the same reflectance and the same transmittance.

We write the incoming plane wave in the complex form

$$E_e = E_e(z,t) = E_0 \exp\left(ikz - i\omega t\right) . \tag{5.1}$$

The wave reflected at one of the mirrors is denoted by E_r, the one transmitted by E_t. We do not consider the phase shifts that may occur upon

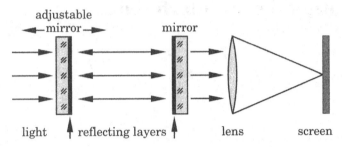

Fig. 5.1. Fabry–Perot interferometer.

reflection and transmission. They do not alter the principle of operation of the instrument. Then, the amplitude reflectance r and the amplitude transmittance t, defined by

$$r = \frac{E_r}{E_e} \quad \text{and} \quad t = \frac{E_t}{E_e}, \tag{5.2}$$

are real numbers between zero and one. Moreover, we suppose that the mirrors are nonabsorptive. Then, by the law of conservation of energy, the following relation holds:

$$r^2 + t^2 = 1. \tag{5.3}$$

With these assumptions, the electric field amplitude immediately behind the left mirror (see Fig. 5.3) after the incoming wave E_e has passed the first mirror is

$$E_t = tE_e(0,t) = tE_{e0}.$$

After passage of the light through the second mirror we have:

$$E_1 = tE_t \exp\left(ikL\right) = E_{e0}t^2 \exp\left(ikL\right). \tag{5.4}$$

Because of the optical path length L between the mirrors a phase factor $\exp\left(ikL\right)$ is introduced. Part of the wave $E_t \exp\left(ikL\right)$ incident on the

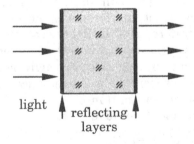

Fig. 5.2. Etalon.

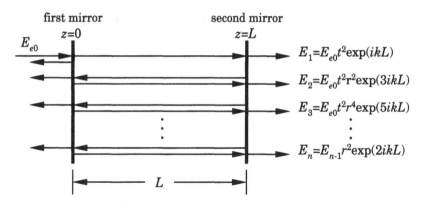

Fig. 5.3. Light waves in a Fabry–Perot interferometer.

second mirror is reflected, giving a wave

$$E_{1r} = trE_{e0}\exp\left(ikL\right)$$

traveling back to the first mirror. After additional reflection at the first mirror and passage through the second mirror, the amplitude

$$E_2 = E_{e0}t^2r^2\exp\left(ik3L\right) \tag{5.5}$$

is obtained. The output wave now has suffered two transmissions (t^2) and two reflections (r^2) and has experienced an additional phase shift $k3L$ because of the path length $3L$. From (5.4) and (5.5) the law for the partial waves being transmitted can immediately be derived. After each round trip the wave amplitude has to be multiplied by $r^2\exp\left(ik2L\right)$, giving

$$E_n = E_{n-1}r^2\exp\left(ik2L\right), \qquad n = 2,3,\ldots, \tag{5.6}$$

or, explicitly written:

$$E_n = E_{e0}t^2\exp\left(ikL\right)r^{2(n-1)}\exp\left[ik2(n-1)L\right]. \tag{5.7}$$

As we have assumed a plane wave, the partial waves E_1, E_2, E_3, ... superimpose behind the second mirror to yield the output wave amplitude E_t:

$$E_t = \sum_{n=1}^{\infty}E_n = E_{e0}t^2\exp\left(ikL\right)\sum_{n=1}^{\infty}r^{2(n-1)}\exp\left[ik2(n-1)L\right]$$

$$= E_{e0}t^2\exp\left(ikL\right)\sum_{n=0}^{\infty}r^{2n}\exp\left(ik2nL\right)$$

$$= E_{e0}t^2\exp\left(ikL\right)\sum_{n=0}^{\infty}[r^2\exp\left(ik2L\right)]^n. \tag{5.8}$$

The sum is a geometric series that can be expressed in closed form:

$$\sum_{n=0}^{\infty} q^n = \frac{1}{1-q} \quad \text{for} \quad |q| < 1.$$ (5.9)

Therefore, with

$$q = r^2 \exp(ik2L),$$ (5.10)

the output amplitude of the instrument is

$$E_t = E_{e0}t^2 \exp(ikL) \frac{1}{1 - r^2 \exp(ik2L)}.$$ (5.11)

As always, we are able to see or measure only the intensity. To simplify the notation, we introduce the phase shift δ for one round trip between the mirrors:

$$\delta = k2L.$$ (5.12)

We then get for the output intensity:

$$
\begin{aligned}
I_t = E_t E_t^* = E_{e0}E_{e0}^* t^4 & \frac{1}{(1 - r^2 \exp[i\delta])(1 - r^2 \exp[-i\delta])} \\
= I_e t^4 & \frac{1}{1 + r^4 - r^2 \exp[-i\delta] - r^2 \exp[i\delta]} \\
= I_e t^4 & \frac{1}{1 + r^4 - 2r^2 \cos\delta},
\end{aligned}
$$ (5.13)

where $I_e = E_e E_e^* = E_{e0}E_{e0}^*$ is the input intensity entering the instrument. This expression is usually given a different form. With the relation $\cos\delta = 1 - 2\sin^2(\delta/2)$ we obtain:

$$
\begin{aligned}
I_t = I_e t^4 & \frac{1}{1 + r^4 - 2r^2(1 - 2\sin^2(\delta/2))} \\
= I_e t^4 & \frac{1}{(1 - r^2)^2 + 4r^2 \sin^2(\delta/2)}.
\end{aligned}
$$ (5.14)

Now, recalling that $r^2 + t^2 = 1$ leads to:

$$
\begin{aligned}
I_t & = \frac{I_e}{1 + (2r/(1 - r^2))^2 \sin^2(\delta/2)} \\
& = \frac{I_e}{1 + K \sin^2(\delta/2)} = \frac{I_e}{1 + K \sin^2 kL}.
\end{aligned}
$$ (5.15)

The quantity

$$K = \left(\frac{2r}{1 - r^2} \right)^2$$ (5.16)

is called finesse coefficient.

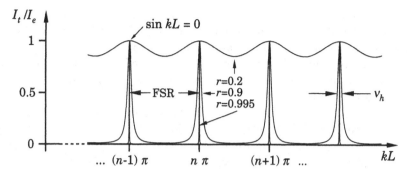

Fig. 5.4. The Airy function for different amplitude reflectances r.

The intensity transmittance T_I of the Fabry–Perot interferometer, consisting of two mirrors a distance L apart, thus is given by

$$T_I = \frac{I_t}{I_e} = \frac{1}{1+K\sin^2 kL}. \qquad (5.17)$$

This function is called the Airy function (Fig. 5.1). We observe that the function is periodic both in L and in k. For instance, if we keep L fixed and alter the wave number $k = 2\pi/\lambda$ or the wavelength λ, then we periodically observe the same intensity transmittance. In particular, because of the roots of the sine function we periodically obtain a transmission of unity, that is, complete transmission for certain wavelengths or frequencies. Therefore the instrument is wavelength selective (frequency selective) and represents an optical filter. Because of the periodicity of the transmitting regions it is a comb filter.

The complete transmission through both mirrors, when the distance between them is $n \cdot \lambda/2$, is highly surprising in view of the high reflectivity of each mirror, 99.9% say. Figure 5.5 demonstrates this peculiar behavior with two experimental arrangements. If just one mirror with a reflectance of 99.9% is placed in front of the detector, then only 0.1% of the incoming light reaches the detector. When a second mirror, again of 99.9% reflectance, is placed in between at a distance of $n \cdot \lambda/2$ from the

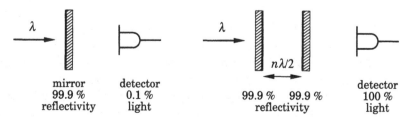

Fig. 5.5. Arrangement for demonstrating the reflection properties of one and of two mirrors.

first one, then not 0.1% of the 0.1% transmitted in the first arrangement is measured, but a full 100% of the incident light is detected, as if neither mirror were present. The actual action as a mirror depends on the experimental arrangement. Thus, an absolute value of reflectance cannot be ascribed to a mirror. A mirror can alter its reflection properties without being altered itself!

Figure 5.1 shows the dependence of the Airy function on the quantity kL for different amplitude reflectances r. For values of r near unity, sharp transmission regions are obtained. The distance between the maxima is of practical importance. The wavelength where complete transmission is observed is not unique, because every wavelength λ_n that obeys the equation

$$n\pi = k_n L = \frac{2\pi}{\lambda_n} L \quad \text{or} \quad n\frac{\lambda_n}{2} = L, \quad n \text{ integer}, \tag{5.18}$$

is completely transmitted. Thus, only the wavelength region between two successive maxima is of interest. This distance is usually given as a frequency difference $\Delta\nu$ and is called the free spectral range (FSR). Its size may be obtained in the following way. Two adjacent maxima of transmission at wave numbers k_n and k_{n+1} obey the relation (5.18) for the integers n and $n+1$, respectively. Subtracting both equations leads to

$$\pi = (k_{n+1} - k_n)L = \Delta k L = \frac{2\pi \Delta\nu}{c} L,$$

where additionally $\Delta\nu = (c/2\pi)\Delta k$ has been used. Thus, the free spectral range is given by

$$\Delta\nu = \frac{c}{2L}. \tag{5.19}$$

Two frequencies whose separation is smaller than $\Delta\nu$ can be distinguished unambiguously (within the resolving power of the instrument). The absolute wavelengths must be determined otherwise, at least to an accuracy of a free spectral range.

A second characteristic number of a Fabry–Perot interferometer is its finesse F. It indicates how many spectral lines are resolvable within a free spectral range, and is defined as

$$F = \frac{\Delta\nu}{\nu_h}, \tag{5.20}$$

$\Delta\nu$ being the free spectral range and ν_h the width at half maximum of the Airy function peaks (see Fig. 5.1). The width ν_h is defined as the full frequency width of a region, where the intensity transmittance T is larger than 1/2. As can be seen by the example with $r = 0.2$ in Fig. 5.1, ν_h and therefore also the finesse F are defined for sufficiently large r only. But just these cases are of interest.

The width ν_h depends on the parameters r and L of the interferometer in the following way. A wave number k'_h is introduced by

$$k'_h L = \frac{2\pi \nu'_h}{c} L = \frac{\pi \nu_h}{c} L, \tag{5.21}$$

where $2\nu'_h = \nu_h$; ν'_h is the frequency width from the maximum to the half maximum. Then, from (5.17), we have (T_h = half of full transmission = 1/2)

$$T_h = \frac{1}{2} = \frac{1}{1 + K \sin^2 k'_h L}. \tag{5.22}$$

For this relation to be valid requires

$$K \sin^2 k'_h L = 1.$$

For $r \lesssim 1$ the transmission regions are small and therefore $k'_h L$ is also small. Then, $\sin(k'_h L)$ can be replaced by $k'_h L$, and we get

$$K(k'_h L)^2 = 1 \quad \text{or} \quad k'_h L = \frac{1}{\sqrt{K}}.$$

Together with (5.21) we have

$$k'_h L = \frac{\pi \nu_h}{c} L = \frac{1}{\sqrt{K}} \tag{5.23}$$

or

$$\nu_h = \frac{c}{\pi L \sqrt{K}} = \frac{c(1 - r^2)}{\pi L 2 r}. \tag{5.24}$$

This is the desired relation for ν_h. Then, using the definition (5.20) together with (5.19) and (5.24), we can express the finesse F by

$$F = \frac{\Delta \nu}{\nu_h} = \frac{c \pi L 2 r}{2 L c (1 - r^2)} = \frac{\pi r}{1 - r^2}. \tag{5.25}$$

From this equation it follows that the number of resolvable spectral lines depends solely on the reflectance r of the mirrors; the larger it is, the higher the reflectance. A Fabry–Perot interferometer with $F = 100$ already is a good instrument; $F = 10000$ is possible at present and is commercially available.

A further important quantity is the resolving power A. As with every spectral instrument, it is defined by

$$A = \frac{\nu}{\nu_h}, \tag{5.26}$$

ν being the (mean) frequency and ν_h the full width at half maximum of a single transmission region, also called the instrumental linewidth. From the definition $F = \Delta \nu / \nu_h$ we have

$$\nu_h = \frac{\Delta \nu}{F}, \tag{5.27}$$

and thus obtain for the resolving power of a Fabry–Perot interferometer:

$$A = \frac{v}{\Delta v}F . \tag{5.28}$$

With this simple instrument, composed of just two plane parallel mirrors, astonishing values for the resolving power are attainable. A typical value for A is 10^8! A comparison with a grating spectrometer puts this figure into perspective. A grating with 1000 lines gives a resolving power of only 10^3 in the first diffraction order. However, the free spectral range of a Fabry–Perot interferometer is much smaller than that of a grating spectrometer.

5.2 Mode Spectrum of a Laser

A laser consists of several parts, one being a resonator to feed back part of the generated light into the amplifying medium. In the simplest case, the resonator consists of two plane parallel mirrors at a distance L. Without the laser material in between this is just the Fabry–Perot arrangement. The reflectance, however, is usually different for both mirrors, near 100% for the first one and significantly less for the second one, the output mirror (about 96% for a He–Ne laser). A resonator of this kind has a longitudinal mode spectrum similar to the line spectrum of a Fabry–Perot instrument. The spectral width of a laser line usually comprises several longitudinal modes of the laser resonator. Those modes that are sufficiently amplified by the laser medium are generated and radiated (Fig. 5.2).

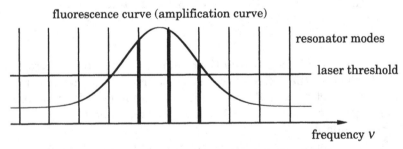

Fig. 5.6. The formation of the spectrum of laser light.

Thus, laser light gets its properties by multiple-beam interference. Several methods are available for measuring the mode spectrum of a laser. We put forward two of them: interference spectroscopy and difference-frequency analysis with fast photodiodes.

5.2.1 Interference Spectroscopy

When high resolving powers are needed in optics, it is usually the Fabry–Perot interferometer that is used. It is best suited for the determination of the mode spectrum of a laser. The arrangement is as follows (Fig. 5.7). The laser light is made slightly divergent by a lens, L_1, and enters the

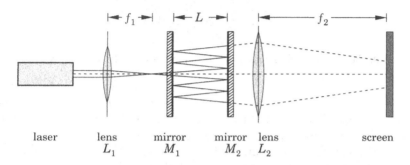

laser   lens   mirror   mirror lens   screen
          L_1 M_1 M_2 L_2

Fig. 5.7. Arrangement for the determination of laser modes with a Fabry–Perot interferometer.

Fabry–Perot interferometer, where the light is reflected back and forth between the two mirrors M_1 and M_2. The output light is collected by a lens, L_2, and directed onto a screen located in the back focal plane of the lens L_2. Thereby, all rays leaving the interferometer at the same angle are collected in one point. Why do we need lens L_1? Most lasers have an extremely small angle of divergence. It then becomes difficult to adjust the instrument to let even one laser mode pass the instrument. However, when we produce a set of oblique rays with lens L_1, different path lengths become available. At an angle of incidence α (Fig. 5.2.1) we have:

$$L_\alpha = \frac{L}{\cos \alpha}.$$
(5.29)

If the angle $\alpha = \alpha_{FSR}$ is such that

$$\frac{L}{\cos \alpha_{FSR}} = L + \frac{\lambda}{2}$$

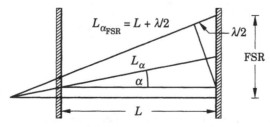

Fig. 5.8. Ring formation at the output of a Fabry–Perot interferometer.

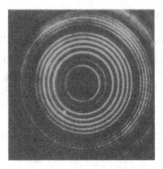

Fig. 5.9. Fabry–Perot ring system with five modes of a 15 mW He–Ne laser.

then the same wavelength λ is transmitted again. A particular wavelength λ from a divergent light beam thus leaves the instrument at discrete angles, for instance, at $\alpha = \alpha_0, \alpha_0 + \alpha_{\mathrm{FSR}}, \ldots$. Because of the rotational symmetry a ring system is obtained on the screen. If the divergent beam contains not just one but several modes, then each of them leads to a separate concentric ring system, $\alpha_\lambda, \alpha_\lambda + \alpha_{\mathrm{FSR}}, \ldots$, shifted by an angle α_λ according to the wavelength (Fig. 5.9).

The meaning of the free spectral range $\Delta\nu$ (or $\Delta\lambda$, or $\Delta\alpha$) is immediately detectable in the ring system. The separation of spectral lines can be determined only within a free spectral range. It is even necessary to know beforehand that the separation is smaller than the free spectral range to arrive at the correct value. The free spectral range can be altered by adjusting the mirror separation L. A smaller L leads to a larger $\Delta\nu$. Absolute wavelengths normally cannot be measured with a Fabry–Perot interferometer. Instead, it is very well suited for measuring difference frequencies, such as the splitting of spectral lines and differences of laser modes (Fig. 5.9) down to the MHz region. There, electronic devices take over for lower frequencies.

A different solution for observing laser modes is depicted in Fig. 5.10: the scanning Fabry–Perot interferometer. The light propagating along

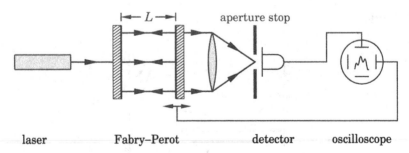

Fig. 5.10. Principle of the scanning Fabry–Perot interferometer.

the optical axis of a Fabry–Perot interferometer is collected and directed onto a photodiode. To shift different wavelengths into the transmission region, the length L between the mirrors is altered, usually piezoelectrically, and the photodiode signal (y-axis) is plotted versus the mirror separation (x-axis) on an oscilloscope. If the wavelength of the incident light is λ, then transmission is obtained at mirror separations of $\ldots, (n-1)\cdot\lambda/2, n\cdot\lambda/2, (n+1)\cdot\lambda/2, \ldots$. Again there is ambiguity because of the finite free spectral range according to $\Delta L = \lambda/2$ (Fig. 5.11). If more lines are present, they are transmitted at the corresponding lengths L_λ and form additional signals on the oscilloscope. Figure 5.12 shows three photographs taken from an oscilloscope screen. Three different He–Ne lasers with two, three, and five longitudinal modes have been analysed.

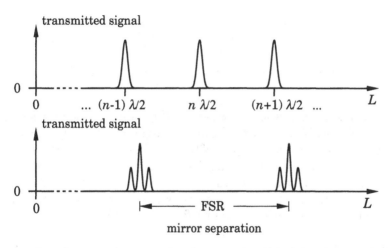

Fig. 5.11. Sketch of signals obtained with a scanning Fabry–Perot interferometer on an oscilloscope for light of one wavelength only (*above*) and for a laser with three modes (*below*). The repetition of the signal groups appears because the length L is varied over more than one free spectral range.

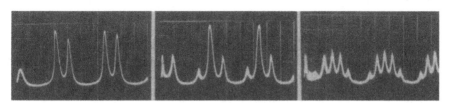

Fig. 5.12. Photographs from an oscilloscope screen showing the mode spectrum of three different He–Ne lasers with two, three, and five modes, respectively.

5.2.2 Difference-Frequency Analysis

Light intensities can be measured with a photodiode. For a large range
of intensities, the number n_i of photoelectrons is proportional to the intensity

$$n_i \propto I = |E|^2. \tag{5.30}$$

After amplification, a photocurrent is obtained that can be transformed
into a voltage U,

$$U \propto n_i \propto I, \tag{5.31}$$

and plotted on the screen of an oscilloscope (Fig. 5.2.2).

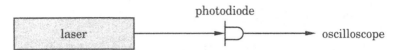

Fig. 5.13. Capturing laser light with a fast photodiode.

Photodiodes with response times in the subnanosecond region are
commercially available. The oscillation of a light wave having a frequency
of about 10^{15} Hz in the visible part of the spectrum cannot be resolved this
way, but sufficiently low difference frequencies from a mixture of waves,
as usually emitted by a laser, can be detected. The superposition of different frequencies leads to beats and thus to a time-dependent intensity
that can be followed with sufficiently fast photodiodes.

Let $E(t)$ be the field amplitude of the superposition of two waves of
circular frequencies ω_1 and ω_2:

$$E(t) = \exp(-i\omega_1 t) + \exp(-i\omega_2 t). \tag{5.32}$$

Then, the dependence of the photocurrent and therefore the voltage U on
time is

$$U(t) \propto I = 2 + 2\cos[(\omega_1 - \omega_2)t]. \tag{5.33}$$

Figure 5.14 shows the time-dependent intensity (short-term intensity)
for two He–Ne lasers of different resonator lengths. From the period of
the oscillation, the mode separation and thus the resonator length can be
obtained.

Today's electronic frequency analysers allow the immediate display of
the frequencies contained in a signal on an oscilloscope. Spectrum analysers are available for frequency ranges reaching some 10 GHz and thus
are suitable for our present purpose. Figure 5.15 shows the screen of a

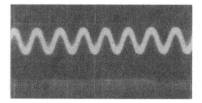 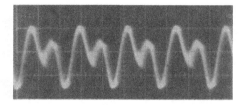

Fig. 5.14. Time-dependent intensity versus time for a 5 mW He–Ne laser (*left*) and a 15 mW He–Ne laser (*right*).

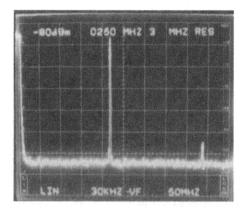

Fig. 5.15. Screen of a spectrum analyzer with an input signal from a He–Ne laser.

spectrum analyser with an input signal derived from the light of a He–Ne laser. The frequency is plotted along the horizontal axis. One horizontal division corresponds to 50 MHz. The amplitudes of the difference frequencies are plotted linearly along the vertical axis in arbitrary units. At the left margin, the zero frequency line is visible, added by the instrument. A strong difference frequency shows up at about $\nu = 220$ MHz, and a weak line at twice the frequency, 440 MHz. The laser thus emits three modes and has a resonator length of $L = c/2\Delta\nu = 68$ cm.

5.3 Dual-Recycling Interferometer

The Michelson interferometer has been used to show that the propagation velocity of light does not depend on the velocity of the light source itself and thereby initiated the theory of special relativity. Light again was used to prove the theory of general relativity through the deflection of light in the gravitational field of the sun. Gravitational fields even can form lenses for the light from stars or distant galaxies. These are stationary gravitational fields. The theory of general relativity, however, also predicts the existence of gravitational waves, that is, of propagating disturbances in

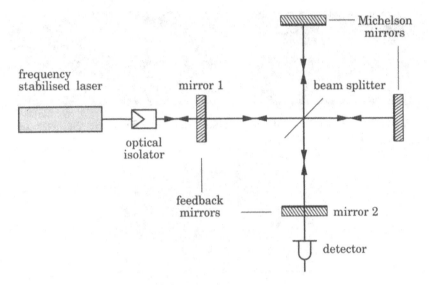

Fig. 5.16. Dual-recycling interferometer.

space–time. A direct measurement on earth has not yet been successfully performed, as the present gravitational wave detectors are not sensitive enough. A gravitational wave should alter the distances between objects, therefore interferometry suggests itself for measuring them, for instance, with the help of the Michelson interferometer. It transforms a distance variation from one arm of the interferometer into a phase shift of a light wave and, with the help of a reference wave from the other arm, into an intensity. In its basic configuration, where the phase shifted wave and the reference wave are superimposed only once, the Michelson interferometer is not sensitive enough. We know from the Fabry–Perot interferometer that the sensitivity can be improved substantially when multiple-beam interference is used. The question is, whether the increased sensitivity of the Fabry–Perot interferometer may be transferred to the Michelson interferometer. This is indeed possible. The Michelson interferometer has to be augmented by two additional mirrors placed at the entrance and exit of the instrument to "recycle" the (monofrequency) laser light to keep it for multiple interference (Fig. 5.16). We then have two crossed, coupled Fabry–Perot interferometers.

At first sight, it does not seem obvious that a mirror at the entrance of the interferometer should increase the sensitivity of the instrument, and even less obvious that a mirror in front of the detector should also increase the sensitivity. To understand this we need a closer look at some properties of the Michelson and the Fabry–Perot interferometer. We start with the Michelson interferometer. Suppose we use a monofrequency light wave of wavelength λ and shift one mirror by $\lambda/4$ from the balanced position. Then, the additional light path amounts to $\lambda/2$, and the detector gets no light, as both waves interfere destructively. What happens to the energy from the two waves? It is reflected back into the laser. The reflection of the incoming wave can be shown experimentally by placing a beam splitter in front of the interferometer to deflect part of the reflected light. Then a maximum in the amount of reflected light is measured when the detector at the output gets no light. A Michelson interferometer in this state of operation acts as a 100% reflector.

This reflector is now used as one mirror of a Fabry–Perot interferometer. The second mirror is placed in front of it and serves as the "recycling" mirror at the input of our Fabry–Perot interferometer (mirror 1 of Fig. 5.16). Control electronics with piezoelements are used to ensure that the Michelson interferometer stays in the state of 100% reflectance for the laser wavelength used. Then the sensitivity of the Fabry–Perot interferometer is transferred to the Michelson interferometer. The same argument can be used a second time for the light that has been generated by the action of a gravitational wave, leading to modulation of the incident laser light.

Systems with "dual recycling", as shown in Fig. 5.16, are presently being developed. Difficulties arise with the long interferometer arms with respect to vibration isolation, active control of the mirrors, control of heat expansion and contraction, etc. Nevertheless, astrophysicists hope to open up a new window to the universe with this instrument.

Problems

5.1 A Fabry–Perot interferometer has a finesse of $F = 10000$ and a mirror separation of $L = 5$ cm. Calculate the mirror recflectance r, the free spectral range $\Delta\nu$, the width ν_h of a spectral line and the resolution A of this instrument at $\lambda = 500$ nm. Can it be used to resolve two spectral lines at 516.7 nm and 516.9 nm unambiguously?

5.2 Calculate the amplitude reflectance R of a Fabry–Perot interferometer in a way analogous to the derivation of the transmittance T given in the text (sum over partial waves). Verify that energy conservation holds: $|T|^2 + |R|^2 = 1$.
Hint: the phase of the partial wave reflected back at the entrance mirror is shifted by π with respect to all remaining partial waves that contribute to the reflected wave.

5.3 Determine the finesse of a Michelson interferometer.

5.4 A Fabry–Perot interferometer with finesse $F = 1000$ is tuned for maximum transmission at the wavelength $\lambda = 632.8$ nm. By what amount has the mirror separation to be changed for the output intensity to drop to one half of its maximum value? How large is the corresponding displacement for a Michelson interferometer?

5.5 A plane, harmonic light wave with intensity I_e is incident on a Fabry–Perot interferometer parallel to the optical axis. Calculate the wave field and the energy density within the resonator for both minimum and maximum transmission T. To do so, represent the field in the resonator cavity as a superposition of a forward and a backward traveling harmonic wave, and use Stokes' formulae $r' = -r$ and $tt' = 1 - r^2$. Here, r and r' are the amplitude reflectances upon reflection of a wave coming from outside the instrument and from within the instrument, respectively; t and t' are the corresponding amplitude transmittances.

5.6 Determine the frequency separation of the longitudinal modes of a He–Ne laser ($\lambda = 632.8$ nm) having a resonator length of 60 cm. Repeat this calculation for a semiconductor laser having a resonator length of 0.5 mm, and an index of refraction of $n = 3.6$, emitting at $\lambda = 1.55$ μm. The light of the He–Ne laser is analyzed by using a fast photodiode with linear characteristic, $U \propto I(t)$; $I(t)$ being the short-term intensity. Assume that in the He–Ne laser three adjacent modes are excited with an amplitude ratio of $1:2:1$. Give the frequencies and relative amplitudes of the harmonic components of the photocurrent.

6. Speckles

When laser light strikes a white wall or just some surface in space, a speckle pattern is observed. An example is given in Fig. 6.1. As a similar phenomenon is not observed with incandescent light or light from a spectral lamp, it must be a special property of laser light. Indeed, it is the special coherence properties of laser light that lead to the appearance of speckle patterns.

Fig. 6.1. A speckle pattern from an argon ion laser, as observed in the laboratory.

The speckle pattern is seen sharply when the eye is focused on any plane in space. Why? The light scattered from the rough surface of the wall sets up a complicated, standing wave field in space because of the high coherence of the light. By interference of the light waves coming from different points on the wall we have statistically distributed field amplitudes and thus also field intensities at different points in space. Hence, the speckle phenomenon is brought about by statistical multiple-beam interference. The mathematical description is somewhat involved. For instance, the central limit theorem of probability theory is needed to arrive at the intensity statistics of a speckle pattern. Speckle sizes, on the other hand, can be roughly estimated by elementary considerations.

6.1 Intensity Statistics

When looking at a speckle pattern, we notice that darker parts obviously occur more often than bright spots. This observation gives rise to the question: what is the probability for a certain value of intensity to be present at any point in the speckle pattern?

In the following, we outline the steps leading to the intensity statistics of a speckle pattern. To this end, we take a coherent scalar light field and assume that the speckle pattern is generated by scattering of the wave at a rough surface. Each point of the surface is a source of spherical waves, whose phase varies statistically along the rough surface. The coherence implies that any two points have a fixed phase difference.

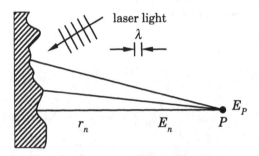

Fig. 6.2. Laser light being scattered at a rough surface.

At a point P in space (Fig. 6.2), the individual field amplitudes,

$$E_n = \frac{A_n}{r_n} \exp(ikr_n),$$ (6.1)

sum to the total amplitude

$$E_p = \sum_n \frac{A_n}{r_n} \exp(ikr_n).$$ (6.2)

The summation can be viewed as a two-dimensional random walk in the complex plane. Figure 6.3 illustrates the erratic motion of the field amplitudes in the complex plane during superposition of the individual partial waves.

For processes of this kind, the joint probability densities can be given for the real and imaginary parts. From them the probability density $p(I)$ for the intensity may be obtained [6.1]. Somewhat lengthy calculations lead to a negative exponential distribution for $p(I)$:

$$p(I) = \frac{1}{\langle I \rangle} \exp\left(-\frac{I}{\langle I \rangle}\right) \qquad \text{for } I \geq 0.$$ (6.3)

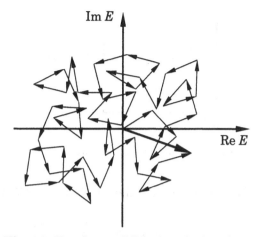

Fig. 6.3. "Random walk" in the complex plane.

This is illustrated in Fig. 6.4. The interesting result, validating our first visual impression, is that intensities near zero occur most often.

For the phase Θ of $E_p = |E_p|e^{i\Theta}$ the probability density

$$p(\Theta) = \frac{1}{2\pi} \qquad \text{for} -\pi \leq \Theta < \pi \tag{6.4}$$

is found. The phase thus is homogeneously distributed (Fig. 6.4).

An important quantity is the standard deviation $\sigma(I)$ of the distribution $p(I)$ in polarized speckle patterns. One obtains

$$\sigma_I^2 = \langle I^2 \rangle - \langle I \rangle^2 = \langle I \rangle^2 \tag{6.5}$$

or

$$\sigma_I = \langle I \rangle. \tag{6.6}$$

The standard deviation thus equals the mean intensity. Expressed differently, the contrast is unity, or the speckle pattern is fully modulated. For

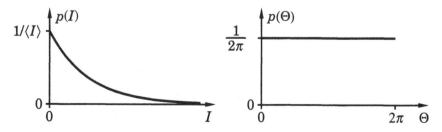

Fig. 6.4. Intensity distribution (*left*) and phase distribution (*right*) in a monofrequency speckle pattern.

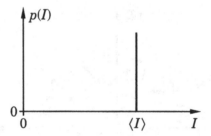

Fig. 6.5. Intensity distribution of scattered white light.

comparison, we give the probability density function $p(I)$ for white light (Fig. 6.5):

$$p(I) = \delta(I - \langle I \rangle), \qquad (\delta = \text{Dirac delta function}). \qquad (6.7)$$

In this case, the same intensity is measured at each point. Quite obviously, the intensity statistics of coherent light (laser light) and of incoherent light (incandescent light) are strongly different.

6.2 Speckle Sizes

In a speckle pattern, "grains" of higher intensity can be spotted. They alter their sizes in a peculiar way, becoming smaller when the eye approaches the scattering surface. Normal objects behave the other way round! This gives rise to the question of their "real" sizes. The answer is at our disposal when we remember Young's interference experiment.

In this context, we change the notation used in Sect. 4.2. The diameter of the area illuminated is called D (instead of d), the distance from this area to the screen is called L (instead of z_L), and the separation of the fringes is d (instead of a) (Fig. 6.6). Then, if we take two extreme points of

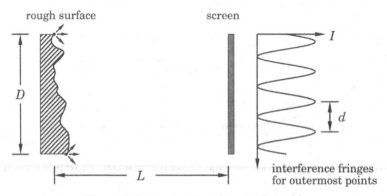

Fig. 6.6. Grain size for "objective" speckles.

the area illuminated, we obtain for the fringe separation d on the screen (refer to (4.43)):

$$d = \frac{\lambda L}{D}. \tag{6.8}$$

Thus, for a given arrangement (λ = const, L = const), we have $d \propto 1/D$; that is, the fringe separation gets larger when the two points approach each other. Hence, the fringe separation d as given in Fig. 6.6 is the smallest speckle size possible. Since light is radiated from the whole area and not just from the two extreme points, this will be an approximation only, so that $d \approx \lambda L/D$. In the case of a spherical area of diameter D, a correction factor of 1.2 (as with the resolving power of the microscope) is obtained:

$$d = 1.2\frac{\lambda L}{D}. \tag{6.9}$$

Speckles of this kind are also called objective speckles, as they are present without further imaging, for instance by the human eye. This notation calls for another type of speckles: "subjective" speckles. Indeed, every imaging system alters the coherent superposition of waves and thus produces its own speckle size. When looking, we use the eye lens to project an image onto the retina, and thus always "see" speckles subjectively. The different quality of objective and subjective speckles is best noticed when projecting objective speckles with large speckle size on a screen and looking at them.

Objective speckles can only be determined instrumentally. On the other hand, subjective speckles can be interpreted as a special kind of objective speckles. Any imaging with coherent light introduces them. To show this, we take an imaging geometry as depicted in Fig. 6.7. The simplest way to estimate the speckle size in this case is as follows. Waves of virtually all directions pass the aperture D of the lens. These are mutually coherent, but, coming from a rough surface, neighbouring waves have

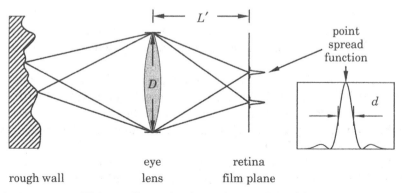

Fig. 6.7. Grain size for "subjective" speckles.

a statistical, although fixed, phase with respect to each other. Therefore the aperture D can be viewed as a rough, radiating surface, and (6.9) can be used as before, giving

$$d = 1.2\frac{\lambda L'}{D}. \tag{6.10}$$

The image distance L' can be expressed by characteristic numbers of the lens. For the case of a distant rough wall, whose image is formed in the back focal plane, we have $L' = f$, f being the focal length of the lens. With the F number of the lens,

$$F = \frac{f}{D}, \tag{6.11}$$

we then get, from (6.10),

$$d = 1.2\lambda F. \tag{6.12}$$

It is seen that d corresponds to the smallest resolvable distance, the resolving power of the optical arrangement. When the wall is not infinitely distant, L' does not equal the focal length f. With the help of the lens equation

$$\frac{1}{Z} + \frac{1}{L'} = \frac{1}{f} \tag{6.13}$$

(Z = object distance, L' = image distance, f = focal length of the lens), and by introducing the magnification $V = L'/Z$, we obtain the image distance:

$$L' = f(V + 1). \tag{6.14}$$

Inserting this expression into (6.10) together with the F number (6.11), we find the speckle size:

$$d = 1.2(1 + V)\lambda F. \tag{6.15}$$

In the case of 1:1 imaging ($V = 1$) we obtain

$$d = 1.2 \cdot 2 \cdot \lambda F.$$

This corresponds to the fact that in the case of 1:1 imaging the image distance L' equals $2f$; that is, the image plane is located two focal lengths away from the lens.

A mixture of objective and subjective speckles is obtained when a coarse speckle pattern is projected onto a screen and is viewed with the eye or is photographed. Then, immersed in the bright objective speckles smaller subjective speckles can be seen. An arrangement for visualizing this effect is given in Fig. 6.8.

The arrangement of Fig. 6.9 may serve for demonstrating the quantitative dependence of speckle size on the geometrical relations. With the help of a ground glass plate, a bright speckle field is projected into space.

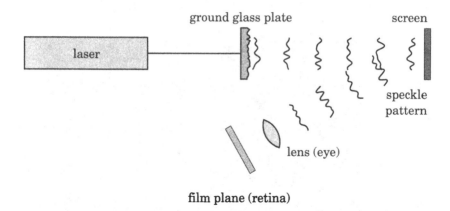

Fig. 6.8. Arrangement for viewing objective and subjective speckles simultaneously.

It is directly captured on photographic film with a camera back that can be moved along the optical axis (alteration of the parameter L in (6.8)). For varying the size of the illuminated area on the ground glass, a lens with a focal length of 11 cm is used that again can be moved along the optical axis (alteration of the parameter D in (6.8)). Figure 6.10 shows four speckle patterns taken with this arrangement for different parameter settings ($a = 11$ cm and 16 cm, $L = 33$ cm and 84 cm). Distinctly different patterns are obtained, demonstrating how the pattern becomes finer with smaller distances between the film plane and the ground glass plate and also with larger areas of the illuminated ground glass. At $a = 11$ cm, the ground glass plate is located in the focal plane of the lens. Then the spot size is smallest and, according to (6.8), the speckles are largest. This observation allows for a quick determination of the focal length of a lens. One example where the speckle size is important is holography, especially when lenses are used for magnification and ground glass plates for diffuse illumination of a scene. Speckles, despite being of size corresponding to the resolving power, lead to a substantial deterioration in a coherent

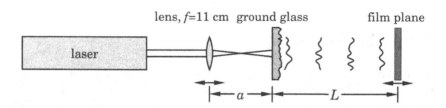

Fig. 6.9. Arrangement for the generation of bright speckles of different size by shifting the lens and the film plane.

	$L = 84\,\text{cm}$	$L = 33\,\text{cm}$
$a = 11\,\text{cm}$		
$a = 16\,\text{cm}$		

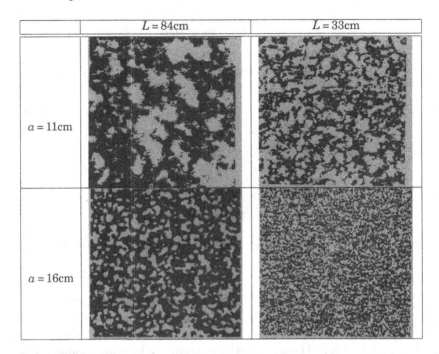

Fig. 6.10. Examples of speckle patterns taken with the arrangement of Fig. 6.9. The distances a and L are defined there.

image. The developement of holographic microscopy, which would have the advantage of a large depth of field, has been hampered by the speckle phenomenon.

Speckles are also formed when coherent light passes a turbulent air layer. Small fluctuations of density and thus of the refractive index in the atmosphere, for instance, deform the wave fronts that reach us from a star and lead to a temporally fluctuating speckle field. This phenomenon is directly visible as scintillation of the stars. This is the reason why the resolving power of conventional optical telescopes typically is limited to about one arc second regardless of the mirror diameter. Planets show almost no scintillation. The reader should be able to find out why.

6.3 Speckle Photography

Usually, the speckle pattern generated upon the diffuse reflection of laser light is a disturbing and therefore unwanted effect. However, as it shifts together with the reflecting object, it can be used for interferometry. The speckle pattern arises as a section cut through a spatial interference pattern. In the main propagation direction of the reflected light the speckle pattern alters slowly. Hence, it is quite insensitive to object motions in

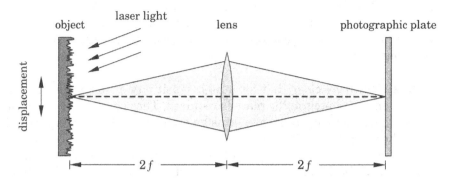

Fig. 6.11. Arrangement for taking a specklegram with 1:1 imaging.

this direction. Motions in the object plane, on the contrary, lead to rapid changes in the speckle field due to its smaller lateral speckle sizes. Therefore, the measurement of small in-plane displacements is the main application of speckle photography.

The experimental arrangement for taking a specklegram is quite simple. Just the photographic image of the object illuminated with laser light has to be taken on a sufficiently fine-grained film (Fig. 6.11). On the photographic plate an image of the object is formed covered with a speckle pattern, the speckle size being given by the aperture of the lens. In the double-exposure technique two exposures are taken, the object being shifted between the first and the second exposure. When the shift is not too large, the photographic plate records two displaced, but otherwise identical, speckle patterns.

The analysis of the specklegrams is done in Fourier space or by spatial correlation. To obtain the spatial Fourier spectrum, the specklegram is illuminated with the unexpanded beam of a laser, such as a He–Ne laser (Fig. 6.12). A lens behind the specklegram performs a Fourier transform

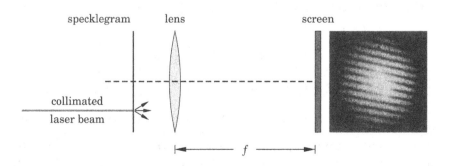

Fig. 6.12. Local optical Fourier transform of the transmission variation on a specklegram.

of the small, sampled area (see the chapter on Fourier optics and the Appendix). In the case of a double-exposure specklegram the spatial Fourier spectrum consists of a fringe system, Young's fringes. The fringe direction and fringe separation indicate the direction and amount of the shift of the object.

It is intuitively clear that for a high sensitivity a fine-grain speckle pattern on the photographic plate is required. The size d of the smallest speckle in an imaging arrangement is given in (6.15). For a lens with an F number of 2, a wavelength of $\lambda = 0.6\,\mu m$, and a 1:1 imaging geometry we obtain a typical speckle size of about $3\,\mu m$. Thus, with the optical Fourier transform method, displacements of $10\,\mu m$ to $30\,\mu m$ can easily be determined. No microscope is involved to reach this resolution.

6.3.1 Double-Exposure Technique

After double exposure, the photographic plate has the following transmission distribution after processing (see the chapter on holography):

$$T(x,y) = a - \frac{bt_B}{2}[I(x,y) + I(x+\Delta x,y)].\qquad(6.16)$$

Here, a and b are the ordinate section and the local slope of the photographic characteristic curve at the working point, respectively, and $t_B/2$ is the exposure time for each of the two exposures. The intensities $I(x,y)$ and $I(x+\Delta x,y)$ denote the nonshifted and shifted speckle patterns, respectively. For the sake of simplicity, we have assumed that the object is shifted in the x-direction only. The displacement Δx of the speckles on the photographic plate is connected with the original displacement Δs by $\Delta x = V\Delta s$, V being the magnification. To determine Δx, the plate is illuminated with a plane wave of amplitude E_0. Then, behind the plate we have the amplitude distribution

$$E(x,y) = E_0 T(x,y).\qquad(6.17)$$

The Fourier transform yields

$$\begin{aligned}
\mathcal{F}[E(x,y)](\nu_x,\nu_y) &= \mathcal{F}[E_0 T(x,y)](\nu_x,\nu_y) \\
&= \mathcal{F}\left[E_0\left(a - \frac{bt_B}{2}\left[I(x,y) + I(x+\Delta x,y)\right]\right)\right](\nu_x,\nu_y) \\
&= E_0 a\delta(\nu_x,\nu_y) \\
&\quad - E_0\frac{bt_B}{2}\mathcal{F}[I(x,y) + I(x+\Delta x,y)](\nu_x,\nu_y),\qquad(6.18)
\end{aligned}$$

where ν_x and ν_y are the spatial frequencies in the Fourier plane (see the chapter on Fourier optics and the Appendix). Making use of the shift theorem of the Fourier transform,

$$\mathcal{F}\left[I(x+\Delta x,y)\right](\nu_x,\nu_y) = \exp(2\pi i\nu_x\Delta x)\mathcal{F}\left[I(x,y)\right](\nu_x,\nu_y),\qquad(6.19)$$

we get

$$\mathcal{F}[E](v_x, v_y) = E_0 a \delta(v_x, v_y) - E_0 \frac{bt_B}{2} \mathcal{F}[I](v_x, v_y)[1 + \exp(2\pi i v_x \Delta x)]. \quad (6.20)$$

When looking at or recording this distribution, we can observe only the intensity. This yields, for $(v_x, v_y) \neq (0,0)$,

$$|\mathcal{F}[E]|^2 = |E_0|^2 b^2 \frac{t_B^2}{4} |\mathcal{F}[I](v_x, v_y)[1 + \exp(2\pi i v_x \Delta x)]|^2$$

$$\propto |\mathcal{F}[I](v_x, v_y)|^2 \cos^2 \pi v_x \Delta x. \quad (6.21)$$

The intensity of the Fourier transform $|\mathcal{F}[I(x,y)](v_x, v_y)|^2$ of a speckle pattern $I(x,y)$ is a speckle pattern. Hence, we have a speckle pattern in the Fourier plane, but now modulated by $\cos^2 \pi v_x \Delta x$; that is, we have a fringe system. From the direction and the distance of the fringes the displacement of the object can be determined.

The fringes are always oriented normal to the direction of the displacement of the object. Thereby, the positive x-direction and the negative x-direction cannot be distinguished. The separation of two fringes, Δv_x, may be obtained from the distance of two maxima of the intensity, for instance. From

$$\cos^2 \pi v_x \Delta x = 1, \quad \text{or} \quad \pi \Delta v_x \Delta x = \pi, \quad (6.22)$$

we find

$$\Delta x = \frac{1}{\Delta v_x}. \quad (6.23)$$

The spatial frequency difference Δv_x is connected with the fringe separation Δu in the back focal plane of the lens used to perform the Fourier transform via

$$\Delta u = \lambda f \Delta v_x, \quad (6.24)$$

f being the focal length of the lens (see the chapter on Fourier optics). By measuring Δu, the displacement Δx of the speckles on the photographic plate can be determined:

$$\Delta x = \frac{1}{\Delta v_x} = \frac{\lambda f}{\Delta u}. \quad (6.25)$$

With the magnification V upon exposure, we obtain the original object displacement

$$\Delta s = \frac{\lambda f}{V \Delta u}. \quad (6.26)$$

Figure 6.13 shows the experimental arrangement for demonstrating the magnitude and direction of displacement of an object: a tilting table. The table carries a metal plate that can be tilted by a screw (Fig. 6.14).

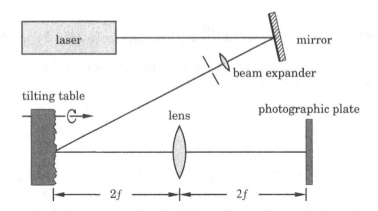

Fig. 6.13. Experimental arrangement for taking a double-exposure specklegram of a tilting table.

The photographic plate is exposed twice. Between the two exposures the metal plate is slightly rotated together with the upper plate of the tilting table by turning the screw. After processing, the photographic plate is illuminated at the four points indicated in Fig. 6.14 with the unexpanded beam of a He–Ne laser. The diffraction pattern (Fourier transform) is projected onto a white screen. A small hole punched into the screen lets the zero diffraction order pass so as not to overexpose the fringe system

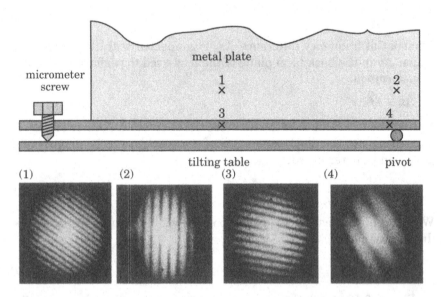

Fig. 6.14. Object used for illustrating the double exposure technique (tilting table with metal plate) and four points where the specklegram has been evaluated to obtain the corresponding fringe patterns.

by the bright central spot (Fig. 6.14). As expected, Young's fringes become denser for larger displacements Δs occurring at larger distances from the turning point of the tilting table. Also, they are always oriented normal to the direction of the displacement.

6.3.2 Time-Average Technique

In the double-exposure technique, two shifted speckle patterns are recorded with the shift possibly being space variant. The technique can be extended to regularly moving, in particular, harmonically oscillating objects. The photographic plate then records a time-dependent speckle pattern. For the sake of simplicity, we again take a one-dimensional motion in the x-direction. After the exposure and processing of the photographic plate, we have a transparency with the transmission distribution

$$T(x,y) = a - b \int_0^{t_B} I(x,y,t)\,dt. \tag{6.27}$$

It is obtained by summing the instantaneous intensities. For an object oscillating harmonically in the x-direction the integral can be solved analytically. Writing for the oscillation

$$s(t) = d \sin \Omega t, \tag{6.28}$$

the transmission distribution (6.27) attains the form

$$T(x,y) = a - b \int_0^{t_B} I(x+d\sin\Omega t,y)\,dt. \tag{6.29}$$

When we illuminate the transparency at the location x with a monofrequency wave of amplitude E_0, we have the amplitude distribution $E(x,y) = E_0 T(x,y)$. As before, we look at the Fourier transform to obtain information about the local motion of the object. With the shift property (see (6.19))

$$\mathcal{F}\left[I(x+d\sin\Omega t,y)\right](v_x,v_y) = \exp\left[2\pi i v_x d \sin \Omega t\right] \mathcal{F}\left[I(x,y)\right](v_x,v_y) \tag{6.30}$$

and an exposure time $t_B = n2\pi/\Omega = nT_S$, T_S being the oscillation period of the object, we get:

$$\begin{aligned}
\mathcal{F}\left[E\right](v_x,v_y) &= \mathcal{F}\left[E_0 T(x,y)\right](v_x,v_y) \\
&= E_0 a \delta(v_x,v_y) - bE_0 \int_0^{t_B} \mathcal{F}\left[I(x+d\sin\Omega t,y)\right](v_x,v_y)\,dt \\
&= E_0 a \delta(v_x,v_y) - bE_0 n \mathcal{F}\left[I(x,y)\right](v_x,v_y) \cdot \\
&\quad \cdot \int_0^{T_S} \exp\left[2\pi i v_x d \sin \Omega t\right]\,dt. \tag{6.31}
\end{aligned}$$

The integral is given essentially by the Bessel function J_0, defined as

$$J_0(\alpha) = \frac{1}{2\pi} \int_0^{2\pi} \exp(i\alpha \sin t) \, dt. \tag{6.32}$$

Then (6.31) yields

$$\begin{aligned}
\mathcal{F}\left[E\right](\nu_x, \nu_y) &= \mathcal{F}\left[E_0 T(x,y)\right](\nu_x, \nu_y) \\
&= E_0 a \delta(\nu_x, \nu_y) \\
&\quad - b E_0 n T_S \mathcal{F}\left[I(x,y)\right](\nu_x, \nu_y) J_0(2\pi \nu_x d).
\end{aligned} \tag{6.33}$$

For $(\nu_x, \nu_y) \neq (0,0)$, we obtain for the intensity:

$$\left|\mathcal{F}\left[E\right]\right|^2 (\nu_x, \nu_y) \propto \left|\mathcal{F}\left[I(x,y)\right]\right|^2 J_0^2(2\pi \nu_x d). \tag{6.34}$$

This is a speckle pattern, $\left|\mathcal{F}\left[I(x,y)\right]\right|^2$, modulated by the square of the Bessel function, J_0^2. Figure 6.15 shows the graph of the function $J_0^2(\alpha)$. The function has roots at $\alpha \approx 2.405$, 5.52, ... They are not equidistant. By determining the roots of (6.34) in the image, the oscillation amplitude

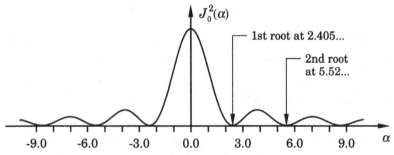

Fig. 6.15. Graph of the function $J_0^2(\alpha)$.

d of the oscillating object can be obtained. From

$$\alpha_0 = 2\pi \nu_{x_0} d = 2.4 \tag{6.35}$$

and with $\nu_{x_0} = u_0/\lambda f$, we find

$$d = \frac{2.4}{2\pi \nu_{x_0}} = \frac{2.4 \lambda f}{2\pi u_0}. \tag{6.36}$$

The quantity u_0 is the distance from the optical axis to the first root of the Bessel function modulating the speckle pattern. When a magnification V is used in taking the specklegram, the oscillation amplitude is determined by

$$d = \frac{1}{V} \frac{2.4 \lambda f}{2\pi u_0}. \tag{6.37}$$

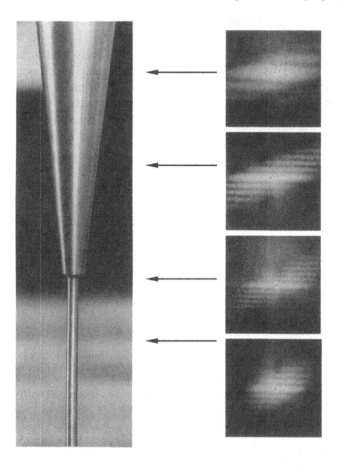

Fig. 6.16. Motion analysis by means of the time-average technique of speckle photography. Shown is the object under investigation, a Mason horn, and the fringe systems corresponding to the locations along the horn as indicated by the arrows.

An example demonstrating this technique is shown in Fig. 6.16. The horn (Mason horn) of an ultrasonic drilling device is set into oscillation at about 20 kHz in the vertical direction. It has a conical shape to transform the low amplitude oscillation generated at the top of the horn into a large amplitude one at the lower end for efficient drilling. To measure the amplitudes along the transformer in order to control the design, speckle photography is chosen because the oscillation amplitudes all along the horn can be measured in one recording. Figure 6.16 shows the diffraction pattern (Fourier transform) from four different locations along the horn. The fringe spacing is different in all four images and, hence, so are the oscillation amplitudes. The amplitude is smallest at the driving side, which is the base of the horn at top of the image.

6.4 Flow Diagnostics

The method of speckle photography originally developed for the determination of in-plane displacements and deformations of solid bodies can also be used for measuring flow velocities in transparent liquids (Fig. 6.17). The liquid is seeded with small particles and a section through the liq-

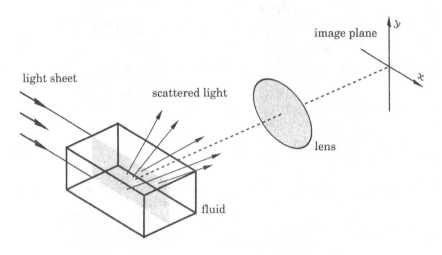

Fig. 6.17. Arrangement for measuring flow velocities in a plane within a fluid by speckle photography.

uid is illuminated with a suitable coherent light wave (ruby laser, argon ion laser, copper-vapor laser). The plane of the section is imaged onto a photographic plate. The seed particles in the liquid act similarly to the diffusively scattering surface of a solid body and form a speckle pattern on the photographic plate. The speckle size again is given by (6.15). Usually, the double-exposure technique is used. When the time interval between the two exposures is suitably chosen and the flow is not too turbulent, then the speckles on the plate have been displaced by only a few diameters. This motion may be different in different areas of the image. In the same way as we have determined the rotation of the tilting table and the metal plate by Fourier transforming local image areas, the fluid flow can be determined with respect to its direction and magnitude by sampling the total image area step by step. With this method it is possible to measure the velocity field simultaneously in one plane. However, just the velocity component normal to the optical axis is obtained. The flow along the optical axis is not accessible this way.

When there are only a few particles in the flow, they are imaged individually in the photographic recording process and do not form a speckle pattern. But in this case the data processing can also be performed via

Young's fringes. This technique is called particle image velocimetry (PIV) for obvious reasons.

In principle, particle image velocimetry does not need coherent light. Nevertheless, because of the necessarily short pulses at high energy needed for the illumination of the scatter particles, lasers serve as the most convenient sources. As their coherence properties are not important, copper-vapor lasers with their small coherence length may also be used. The images containing the particle pairs may be processed directly by scanning them into a computer. This is done in small patches of the size wanted for velocity vector determination. They correspond to the laser beam diameter when performing the Fourier transform by diffraction. Each image section is autocorrelated. This takes the place of the Fourier transform of Young's fringes. Ideally, the autocorrelation yields three points on a line perpendicular to Young's fringes (see the chapter on Fourier optics). The center point corresponds to the zeroth diffraction order when the transparency with Young's fringes is illuminated by laser light. The distance of the outer points from the central one is a measure of the velocity. The line connecting the points determines the flow direction except for the sign. By moving the film between the exposures, this ambiguity may be removed. This way, the determination of the velocity vector may be automated. Sampling the total image of a PIV transparency gives the two-dimensional velocity field in the plane of the light section. Due to the great importance for fluid-flow diagnostics complete devices are commercially available.

As an extension of the method, several light sheets may be used simultaneously in an approach to three-dimensional velocity-field determination. Unfortunately, again just the velocity components in the plane are obtainable. A complete, three-dimensional determination of fluid flows seems to require holographic methods (see the chapter on holography).

Fluid flows are dynamical systems. The measurement of their three-dimensional evolution in time presents a challenge to the experimentalist. The PIV method can be extended also in this direction by taking fast sequences of images. For a simple plane, this has been realized to visualize the fluid flow around a collapsing cavitation bubble [6.2]. Holographic cinematography (see Sect. 7.4) allows the recording of the complete spatiotemporal dynamics. At present, however, such systems are complex and expensive.

In this context a further method, laser Doppler anemometry (LDA), will be discussed [6.3]. This method gives the velocity vector (in its simplest form it is a one-velocity component) at a point in the liquid. In view of the above methods, this seems to be a strong limitation. The method, however, works quasi-continuously and delivers the data immediately without the necessity of first processing a photographic or holographic plate. This is of great advantage when doing measurements, for exam-

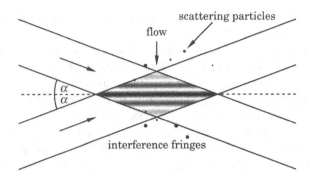

Fig. 6.18. Standing waves of the electrical field of two crossed laser beams in laser Doppler anemometry. Maxima and minima of the electric field having the distance d both yield an intensity maximum upon scattering.

ple, in wind and water tunnels that are expensive to run and where long stand-by times for data to be processed are not tolerable.

In LDA, two laser beams are crossed at a small angle (Fig. 6.18). This sets up an interference field – a standing wave field of the electrical field strength. It oscillates at the high frequencies of light that cannot be resolved. A screen placed across the interference field would make the standing wave field visible as an interference pattern. The distance d between the fringes is

$$d = \frac{\lambda}{2\sin\alpha}, \tag{6.38}$$

where 2α is the angle between the two laser beams and λ is the wavelength of the laser light. In the limit of counterpropagating waves ($\alpha=90°$) a fringe spacing of $\lambda/2$, the smallest possible one, is obtained. In the limit of parallel laser beams ($\alpha = 0$) the fringe spacing becomes infinitely large.

The standing wave field can be used to measure the flow velocity in a gas or liquid in the following way. The medium is seeded with small scatter particles. When they cross the standing wave field, light is scattered with a characteristic modulation of the intensity $I(t)$. Figure 6.19 shows a typical scattering signal. The time t_s between successive intensity maxima is given by the time of flight of the particle from one field maximum to the next minimum in the interference field separated by the distance d. The time of flight t_s of the particle is connected with its velocity component $u_\perp$ perpendicular to the standing wave field and with the fringe spacing d by

$$t_s = \frac{d}{u_\perp} = \frac{\lambda}{2u_\perp \sin\alpha}. \tag{6.39}$$

Since the parameters α and λ are fixed in a given arrangement, t_s has to be measured to yield the velocity $u_\perp$.

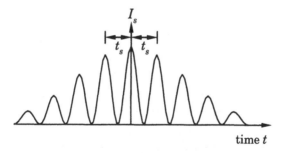

Fig. 6.19. A typical scattering signal (burst) recorded when a small particle crosses the standing wave field of two coherent laser beams.

Only the velocity component perpendicular to the standing wave field can be measured. If three pairs of crossed beams of different wavelengths (for separating the scattering signals) are used, all three velocity components, that is, the complete velocity vector, can be determined. The time t_s is usually obtained by autocorrelating the scattering intensity $I(t)$ and averaging over several scattering events. LDA devices are commercially available and widely used in fluid-flow experiments as they allow continuous monitoring of fluid velocities in real time without disturbing the fluid. Applications include the diagnostics of the flow around airfoils and cars in wind tunnels.

6.5 Stellar Speckle Interferometry

The light of the stars that reaches us on earth shows coherence properties. The Michelson stellar interferometer (see Sect. 4.5) takes advantage of this fact for determining star diameters and distances of binary stars. The coherence of star light, on the other hand, having passed through the atmosphere, leads to a temporally varying speckle field (scintillation). It is very disturbing for astronomical observations. The almost plane wave fronts that reach us from a star are deflected and deformed by the inhomogeneities of the atmospheric layer, similar to the laser light scattered at a rough surface. Imaging a star (point light source) in a telescope yields a speckle pattern in the image plane that typically corresponds to an angular diameter of about 1 arc second. The individual speckles are of size corresponding to the resolving power of the telescope (Fig. 6.20). The speckle pattern quickly varies with time so that at longer exposure times the patterns add up to a bright extended spot. Thus, the theoretical resolving power of a telescope (for instance 0.027 arc seconds at $\lambda = 550\,\mathrm{nm}$ for the 5 m mirror of the Mt. Palomar observatory) cannot be reached.

In some cases, the remedy is available in the form of stellar speckle interferometry. The object in question, for instance a binary star, is taken at

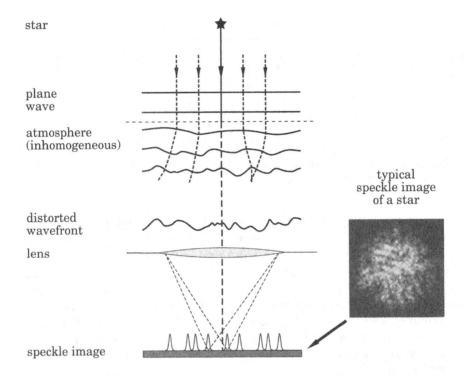

star

plane
wave

atmosphere
(inhomogeneous)

distorted
wavefront

lens

speckle image

typical
speckle image
of a star

Fig. 6.20. Deformation of a plane wave front of star light by inhomogeneities in
the atmosphere and origin of the speckle image in a telescope.

a sufficiently short exposure time (about 10 ms in the visible), so that the
state of the atmosphere has not altered during exposure and the speckle
pattern is not blurred. All waves radiated from different points of the
object into a certain solid angle, the isoplanatic region, mainly pass the
same inhomogeneities and thus lead to the same speckle pattern. In the
image plane of the telescope, speckle patterns from incoherent sources superimpose with their intensity. In the case of a binary star, the complete
intensity distribution is then simply composed of two slightly shifted identical speckle patterns, as with the double-exposure technique of speckle
photography. Since only the distance between the two stars is of interest,
the telescopic image may be processed, as in the case of a double-exposure
specklegram.

Complicated objects cannot be reconstructed this way. However, we
can do better. If $O(x,y)$ denotes the (unknown, undisturbed) intensity
distribution of the object and $P(x,y)$ is the speckle pattern for one object

point (the point spread function), then the actual intensity distribution $I(x,y)$ measured is given by the convolution of O and P (see the Appendix):

$$I(x,y) = (O*P)(x,y) = \int\limits_{-\infty}^{+\infty}\int\limits_{-\infty}^{+\infty} O(\xi,\eta)P(x-\xi,y-\eta)\,d\xi\,d\eta. \qquad (6.40)$$

The task of stellar speckle interferometry is obvious: determine the object intensity O from the knowledge of I and of the independently to be measured speckle pattern P.

The task looks involved in view of the integral. However, it becomes quite simple if we look at the Fourier transform of (6.40):

$$\tilde{I}(u,v) = \mathcal{F}[I](u,v) = \mathcal{F}[O](u,v)\cdot\mathcal{F}[P](u,v) = \tilde{O}(u,v)\tilde{P}(u,v). \qquad (6.41)$$

The convolution then is replaced by a multiplication. The Fourier transform nowadays can be done quickly on a computer.

To arrive at O, we need the speckle pattern P. It can be determined directly only in special cases, that is, in those cases, where in close neighborhood of the object to be investigated (within the isoplanatic region) a (pointlike) reference star is present, whose speckle pattern P is recorded simultaneously. This speckle pattern then is used to invert (6.40). Actually, several hundred speckle recordings are averaged to improve the signal-to-noise ratio.

If no suitable reference star can be found in the close neighborhood to the object, a method proposed first by A. Labeyrie [6.4] may be used. The idea is to suppress the speckles by averaging over the correlation function of many recordings. As the speckle pattern changes from one to the next recording, it does not make sense to average the measured intensities directly. Instead, for each image the autocorrelation function is calculated:

$$A(x,y) = \int\limits_{-\infty}^{+\infty}\int\limits_{-\infty}^{+\infty} I(\xi,\eta)I(\xi+x,\eta+y)\,d\xi\,d\eta, \qquad (6.42)$$

and this function is averaged over several hundred recordings. The details of a recorded image are contained in the central maximum of the autocorrelation function. *Labeyrie* showed that this information is not lost upon averaging. The statistically appearing speckles, on the other hand, average out to a constant background as they do in a long exposure in a normal telescopic image.

The Fourier transform of the autocorrelation A of a function equals its power spectrum. Therefore, we have

$$\mathcal{F}[A](u,v) = \tilde{A}(u,v) = \tilde{I}(u,v)\tilde{I}^*(u,v) = |\tilde{I}(u,v)|^2. \qquad (6.43)$$

Squaring (6.41) yields:

$$|\tilde{I}(u,v)|^2 = |\tilde{O}(u,v)|^2\,|\tilde{P}(u,v)|^2. \qquad (6.44)$$

Since the averaging of the autocorrelation function over many images commutes with the Fourier transform, we get

$$\langle|\tilde{I}(u,v)|^2\rangle = |\tilde{O}(u,v)|^2 \langle|\tilde{P}(u,v)|^2\rangle \tag{6.45}$$

or

$$|\tilde{O}(u,v)| = \left(\frac{\langle|\tilde{I}(u,v)|^2\rangle}{\langle|\tilde{P}(u,v)|^2\rangle}\right)^{1/2}. \tag{6.46}$$

Again, information is needed about the speckle image P of a point object, but this time only the quantity $\langle|\tilde{P}(u,v)|^2\rangle$, that is, the Fourier transform of the averaged autocorrelation function of the speckle image P, is required. Practically, it contains the information about the degree of inhomogeneity of the atmosphere at the time of observation. Experience has shown that this quantity can also be obtained to a good approximation with reference stars located farther away; that is, not in the isoplanatic region.

According to (6.46), only the modulus of the Fourier transform of O is calculated, the phase gets lost. This means that the image geometry cannot be fully retrieved with the aid of the autocorrelation function. This drawback is overcome with the speckle-masking method [6.5]. In this method the triple correlation of the speckle images and its Fourier transform, the bispectrum, are calculated to get the phase information for $\tilde{O}$. The interested reader may consult the literature [6.6].

From the Fourier spectrum $\tilde{O}$, back transformation yields the object intensity O. As an example, Fig. 6.21 shows the speckle image of a binary star and the corresponding reconstructed object image. The improvement in resolution by this ingenious method is really convincing.

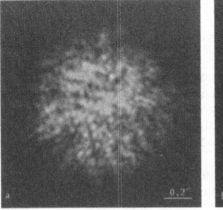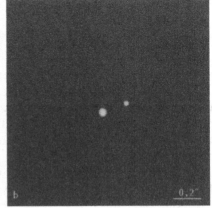

Fig. 6.21. Specklegram of the spectroscopic binary star Ψ Sagitarii (*left*) and object image (*right*) reconstructed by the speckle-masking method (from Ref. 6.5).

Problems

6.1 Consider the probability distribution $p(I)$ of the speckle intensity for a linearly polarized, monofrequency wave field, (6.3). Calculate the corresponding distribution of the field amplitudes, $p(|E|)$. Express the average of the field amplitude, $\langle|E|\rangle$, in terms of the root-mean-square field amplitude, $\overline{E} = \sqrt{\langle I \rangle}$.

6.2 In the arrangement of Fig. 6.8 a He–Ne laser beam ($\lambda = 633$ nm), expanded by a pinhole to a diameter of $d = 2$ mm, illuminates a ground glass plate. At a distance $L_1 = 1$ m behind the plate, an observation screen is placed. An observer is looking at the screen from a distance of $L_2 = 2.5$ m. Calculate the minimum speckle grain size (a) of the objective speckles on the observation screen, (b) of the objective speckle grain images on the observer's retina, (c) of the subjective speckles formed on his or her retina (diameter of pupil = 5 mm, distance between the eye lens and the retina = 1.7 cm). How does the value obtained in (c) depend on the distance L_2? Argue why a short-sighted observer without glasses can see the subjective speckles sharply defined.

6.3 In an arrangement for double-exposure speckle photography the scattering object, illuminated by laser light ($\lambda = 488$ nm), is taken twice on a photographic plate with magnification $V = 1$. The lens used has an F number of $F = f/D = 2$. If we assume that a displacement of the two superimposed speckle patterns by twice the minimum speckle grain size is still detectable, what is the minimum object shift that can be measured with this arrangement? Is it possible to improve the sensitivity by taking a magnified image of the object with the same lens?

6.4 A time-averaged specklegram is taken from an oscillating object, the oscillation being triangular; that is, it is given by the function

$$f(t) = \begin{cases} (2A/T)(t - (m+1/4)T) & \text{for } mT \leq t \leq (m+1/2)T, \\ -(2A/T)(t - (m+3/4)T) & \text{for } (m+1/2)T \leq t \leq (m+1)T, \end{cases}$$

for any integer m. Describe the interference pattern that appears upon illumination of the processed photographic plate by an unexpanded laser beam as described in the text.

6.5 In the measuring volume of a laser Doppler anemometer two argon-ion laser beams ($\lambda = 514$ nm) intersect at an angle of $\gamma = 12°$. The viewing direction of the anemometer's detector coincides with the bisector of the angle γ in the plane spanned by the two light rays. Upon passage of a scattering particle, a burst with a mean frequency of 420 kHz is detected. Give the particle's velocity perpendicular to the line of sight of the detector.

7. Holography

In 1948, *Dennis Gabor* (1900–1972) presented a method for three-dimensional imaging of objects: holography [7.1]. Holography requires coherent light and thus did not become important until after the invention of the laser. It soon found application in holographic interferometry, as now arbitrarily shaped diffusely scattering objects could be treated interferometrically and the same object could be compared with itself at a later time. Today, holography has expanded into various fields. Holograms and their images can now even be handled digitally. In optical data processing, for instance, more and more digital holograms and holographic optical elements are used. That way, data channels can be fanned out and optical connections can be realized on and between processor boards. It has become a customary practice to insert holograms onto credit cards to make forgery more difficult. Art, too, has discovered holography as an element of design.

7.1 Principle of Holography

In usual photography, only the intensity of the light field of an object is recorded. Thereby, the object has to be imaged, and usually incoherent light is used. With coherent light, depending on the stop, differently sized speckles appear in the image. Upon recording, the phase information of the light wave is lost. Therefore, it cannot be retrieved. In holography, on the contrary, the amplitude and phase of a light wave are recorded by adding a reference wave. Also, the object need not be imaged. However, coherent light is required for establishing a fixed phase relation between the reference and object wave. The phase then is coded in the interference pattern, that means, in the position of the interference fringes in a manner similar to Young's interference experiment. How can we reconstruct the three-dimensional image of the object from the interference pattern of the hologram? It was *Gabor* who found the simple solution: just illuminate the hologram with the reference wave used for the recording.

7.1.1 Hologram Recording

The basic optical arrangement for recording a hologram is sketched in Fig. 7.1. The object to be recorded holographically is illuminated with coherent light. The light scattered at the object is recorded, with a reference wave superimposed, on a fine-grain photographic plate, the later hologram.

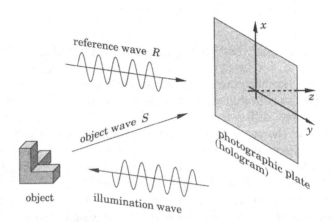

Fig. 7.1. Basic arrangement for recording a hologram.

A theoretical description, because of the coherence of the light used, has to proceed from the superposition of the object wave S and the reference wave R at the position of the holographic plate. Subsequently, the intensity is determined from the sum of the wave amplitudes. This intensity is recorded.

At a point (x,y) of the holographic plate, we have the electric field

$$E(x,y,t) = S(x,y,t) + R(x,y,t). \tag{7.1}$$

When using monofrequency light, $R(x,y,t) = R(x,y)\exp(-i\omega t)$, as the reference wave, and an object that does not move during exposure, that is, $S(x,y,t) = S(x,y)\exp(-i\omega t)$, we immediately get for the intensity:

$$
\begin{aligned}
I(x,y) &= |R(x,y) + S(x,y)|^2 = (R+S)(R+S)^* \\
&= RR^* + SS^* + R^*S + RS^*.
\end{aligned}
\tag{7.2}
$$

The photographic material responds to the total energy per area received during the exposure time t_B and converts it into an optical density and a change of the refractive index. The exposure B (energy/area) is given by

$$B(x,y) = \int_0^{t_B} I(x,y,t)\,\mathrm{d}t = \int_0^{t_B} E(x,y,t)E^*(x,y,t)\,\mathrm{d}t. \tag{7.3}$$

The exposure leads to a complex amplitude transmittanceof the negative. The complex amplitude transmittance τ is given by

$$\tau = \frac{E_t}{E_e} = T\exp(i\varphi) = |\tau|\exp(i\varphi) . \tag{7.4}$$

This definition is analogous to the definition (5.2) of the amplitude transmittance of a mirror we put forward in connection with the Fabry–Perot interferometer. Here, E_t is the outgoing light wave immediately behind the photographic plate, and E_e is the incident light wave (Fig. 7.2). The

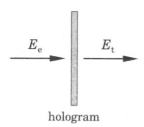

hologram

Fig. 7.2. A hologram transforms an input wave E_e to an output wave E_t.

complex amplitude transmittance is usually a function of the position on the photographic plate:

$$\tau = \tau(x,y) = T(x,y)\exp\left[i\varphi(x,y)\right] . \tag{7.5}$$

Two limiting cases are distinguished in the recording of a hologram: the amplitude hologram ($\varphi = $ const) and the phase hologram ($T = $ const). In both cases, the complete information contained in the wave field is recorded. The phase hologram, however, has the advantage that it absorbs almost no light and thus yields a brighter image upon image reconstruction.

The usual processing of a photographic plate yields a real amplitude transmittance T which depends on the exposure B as given in Fig. 7.3; that is, it gives an amplitude hologram.

The holographic recording is carried out in the linear part of the characteristic near the working point B_A of the exposure. There, the characteristic can be approximated by a straight line, $T = a - bB$. When the intensity does not vary with time, then $B = It_B$ and we obtain

$$T = a - bIt_B . \tag{7.6}$$

The constants a and b depend on the photographic material and on the details of processing.

In the recording of an amplitude hologram, the photographic plate stores the transmittance distribution

$$T(x,y) = a - bt_B(RR^* + SS^* + R^*S + RS^*)(x,y) . \tag{7.7}$$

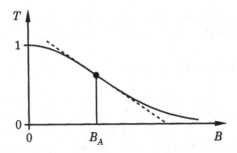

Fig. 7.3. Amplitude transmittance of a holographic plate versus exposure B.

When producing a phase hologram by special processing, we obtain a phase transmittance curve $\varphi(B)$, whose form is sketched in Fig. 7.4. In the neighborhood of the working point B'_A we again can assume a linear relationship

$$\varphi(I) = a' + b' t_B I. \tag{7.8}$$

The transmittance of the photographic plate then reads:

$$\tau = \exp\left[i\varphi(I)\right]. \tag{7.9}$$

It depends nonlinearly on the intensity I. When $\varphi(I) \ll \pi/2$, the exponential function in (7.9) can be expanded to yield, in a linear approximation,

$$\tau = \exp[i\varphi(I)] \approx 1 + i\varphi(I) = (1 + ia') + ib' t_B (RR^* + SS^* + R^* S + RS^*). \tag{7.10}$$

An expression, similar to (7.7) for the amplitude hologram, is obtained for the phase hologram, except for different constants.

7.1.2 Image Reconstruction

To obtain the image of the object (image reconstruction), the hologram is illuminated with the former reference wave as the reconstruction wave. In the reconstruction, the replica of the reference wave $R(x,y) \exp(-i\omega t)$

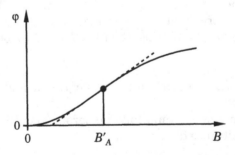

Fig. 7.4. Phase of transmittance for a suitably developed holographic plate.

is modulated with the complex transmittance $\tau(x,y)$. In the case of an amplitude hologram we get the complex amplitude

$$E_t(x,y) = T(x,y)E_e = T(x,y)R$$
$$= Ra - bt_BR(RR^* + SS^* + R^*S + RS^*). \tag{7.11}$$

Four terms are obtained, which have the following meaning:

1. $(a - bt_B|R|^2)R \propto R$ zeroth diffraction order, the reference wave is multiplied by a constant factor,
2. $bt_B|S|^2R$ broadened zeroth diffraction order, modulated by $|S|^2$ (usually a speckle pattern),
3. $bt_B|R|^2S \propto S$ direct image (virtual),
4. $bt_BR^2S^*$ conjugate image (real).

The third term reproduces the original wave S, except for a constant factor, when a plane reference wave is used. The object wave S thus has been reconstructed via the hologram recording and illumination of the hologram with a replica of the reference wave. To obtain the object wave S undisturbed, provision must be made so that the other three waves do not interfere. This problem caused some trouble in the early days of holography. *Gabor*, the inventor of holography, had at his disposal light of limited coherence only. Nevertheless, he was able to record holograms, albeit in-line holograms only, where the object wave and the plane reference wave in effect have almost the same propagation direction perpendicular to the holographic plate (see Sect. 7.3.1). Nowadays, any angle between the object and reference wave can be realized with suitable lasers. A prerequisite is a sufficiently large coherence length or a suitable coherence function of the light used.

The fourth term also gives an image, the conjugate image, closely related to the original image of the object. It is real and pseudoscopic; that is, the object can only be looked at from the back to see the front! A pseudoscopic image has peculiar properties of parallax. If we try to look from above onto a plane of the object, the plane tilts, against expectation, into the opposite direction and is seen at a smaller angle. Moreover, when the viewing direction is altered, parts of the object obviously lying in greater depth can hide parts of the object in the foreground. The reason is that the object parts now seen at greater depth belong to object parts originally lying closer to the observer. A pseudoscopic image appears to be turned inside out.

When the image is reconstructed from a phase hologram, we get

$$E_t(x,y) = \tau(x,y)R = \exp\left[i\varphi(I(x,y))\right]R \approx (1 + i\varphi(I(x,y))R$$
$$\approx (1 + ia')R + ib't_BR(RR^* + SS^* + R^*S + RS^*). \tag{7.12}$$

As before, we obtain four terms. As a result of the additional approximation $\exp(i\varphi) \approx 1 + i\varphi$, the nonlinear effects are more pronounced with phase holograms than with amplitude holograms. Moreover, as they are brighter than amplitude holograms, the nonlinear effects become even more enhanced in the image. This may lead to additional images in higher diffraction orders.

7.1.3 Location of the Images

The images generated upon reconstruction are found in the directions indicated in Fig. 7.5. When the reference and reconstruction waves are incident perpendicularly onto the hologram ($\alpha = 0$), the geometry is substantially simpler, as depicted in Fig. 7.6.

The construction shows that the direct image is a virtual image. It is viewed by looking through the hologram as if through a window and is located behind the hologram. The conjugate image, on the other hand, is real. It is located in front of the hologram and can be intercepted with a ground glass plate.

The directions given for the different images in the two figures directly follow from the mathematical operations associated with them. For simplicity, we assume that S is a plane wave striking the hologram at an angle γ. We then have for S^*, when we omit the factor $\exp(-i\omega t)$ common to all terms,

$$S = \exp(ikx\sin\gamma) \rightarrow S^* = \exp(-ikx\sin\gamma) = \exp[ikx\sin(-\gamma)]. \quad (7.13)$$

Thus, the wave S^* can be interpreted as a wave leaving the hologram at the angle $-\gamma$.

Let the reference and reconstruction wave R also be a plane wave, $R = \exp(ikx\sin\alpha)$, striking the holographic plate at the angle α.

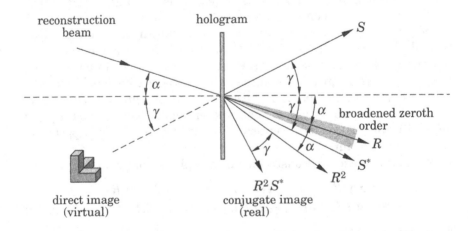

Fig. 7.5. Location of the images upon reconstruction for small angles of incidence, α and γ (α, γ exaggerated).

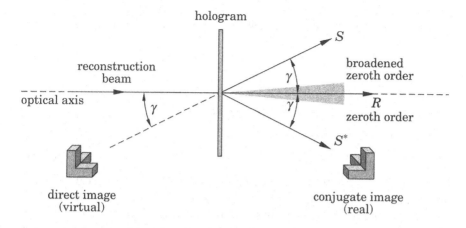

Fig. 7.6. Location of the images when the reference and reconstruction waves are perpendicular to the hologram.

For R^2 we then have:

$$R^2 = \exp(ikx2\sin\alpha) = \exp(ikx\sin\alpha') .$$ (7.14)

For small angles of incidence, α, we have $\alpha' = 2\alpha$. This case is assumed in Fig. 7.5. There, for better demonstration only, the angles α and γ are exaggerated.

Using (7.13) and (7.14), we immediately obtain the result for R^2S^*:

$$R^2S^* = \exp(ikx2\sin\alpha)\exp\left[ikx\sin(-\gamma)\right]$$
$$= \exp\left[ikx(2\sin\alpha + \sin(-\gamma))\right] .$$ (7.15)

Writing the expression R^2S^* as a plane wave leaving the hologram at the angle β, $R^2S^* = \exp(ikx\sin\beta)$, we obtain the relation for the wave leading to the conjugate image:

$$\sin\beta = 2\sin\alpha + \sin(-\gamma) .$$ (7.16)

With increasing α or γ, the conjugate image becomes more and more distorted until, for

$$\sin\beta = 2\sin\alpha + \sin(-\gamma) > 1 ,$$ (7.17)

an angle β does no longer exist to satisfy the relation (7.16). Then, the corresponding parts of the conjugate image disappear.

7.1.4 Phase Conjugation

In the preceding section, we repeatedly used complex conjugate amplitudes (R^*, S^*). To them, we have assigned light waves. For instance, the wave S^* leads to the conjugate image. This is an example of a phase conjugate wave. We give its definition for a plane wave propagating in the z-direction:

$$E(z,t) = E_0 \exp\left[i(kz - \omega t)\right]$$
$$= E_0 \exp(ikz)\exp(-i\omega t). \tag{7.18}$$

The corresponding phase conjugate wave is given by

$$E_{pc}(z,t) = E_0^* \exp(-ikz)\exp(-i\omega t)$$
$$= E_0^* \exp\left[i(-kz - \omega t)\right]. \tag{7.19}$$

It differs from the complex conjugate wave, abbreviated by c.c., in that the complex conjugate is only taken for the spatial part. Mathematically, a phase conjugate wave is thus given by a simple operation. Physically, a phase conjugate plane wave is just the plane wave propagating in the opposite direction. More generally, to obtain the phase conjugate wave, the wave vector k has to be replaced by $-k$.

Phase conjugate waves have become of importance, since they can be obtained almost in real time. A realization, called a phase conjugating mirror (Fig. 7.7, see also Fig. 2.3), can be achieved by four-wave mixing in a nonlinear crystal such as $LiNbO_3$, $BaTiO_3$, or $Bi_{12}SiO_{20}$ (BSO). For these mirrors, the usual law of reflection is no longer valid.

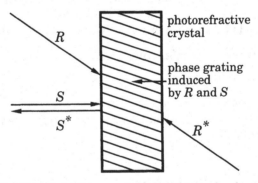

Fig. 7.7. Realization of a phase conjugating mirror by four-wave mixing in a photorefractive crystal.

When a wave front is distorted, for instance by inhomogeneities of the refractive index, then the phase conjugate wave encounters the same inhomogeneities while propagating in the opposite direction (if these are varying sufficiently slowly in time). Hence, it leaves the distorting region with its wave form restored. In this way, phase conjugating mirrors may correct aberrations (Fig. 7.8).

Holography presents a simple and convenient means for generating phase conjugate waves. Choosing a plane wave as the reference wave R,

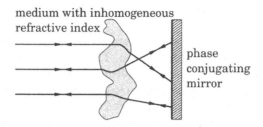

Fig. 7.8. Aberration correction by a phase conjugating mirror.

we can easily obtain the phase conjugate wave R^* as the wave propagating in the opposite direction. Illuminating the hologram with R^* as the reconstruction wave, we have

$$E_t = T(x,y)R^* = R^*a - bt_\mathrm{B}R^*(RR^* + SS^* + R^*S + RS^*). \qquad (7.20)$$

Now the last term, $\propto |R|^2 S^*$, reconstructs the conjugate wave S^* undistorted. The simplest way to obtain the conjugate reference wave is to turn the hologram around by 180° in the arrangement where it has been recorded (Fig. 7.9).

A comparison with the recording geometry shows that the wave S^* leaves the hologram and retraces S to the former location of the object. It seems as if an image of the object is fully recovered in three dimensions. This is only partly true. When the object is no longer in place, the light can of course no longer be scattered. Instead, the light propagates farther into the space formerly occupied by the object and may disturb other parts of the image.

The real image obtained with the phase conjugate reference wave is used, for instance, in particle-size analysis of sprays from injection nozzles and of bubbles in water. A camera is translated through the conjugate real image to feed it into a computer plane by plane for processing and generating a three-dimensional image of the drop or bubble distribution [7.2].

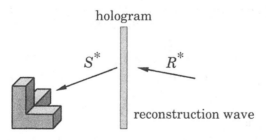

Fig. 7.9. Reconstruction geometry with a phase conjugate reference wave R^* as reconstruction wave.

7.2 The Imaging Equations of Holography

Up to now, we have used the same geometry in reconstructing the images from a hologram as in recording them. The sole exception has been the reconstruction with the conjugate reference wave just discussed. Surely, an important question will be, how closely the recording geometry has to be met to get a good image upon reconstruction. It can be expected that the image does not disappear immediately when the reference wave is slightly modified, for instance, when the recording angle is not precisely reproduced by the reconstruction wave. The laws effective in these circumstances are given by the holographic imaging equations.

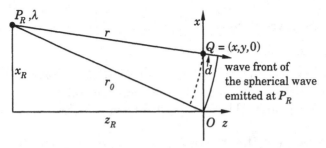

Fig. 7.10. Phase variation of a spherical wave in a plane.

In deriving the imaging equations, we consider the hologram of a point source. The image of an extended object can be obtained with respect to the location and distortion by identifying the images of several of its points. Fundamental for the derivation of the imaging equations of holography is the variation of the phase of a spherical wave in the plane of the hologram (the $(x,y,0)$-plane) in its dependence on the location of its origin P_R in space (Fig. 7.10). The point P_R later on will be the origin of the reference wave. Its coordinates are denoted by (x_R, y_R, z_R). A plane wave is obtained in the limit $z_R \rightarrow -\infty$ with x_R/z_R, $y_R/z_R = \text{const}$. The phase variation on the (holographic) plane is effected by the different path lengths from P_R to the different points on the plane. Let the phase at the origin O of the coordinate system be the phase for reference. Then, the phase at a point $Q = (x,y,0)$ is given by the path difference $d = r - r_0$ as

$$\varphi_R(x,y,0) = kd = k(\overline{P_R Q} - \overline{P_R O})$$

$$= k\left(\sqrt{(x-x_R)^2 + (y-y_R)^2 + z_R^2} - \sqrt{x_R^2 + y_R^2 + z_R^2}\right)$$

$$= k|z_R|\left(\sqrt{1 + \frac{(x-x_R)^2 + (y-y_R)^2}{z_R^2}} - \sqrt{1 + \frac{x_R^2 + y_R^2}{z_R^2}}\right). \quad (7.21)$$

For paraxial rays the approximation $|x_R|, |y_R|, |x|, |y| \ll |z_R|$ is valid. Then the root can be expanded into a series. Setting z_R instead of $|z_R|$ (as the

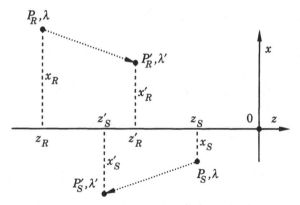

Fig. 7.11. The two spherical waves originating in P_R and P_S during recording and the image point P_S' upon reconstruction with a spherical reconstruction wave with origin P_R'.

result is not altered) we have

$$\varphi_R(x,y,0) = k\frac{1}{2}\frac{x^2+y^2-2xx_R-2yy_R}{z_R} - k\frac{1}{8z_R^3}(x^4+y^4+2x^2y^2-4x^3x_R$$
$$-4y^3y_R-4x^2yy_R-4xy^2x_R+6x^2x_R^2+6y^2y_R^2+2x^2y_R^2$$
$$+2y^2x_R^2+8xyx_Ry_R-4xx_R^3-4yy_R^3-4xx_Ry_R^2-4yx_R^2y_R)$$
$$+\text{higher order terms } (\mathcal{O}(z_R^{-5})). \tag{7.22}$$

To a first approximation, the phase in the plane $z=0$ of a spherical wave originating from the point (x_R,y_R,z_R) varies as

$$\varphi_R(x,y,0) = k\frac{1}{2}\frac{x^2+y^2-2xx_R-2yy_R}{z_R}. \tag{7.23}$$

The additional terms appearing in (7.22) are aberration terms.

We take the spherical wave with the origin P_R as the reference wave to record a hologram of an object wave. Our object simply is another spherical wavewith the origin P_S (Fig. 7.11).

The phase variation in the $(x,y,0)$-plane of the spherical wave with the origin P_S is given to a first approximation, similarly to (7.22), by

$$\varphi_S(x,y,0) = k\frac{1}{2}\frac{x^2+y^2-2xx_S-2yy_S}{z_S}. \tag{7.24}$$

The hologram of a point object thus recorded will now not be illuminated with a spherical wave with the origin P_R and wave number k (wavelength λ) but with a spherical wave with the origin P_R' and wave number k' (wavelength λ'). Again, we can immediately give the phase variation of this wave in the $(x,y,0)$ plane to a first approximation:

$$\varphi_R'(x,y,0) = k'\frac{1}{2}\frac{x^2+y^2-2xx_R'-2yy_R'}{z_R'}. \tag{7.25}$$

We ask where the point P_S can be found now. This new location is called P_S'. First of all, we know that there will be two images, the direct image $P_{S,d}'$ and the conjugate image $P_{S,c}'$ (sometimes only one may exist, however). Their location is obtained by looking at the reconstruction terms. Connected with the image point $P_{S,d}'$ is a light wave S_d' in the plane of the hologram. It is given by the term

$$S_d' \propto R'R^*S \tag{7.26}$$

appearing in the reconstruction for the direct, usually virtual, image. Upon reconstruction with the identical reference wave R, this term reads $S_d \propto RR^*S = |R|^2 S$. The conjugate, usually real, image is represented by the following expression:

$$S_c' \propto R'RS^*. \tag{7.27}$$

The phase $\varphi_{S,d}'(x,y,0)$ of the wave S_d' is given by the sum of the phases of the waves R', R^*, and S, the phase $\varphi_{S,c}'(x,y,0)$ of the wave S_c' by the sum of the phases of R', R, and S^*:

$$\varphi_{S,d}'(x,y,0) = \varphi_R' - \varphi_R + \varphi_S, \tag{7.28}$$

$$\varphi_{S,c}'(x,y,0) = \varphi_R' + \varphi_R - \varphi_S. \tag{7.29}$$

The imaging equations, that is, the relation between the coordinates of the four points P_R, P_R', P_S, and P_S', are obtained from these equations by inserting the expressions for φ_R', φ_R, and φ_S, rearranging the terms as a phase variation of a spherical wave, and comparing the coefficients. For the phase of the direct image we obtain:

$$
\begin{aligned}
\varphi_{S,d}' &= \varphi_R' - \varphi_R + \varphi_S \\
&= k'\frac{1}{2}\frac{x^2+y^2-2xx_R'-2yy_R'}{z_R'} \\
&\quad -k\frac{1}{2}\frac{x^2+y^2-2xx_R-2yy_R}{z_R} \\
&\quad +k\frac{1}{2}\frac{x^2+y^2-2xx_S-2yy_S}{z_S}
\end{aligned}
\tag{7.30}
$$

or, with $m = k/k' = \lambda'/\lambda$,

$$
\begin{aligned}
\varphi_{S,d}' &= k'\frac{1}{2}\left[(x^2+y^2)\left(\frac{1}{z_R'}-\frac{m}{z_R}+\frac{m}{z_S}\right) - 2x\left(\frac{x_R'}{z_R'}-\frac{mx_R}{z_R}+\frac{mx_S}{z_S}\right)\right. \\
&\quad \left. -2y\left(\frac{y_R'}{z_R'}-\frac{my_R}{z_R}+\frac{my_S}{z_S}\right)\right] \\
&= k'\frac{1}{2}\frac{x^2+y^2-2xx_{S,d}'-2yy_{S,d}'}{z_{S,d}'}.
\end{aligned}
\tag{7.31}
$$

In the last step, we have written the equation as a phase variation of a spherical wave with origin $(x'_{S,d}, y'_{S,d}, z'_{S,d})$. Since the equation has to be valid for any (x,y), a comparison of the coefficients yields:

$$\frac{1}{z'_{S,d}} = \frac{1}{z'_R} - \frac{m}{z_R} + \frac{m}{z_S}, \tag{7.32}$$

$$\frac{x'_{S,d}}{z'_{S,d}} = \frac{x'_R}{z'_R} - \frac{mx_R}{z_R} + \frac{mx_S}{z_S}, \tag{7.33}$$

$$\frac{y'_{S,d}}{z'_{S,d}} = \frac{y'_R}{z'_R} - \frac{my_R}{z_R} + \frac{my_S}{z_S}. \tag{7.34}$$

These are the holographic imaging equations for the direct image to a first approximation. Similarly, we obtain for the conjugate image:

$$\frac{1}{z'_{S,c}} = \frac{1}{z'_R} + \frac{m}{z_R} - \frac{m}{z_S}, \tag{7.35}$$

$$\frac{x'_{S,c}}{z'_{S,c}} = \frac{x'_R}{z_{R'}} + \frac{mx_R}{z_R} - \frac{mx_S}{z_S}, \tag{7.36}$$

$$\frac{y'_{S,c}}{z'_{S,c}} = \frac{y'_R}{z'_R} + \frac{my_R}{z_R} - \frac{my_S}{z_S}. \tag{7.37}$$

As can be noticed, the conjugate image point results from the direct image point when the signs of the two last terms are changed in each line.

We now discuss the imaging equations for special cases.

When using the same light wave for reconstruction as for recording, that is, when $R' = R$ and $\lambda' = \lambda$ $(m = 1)$, then

$$(x'_{S,d}, y'_{S,d}, z'_{S,d}) = (x_S, y_S, z_S) \tag{7.38}$$

is obtained for the direct image point. In this case the direct image is located at the site of the original object, as it should be.

In the case of plane waves along the optical axis, that is, z_R, $z'_R \to -\infty$ and x_R/z_R, y_R/z_R, x'_R/z'_R, $y'_R/z'_R = 0$, we get:

$$(x'_{S,d}, y'_{S,d}, z'_{S,d}) = \left(x_S, y_S, \frac{1}{m}z_S\right) = \left(x_S, y_S, \frac{\lambda}{\lambda'}z_S\right). \tag{7.39}$$

This case corresponds to an illumination of the holographic plate with plane reference and reconstruction waves, incident along the normal, possibly with different wavelengths. According to (7.39), the object then is shifted along the z-axis and is expanded or contracted depending on the ratio of the wavelengths, λ'/λ. The lateral sizes stay the same!

When the same wavelength is used for the reconstruction, the direct image is distortion free when the recording geometry is used; a different wavelength always yields distortion. The distortions could be avoided by adjusting all dimensions including the hologram size according to m. Unfortunately, this method normally cannot be realized. Moreover, a modification of the hologram size is not included in the imaging equations given above.

The aberrations have not been discussed further. They are of importance, when the quality of an image point has to be judged, that is, how a point is spread out in the imaging process. As can be guessed, this needs some effort and is usually done with the help of a computer for a given case. The problem of image quality arises in the holographic recording and reconstruction of small particles. Numerical studies show that the resolution usually drops quickly when the hologram is illuminated with light with a wavelength different from the recording one. Then the sizes of small particles cannot faithfully be recovered.

7.3 Holographic Arrangements

As the wavelength of light is very small in the visible, interference phenomena are very sensitive to displacements. Therefore, a holographic arrangement demands a high mechanical stability, in particular when exposure times of the order of minutes are unavoidable. No special precautions, on the other hand, are necessary, when short, coherent light pulses are used, which nowadays are easily available with Q-switched lasers.

7.3.1 In-line Holograms

The simplest way to generate a hologram consists in illuminating a holographic plate with an expanded laser beam through the (sufficiently transparent) object (Fig. 7.12).

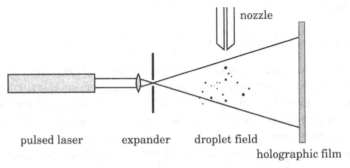

Fig. 7.12. Recording an in-line hologram of a spray from a nozzle.

This type of recording is used in particle diagnostics for investigating sprays of all kinds, bubble distributions in liquids (cavitation, traces of particles in a bubble chamber), rain drops (meteorology), plankton (marine biology), etc. The method poses the least requirements concerning the coherence of the light used and was already introduced by *Gabor* [7.1]. The reference wave is given by the light passing the particle cloud undisturbed, the object wave is formed by the light scattered at the particles.

7.3.2 Reflection Holograms

A typical arrangement for recording holograms of reflecting objects is sketched in Fig. 7.13. The coherent light is provided by a laser such as

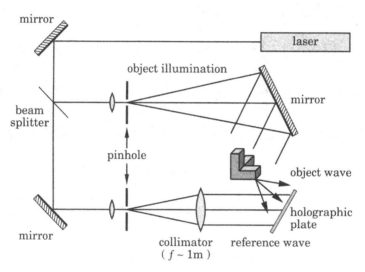

Fig. 7.13. Example of an arrangement for recording reflection holograms.

a He–Ne laser. The light wave is divided into two by a beam splitter to obtain a reference wave and a wave for illuminating the object, yielding the object wave. Both waves have to be expanded to illuminate a larger area. This is done by means of microscope objectives with an aperture stop (pinhole) in the back focal plane acting as a spatial frequency filter. The pinhole is inserted to obtain a pure, interference-free light wave. Its operation is discussed more closely in the chapter on Fourier optics. The object wave illuminates a scene via a large mirror. Scattered light from the objects of the scene reaches the holographic plate and interferes with the reference wave. The interference pattern is recorded and yields the hologram. When setting up the experiment, one should ensure that the path lengths of the object and reference waves from the beam splitter to the holographic plate should match as closely as possible. Due to the usually limited coherence length, even of laser light (compare Sect. 4.3), the adjustment is necessary to obtain an interference pattern of high contrast on the holographic plate. Illumination of the hologram with the reference wave as the reconstruction wave yields a three-dimensional image of the objects. If this is done in the same arrangement (Fig. 7.13), the holographic plate may be developed in place, otherwise it has to be brought back to its former location after processing. To view the image, either the object wave is blocked or the beam splitter is removed. Figure 7.14 shows two different views of a reconstructed image that illustrate the three-dimensionality. The real impression, however, is only given by a view through the hologram itself.

Fig. 7.14. Two different views of a reconstructed image. The left image shows a photograph taken with a small F number and focusing to the foreground. The right image was taken with a large F number and therefore a large depth of field. The focusing is on the center of the image. Speckles appear in this image due to the large F number.

7.3.3 Transmission Holograms

Transmissive objects are recorded with backlight and a separate reference wave. Figure 7.15 shows an arrangement for recording transmission holograms in an experiment for investigating laser-induced breakdown of liquids. This experiment is an interesting application of transmission holography, which makes use of the different properties of coherent and incoherent light. When intense laser light is focused into air or water, dielectric breakdown is observed. It is very similar to the electric spark discharge between two electrodes when a high voltage is applied across them. During discharge or breakdown intense white light and a shock wave are emitted. In water, a bubble is additionally formed after the breakdown, which initially expands at high speed. If we want to investigate the physical phenomenon of breakdown with bubble and shock-wave

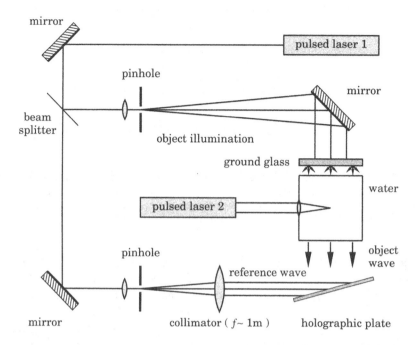

Fig. 7.15. Example of an arrangement for recording transmission holograms.

formation, the intense white light is very disturbing as it swamps photographic and interferometric recordings. However, when a hologram is
taken with a short exposure time in the range of a few nanoseconds, then
both the fast dynamics is stopped in three dimensions and the bright
white light emitted during breakdown is suppressed. As it is incoherent,
the white light illuminates the holographic plate homogeneously and is
not reproduced when the image is reconstructed. This makes possible the
undisturbed recording of both the bubble generated and the shock wave
emitted (Fig. 7.16).

7.3.4 White-Light Holograms

As holography essentially makes use of coherent light, we so far have
supposed the necessary coherence. But even today, lasers are not readily
available everywhere. Thus, the question may be posed about which coherence properties are really needed in a given arrangement or whether
arrangements can be conceived that work with reduced coherence. It has
been found that, indeed, good image reconstructions can be obtained even
with white light, when the holograms are produced in a suitable way.

   To begin, we attempt to reconstruct the image stored in a usual hologram by illuminating the hologram with white light (Fig. 7.17). Can we

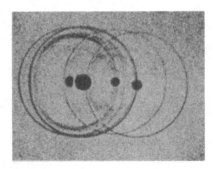

Fig. 7.16. Example of light-induced breakdown by a giant laser pulse focused into water. Image reconstruction is from a transmission hologram. The bubbles generated and the corresponding shock waves are visible without any sign of the bright white light emitted during breakdown.

expect to see anything? We know that white light is composed of a spectrum of frequencies. Each of them is diffracted in a different way by the hologram and thus reconstructs its own image. These images superimpose in space, and the original image is unrecognizable in most of the cases. If we look at a single image point, we may detect that it is spread out to a small white-light spectrum. With the help of color filters, the image quality may be improved. The image, however, is then quite dark.

There exists a different way to produce holograms that can be viewed in white light: white-light holograms. They are obtained when in the recording of the hologram, the coherent reference and object waves strike the photographic plate from opposite sides (Fig. 7.18). Then, interference fringes are formed as layers in the photographic emulsion that essentially run parallel to the surface of the plate. In a photographic emulsion of about 7 μm thickness, about 15 parallel interference layers are present. When the hologram produced this way is illuminated with white light in the direction of R^*, the different layers reflect the light, which con-

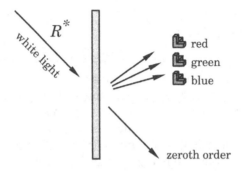

Fig. 7.17. Superposition of multiple images, when the reconstruction is with white light instead of the conjugate reference wave R^*.

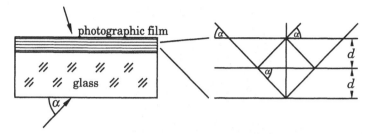

Fig. 7.18. Recording geometry for a white-light hologram (*left*) and Bragg reflection (*right*).

structively interferes in certain directions (Bragg reflection). If d is the separation of the interference layers and α is the viewing angle, then the Bragg condition reads:

$$2d\sin\alpha = n\lambda, \quad n = 1, 2 \ldots . \tag{7.40}$$

For $n = 1$, a constructive interference of the partial waves reflected is obtained only at a wavelength λ_α according to the angle α (Fig. 7.18). Therefore, when the hologram is viewed at an angle α, the image is seen only at the wavelength λ_α, without any disturbing interference from waves of other wavelengths. The parallel interference layers modulated with the image information act as an interference filter for the viewing wavelength.

A simple arrangement for taking a white-light hologram is shown in Fig. 7.19. For recording, a coherent light source (for instance, a He–Ne laser) is required as before. The laser beam is expanded, collimated, and, via a large mirror, directed through the photographic plate onto the object. The reference wave corresponds to the light coming directly from the laser, the object wave is given by the light scattered back from the object. Strongly reflecting objects will give the best results because they provide enough light by back scattering.

The distance between the object and the holographic plate has to be kept small, since the spatial coherence condition also has to be obeyed.

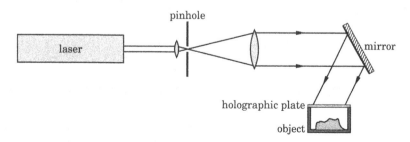

Fig. 7.19. Arrangement for recording a white-light hologram.

Fig. 7.20. Example of a photographic image from a white-light hologram viewed in white light.

Without special precautions in the developing of the photographic plate, the emulsion will shrink during processing. Consequently, a hologram taken in red light and illuminated with white light yields the best images in the green. Figure 7.20 shows a photographic image of a horseman taken in reflection from a white-light hologram illuminated with white light.

7.3.5 Rainbow Holograms

There exists a further method to circumvent the disturbing superposition of differently colored images. It sacrifices the degree of freedom of vertical parallax in exchange for a degree of freedom of wavelength. This will be made clear in the following. The exchange needs the help of phase conjugate waves, which are easily realizable with holograms. The method consists in recording a hologram, in a normal or white-light geometry, of the conjugate image of a primary (master) hologram masked by a long slit. Figure 7.21 explains how the color separation in the eye is achieved when viewing the hologram. The images again are reconstructed with conjugate waves to yield a real image for each color contained in the white light. These images superimpose in space as before. However, upon

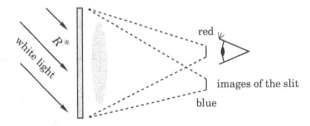

Fig. 7.21. Reconstructed images from a rainbow hologram. The shaded area in front of the hologram indicates the superimposed colored images.

propagation each color converges to a different slit, since the object image has been projected into space only from a small slit area of the primary hologram. Surely, these slit images also superimpose, but the shift from red to blue can be made sufficiently large so that sharp images in quite pure spectral colors appear.

An arrangement for recording rainbow holograms is given in Fig. 7.22. The master hologram is illuminated with the conjugate reference wave

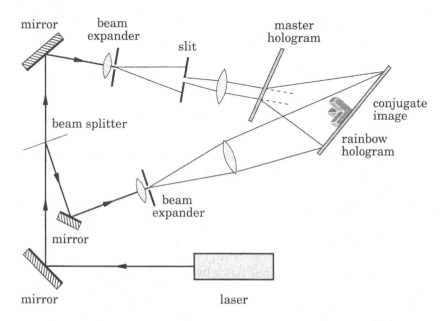

Fig. 7.22. Example of an arrangement for recording a rainbow hologram.

to act as a small slit. In the image reconstruction from the rainbow holo-gram, the white light should impinge obliquely from above. To achieve this, the primary hologram along with the slit and the photographic plate for the rainbow hologram are turned by 90 degrees. Then the arrange-ment becomes simpler, since oblique incidence from the side is realized more easily. The conjugate image of the primary hologram should have a distance of about 30–50 cm from the hologram, because later the eye is in the position of the reconstructed slit of the primary hologram and there-fore should be able to focus onto the object image. The photographic plate for the rainbow hologram is placed near to the reconstructed conjugate object so that the complete image can be seen for all colors.

Then, as shown in Fig. 7.21, the rainbow hologram may be viewed in white light coming from the direction of the conjugate reference wave. In sun light, for instance, the object appears in bright, clear spectral color with full horizontal parallax. When, however, the eye is moved vertically up and down, the color changes from brilliant blue to brilliant red; that is, there is no vertical parallax. Instead, a different wavelength out of the white light is chosen for the reconstruction of the image. When the eye is not placed exactly in the slit plane, but in front of or behind it, the colors start to superimpose. This happens in such a way that part of the image starts to become differently colored, keeping its sharpness. It is possible to view the object in all spectral colors from red to blue, vertically coloring the object. In all cases, despite moving one's head up and down, it is not possible to look from above or below at the object: the vertical parallax has been lost. But it is always possible to look from the left and the right around the object (as far as the hologram allows): the horizontal parallax has been kept.

7.4 Holographic Cinematography

Single images are not sufficient to represent motion. The problem of reproducing motion is solved by movie and television techniques, but just for two-dimensional images only. Are three-dimensional movies possible with holography? That is indeed the case. However, the method is involved and contains many restrictions. Thus, the holographic movie for entertainment has not made its way. Science, on the other hand, has developed some applications, for instance, for three-dimensional motion analysis of particle clouds (drops in air, bubbles in water) and for the visualization of three-dimensional flows by following scattering particles in fluids. This opens up new possibilities in experimental hydrodynamics to investigate turbulence and coherent flow structures [7.3].

An arrangement for taking a series of holograms in rapid succession is shown in Fig. 7.23. For that a series of coherent light pulses is needed. Short coherent pulses of high energy per pulse are hard to get. Therefore a copper vapor laser is used that emits a series of light pulses at a rate of up to 20 kHz with a duration of 35 ns and an energy per pulse of about 1 mJ. As the coherence length of these pulses is short (a few millimeters only) they are used to pump a dye-laser (in an oscillator-amplifier configuration) whose coherence can be controlled by dispersive elements inside the cavity. The dye laser thus acts as a coherence transformer for the pulses of the copper vapor laser. The pulses from the dye laser have a duration of 15 ns and an energy of about 50 µJ, with a coherence length of several centimeters. The energy is sufficient to expose a hologram of about 1 cm^2 in size.

To obtain a specified number of pulses an electrically activated shutter opens for a prescribed time interval equal to the time of one revolution of the rotating holographic plate on which the holograms with a diameter of up to 1 cm are recorded. The short pulse duration of 15 ns prevents the blurring of the interference fringes despite of the high tangential velocity

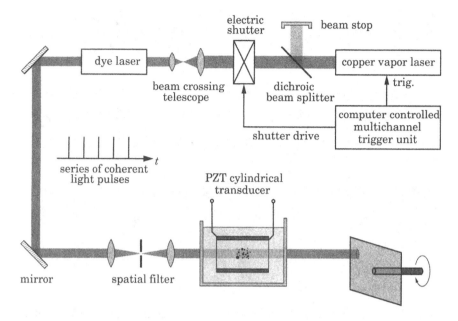

Fig. 7.23. Arragement for high-speed holographic cinematography with a copper-vapor laser pumped dye-laser in an in-line geometry with a rotating holographic plate [7.3].

of up to 70 m/s of the rotating holographic plate, necessary to obtain high framing rates. Series of up to 40 holograms at rates of up to 10000 frames per second have been taken [7.3]. With a different arrangement using an argon-ion laser holographic series have been recorded at a rate of up to 300000 holograms per second [7.4].

Figure 7.24 shows twelve reconstructed images from twelve successive holograms out of a series of holograms taken at 69300 frames per second of oscillating bubbles in water driven by an acoustic field at a frequency of 23.1 kHz. They are arranged in a way so that each row contains images corresponding to a fixed phase of the sound field. The subharmonic dynamics of the bubbles is clearly noticed [7.5, 7.6].

7.5 Digital Holography

The holographic recording process can be simulated on a computer. In this digital or synthetic holography the computer calculates the transmittance of the holographic plate. A suitable plotting method serves for the output and the transfer onto photographic film. Illumination of the digital hologram with coherent light according to the simulated record-

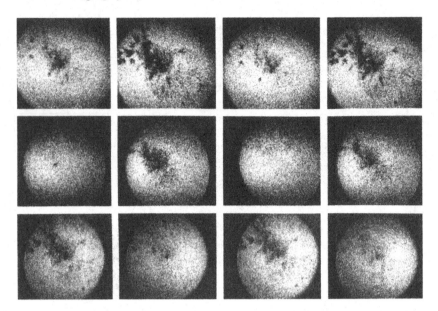

Fig. 7.24. Reconstructed images from a holographic series taken at 69300 frames per second of bubbles inside a piezoelectric cylinder driven at 23.1 kHz [7.5]

ing geometry reconstructs the object, which need not exist in reality. This method has the obvious advantage that holograms can be calculated and generated from objects for which only a mathematical description exists, for instance, aspheric lenses or lenses with multiple focal points.

The generation of a digital hologram of an object calls for

- a mathematical model (algorithm) as simple and time effective as possible for calculating the diffraction patterns,
- a powerful digital computer,
- a fast, high-definition plotter for the graphical output of the calculated diffraction pattern.

Here, we only want to present the direct digital simulation of the holographic recording process considered so far and a digitally possible extension of it. Numerous further methods exist that may be looked up in [7.7].

7.5.1 Direct Simulation

In the numerical calculation, the object is represented by a number L of radiating points $P_l, l = 1, \ldots, L$, with coordinates (x_l, y_l, z_l). The holographic plate, placed in the plane $z = 0$, is represented by a grid of sample points $\{(x_m, y_n, 0); m = 1, \ldots, M; n = 1, \ldots, N\}$ as illustrated in Fig. 7.25.

Let each point P_l be a source of a spherical wave yielding the complex amplitude E_{lmn} at the location (x_m, y_n) of the holographic plate (the

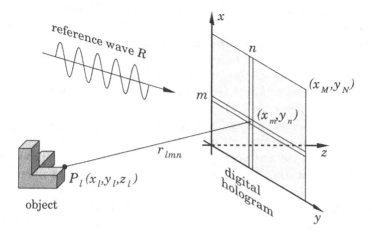

Fig. 7.25. Sketch for illustrating the calculation of a digital hologram.

z-coordinate is omitted for simplicity because of $z = 0$ in the plane of the hologram):

$$E_{lmn} = A_l \exp(i\varphi_l) \frac{\exp(ikr_{lmn})}{r_{lmn}}. \tag{7.41}$$

Here, A_l and φ_l are the amplitude and phase, respectively, of the wave originating from the point P_l. The quantity r_{lmn} is the distance of P_l from the grid point (x_m, y_n) of the hologram and is given by

$$r_{lmn} = \sqrt{(x_l - x_m)^2 + (y_l - y_n)^2 + z_l^2}. \tag{7.42}$$

To get the total amplitude of the object wave at the point (x_m, y_n), all partial waves from the individual object points have to be summed:

$$E_{Smn} = \sum_{l=1}^{L} E_{lmn}. \tag{7.43}$$

Including the reference wave E_R leads to the intensity $I_{mn} = I(x_m, y_n)$ at the point (x_m, y_n):

$$I_{mn} = |E_{Rmn} + E_{Smn}|^2 = |E_{mn}|^2. \tag{7.44}$$

Pausing here to calculate the number of operations necessary for a modest example, we immediately notice that with the normal resolution of a holographic plate (3000 lines per millimeter) and an even fair number of object points (one million), that is, approximately the number of pixels on a computer screen, the computing time on present-day (2002) workstations becomes unrealistically large, unless massively parallel computers with thousands of processors are used.

As an example, we consider the calculation of a digital hologram with a grid size of $2000 \times 2000 = 4$ million points and an object of 1000 points. Then, according to (7.44), four million complex additions and multiplications have to be performed to arrive at the intensity for each hologram point. Moreover, to obtain the amplitudes E_{Smn} needed in (7.44) for each m and n, four billion complex additions are necessary according to (7.43). For this, the expressions (7.41) and (7.42), with their complex exponential function and square root, have to be evaluated, again four billion times. Just supposing that a complex addition needs 10 ns, four billion additions already yield 40 seconds, that is almost a minute of computing time. As multiplication and the calculation of the complex exponential function and the square root take substantially longer, the difficulties digital holography encounters at present are obvious.

The computational effort necessary can be drastically reduced, when the following simplifying assumptions are made:

1. Each object point emits a light wave of the same amplitude A_l:

$$A_l = \text{const}, \qquad \text{for all } l. \tag{7.45}$$

2. Each object wave is emitted with the same phase φ_l set to zero:

$$\varphi_l = 0, \qquad \text{for all } l. \tag{7.46}$$

3. Let the object be far away from the hologram and its size be small compared to its distance from the hologram. This means that the amplitude of the waves originating in P_l can be considered to be of equal magnitude at each point (x_m, y_n). Thus,

$$\frac{A_l \exp(i\varphi_l)}{r_{lmn}} = \frac{A_l}{r_{lmn}} = \text{const}, \qquad \text{for all } l, m, n.$$

Without further restriction we set

$$\frac{A_l \exp(i\varphi_l)}{r_{lmn}} = 1, \qquad \text{for all } l, m, n. \tag{7.47}$$

4. This leaves the expression $E_{lmn} = \exp(ikr_{lmn})$. Since we have not yet fixed λ nor $k = 2\pi/\lambda$, we choose

$$k = 1, \qquad \text{or} \qquad \lambda = 2\pi. \tag{7.48}$$

This finally leads to

$$E_{lmn} = \exp(ir_{lmn}). \tag{7.49}$$

5. As a further simplification we assume that the reference wave E_R strikes the holographic plate at normal incidence. Moreover, to avoid a phase factor, we set E_R real:

$$E_{Rmn} = E_R = \text{const}, \quad \text{for all } m, n; \quad E_R \text{ real}. \tag{7.50}$$

At first sight this seems all we can do.

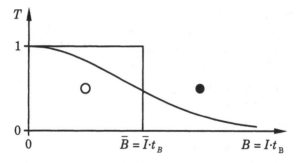

Fig. 7.26. Comparison of the smooth characteristic of a photographic emulsion and a binary characteristic whose threshold is placed at the working point $\bar{B}$.

Now, having calculated the intensity values I_{mn}, we consider their conversion into a real hologram. In an exact simulation of the analogue recording process the intensity distribution should be transferred into a grey pattern according to the characteristic of the photographic plate. Printing grey scales at high resolution, however, is difficult. On the other hand, it is easy to plot a black dot or none, for instance on a laser printer. Therefore, binary plots are commonly used.

In Fig. 7.26, the smooth characteristic of a photographic emulsion is compared with a binary one. The comparison suggests we should plot a black dot when the intensity I_{mn} at the hologram point considered is larger than the average intensity $\bar{I}$. The average depends on the intensity of the reference wave as known from usual holography.

Now we geometrically illustrate the summation of E_{lmn} in the complex plane (Fig. 7.27). If we choose the real amplitude E_R of the reference wave to be large compared to the amplitude of the object wave, it seems logical to choose for the average intensity

$$\bar{I} = |E_R|^2 = E_R^2 . \tag{7.51}$$

When the sum vector $E_{mn} = E_{Smn} + E_R$ points to the outside of the circle of radius $E_R = \sqrt{\bar{I}}$, a black dot is to be plotted. With E_R large, the

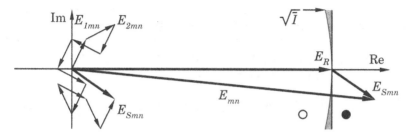

Fig. 7.27. Illustration of the summation of the object waves E_{lmn} and the reference wave E_R. A black dot is plotted whenever the sum vector E_{mn} points to the right of the vertical $\mathrm{Re}\{E\} = E_R$.

shaded area in Fig. 7.27 gets small and hence the circle may be replaced by a straight line without introducing a large error. This tremendously simplifies the decision of whether a point should be blackened or not. Now, the knowledge of E_{Smn} instead of E_{mn} suffices to arrive at this decision:

$$\text{Re}(E_{Smn}) > 0 \rightarrow \text{black dot,}$$
$$\text{else} \rightarrow \text{none.} \tag{7.52}$$

Surprisingly, this simplification has the effect that the reference wave disappears from the calculations. Thus, no intensity has to be calculated. Moreover, we can dispense with the calculation of complex values, as for evaluating (7.52), we only need the real part of E_{Smn}:

$$\text{Re}(E_{Smn}) = \sum_{l=1}^{L} \cos(r_{lmn}). \tag{7.53}$$

The calculation of the hologram now proceeds as follows: calculate $\cos(r_{lmn})$ for all l, then (7.53) for all m and n, and arrange the plot according to the condition (7.52).

For storing the intensity distribution of the hologram in digital form without losing information about the object, according to the sampling theorem certain conditions must be fulfilled with respect to the discretization of the holographic plate [7.8]. This implies that only a certain range of distances of the object from the holographic plane is allowed. The total information of a signal (here the intensity distribution) is retained when the sampling is done at a frequency (here the grid point density) of at least twice the maximum frequency contained in the signal. Here, the frequencies always are the spatial frequencies, which are more closely discussed in the chapter on Fourier optics.

In our case, the sampling frequency is given; it corresponds to the separation of two adjacent grid points. Hence, the z-coordinate of an object point has to be so large that the corresponding interference fringes on the plate are sampled with at least two points per fringe spacing.

As an example, we consider the hologram of a point source on the optical axis taken with a reference beam normal to the plate. This renders a radially symmetric intensity or optical density distribution as depicted in Fig. 7.28 for two cases. Let a be the radius of the holographic plate and b the local fringe spacing between the maxima of intensity at the edge of the plate. Then, the smallest allowed distance $z_{\min}$ of the object point from the holographic plate is given by

$$z_{\min} = \sqrt{\frac{(-2ab + b^2 - \lambda^2)^2}{4\lambda^2} - a^2} \approx a\sqrt{\frac{b^2}{\lambda^2} - 1}. \tag{7.54}$$

If we choose the distance between two grid points in the hologram to be unity, b has to be taken as 2. For a hologram size of $a = 2500$ and with $\lambda = 1$ we get $z_{\min} = 4332$. Previously, we had set $\lambda = 2\pi$. Then $z_{\min} = 0$,

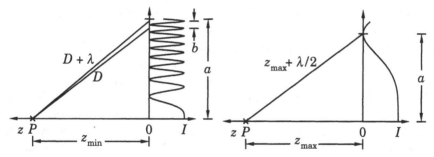

Fig. 7.28. Calculating the minimum distance (*left*) and the maximum distance (*right*) of an object point from the holographic plate.

because even at arbitrarily oblique incidence of the wave from P onto the plate the fringe spacing stays larger than 2. For $\lambda \geq 2$ no restrictions apply for $z_{\min}$.

On the other hand, the object should not be placed too far away from the hologram plane. Otherwise, no modulation may be obtained on the hologram as the central black or white disk covers the total hologram (see Fig. 7.28). The first minimum of the intensity distribution should be within the hologram area. This requirement yields

$$z_{\max} = \frac{4a^2 - \lambda^2}{4\lambda} \tag{7.55}$$

as the maximum allowed distance of an object point. If we again take $a = 2500$, $\lambda = 1$, we get $z_{\max} = 6.25 \cdot 10^6$. In the limit $a \gg \lambda$, (7.55) simplifies to

$$z_{\max} \approx \frac{a^2}{\lambda}. \tag{7.56}$$

The hologram of a single point calculated in the way described and plotted digitally is a Fresnel zone plate (Fig. 7.29). It shows a ring system. The thickness of each zone corresponds to a path difference of $\lambda/2$, taken from

Fig. 7.29. Digital binary hologram of a single point (Fresnel zone plate).

the edges of a zone to the object point. Fresnel zone plates are used in X-ray microscopy for imaging purposes, since lenses are lacking because of too low an index of refraction for transparent media [7.9].

Figures 7.30 and 7.31 show two digital holograms for an object in space, a collection of points forming the word HOLOGRAM.

The letter groups HOLO and GRAM are located at different depths. The hologram patterns were plotted on a laser printer with a resolution of 600 dots per 2.54 cm (600 dpi). The letters are arranged above and parallel to the long hologram side in Fig. 7.31, the side with the coarser interference pattern. The object consists of a total of $L = 130$ points, the hologram area is sampled by $M \times N = 5000 \times 3200 = 16$ million points. The computing time was about 6 hours on a low-cost workstation. One object point required about 11 s per hologram size of $1000 \times 1000 = 10^6$ points. The hologram, when reduced to 2 cm $\times$ 3 cm on a high-definition emulsion (copy film), gives a good reconstruction of the object when being illuminated with He–Ne laser light. Figure 7.32 shows two photographs of the reconstructed word from two views and focused onto two planes at different depths. The spatial extent of the letter arrangement is clearly visible.

The reader wishing to dig deeper into the subject of generating holograms digitally may consult Ref. [7.10]. There speed-up in hologram calculation on the basis of interference of "complex textures" is described, i. e. of holograms of single graphics primitives (points, lines and curves).

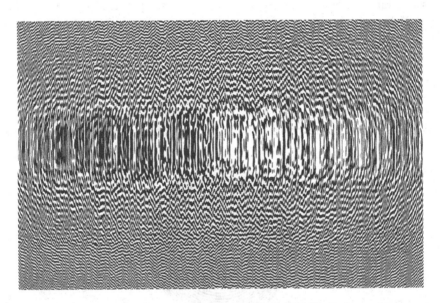

Fig. 7.30. Digital hologram of the letters HOLOGRAM. The holographic area is located in front of the letters.

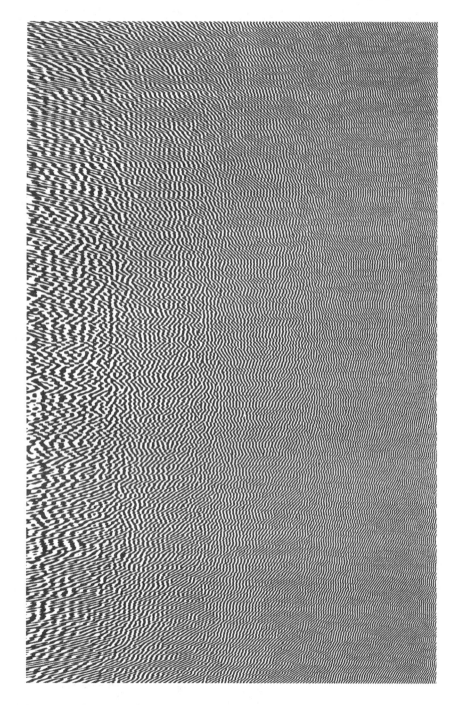

Fig. 7.31. Digital hologram of the letters HOLOGRAM. Turn the page 90 degrees to the right and the letters are located above and along the the upper edge of the hologram area (5000×3200 points).

Fig. 7.32. Reconstruction of the image of the digital hologram of Fig. 7.31 with He–Ne laser light. Two different views and focal settings are used.

7.5.2 Simulation with Square Light Waves

We have pushed the simplification so far that finally the reference wave has disappeared from the calculations. According to (7.53), only a sum of cosine functions has to be formed to obtain the resulting field strength at a point on the holographic plate. The calculation of the cosine function is time consuming, and one may think about replacing it in a further simplification. Proceeding with boldness, we can indeed do this. Why does the cosine function enter the sum? It enters because we assumed harmonic waves for the light. But is holography really bound to harmonic waves? If not what would holography look like with other waveforms such as square light waves? The numeric experimental experience shows that holography with square light waves is feasible.

We define a square wave with period 2π by

$$\text{rect}(r) = \begin{cases} +1 & \text{if} \quad \text{mod}(r, 2\pi) < \pi, \\ -1 & \text{if} \quad \text{mod}(r, 2\pi) \geq \pi. \end{cases} \tag{7.57}$$

If we shift the phase of the cosine by $-\pi/2$, that is, if we consider a sine function instead of a cosine function, the positive and negative sections of the sine wave and the square wave coincide (Fig. 7.33). Then, the real part of E_{Smn} is given as a sum of square waves instead of sine waves:

$$\text{Re}\left(E_{Smn}^{\text{rect}}\right) = \sum_l \text{Re}\left(E_{lmn}^{\text{rect}}\right) = \sum_l \text{rect}(r_{lmn}). \tag{7.58}$$

The corresponding object wave has been denoted by E_{Smn}^{rect} as it differs from the one calculated with sine waves. Instead of a cosine or sine wave we only need a modulo function. For this change of operation we get a fourfold to fivefold decrease in the computing time. Moreover, the still

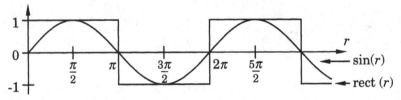

Fig. 7.33. Comparison of sine and square waves.

time-consuming modulo function can be avoided when a suitable wavelength is chosen for the square wave. It just has to be decided whether r_{lmn} falls into the region of an odd or an even multiple of π. With $\lambda = 2$, the square wave attains the value $+1$, when the integer part of r_{lmn} is even, and -1 if it is odd. In the computer code, this requirement can be met by letting r_{lmn} be an integer variable. Thus the noninteger part is cut. Then, after r_{lmn} is calculated, only the least significant bit of the variable has to be checked to decide whether an odd or an even integer is present. The calculation of the distances r_{lmn} is left. No substantial simplification for saving computer time could be found here. The calculation of the square root usually is so fast that an approximation by a Taylor expansion is not rewarding. Furthermore, fast algorithms exist for the calculation of integer square roots of integers.

7.5.3 Digital hologram recording and reconstruction

The handling of holographic films and plates is time consuming. With the advent of CCD chips and cameras the question can be posed, whether these devices are capable of recording a hologram. In principle it should be possible, because a hologram is just an interference pattern. The problem occurs in the resolution, i. e. the number of picture elements (pixels) and their individual size. A holographic plate has a resolution of about 3000 lines per millimeter. CCD chips are available for about 5000×5000 pixels with pixel sizes down to about $7 \, \mu m$. As we have seen in the calculation of digital holograms (Fig. 7.31) modest holograms can be rendered with this number of pixels. The spatial frequency in the interference pattern has to be sufficiently low, of course, to meet the restrictions imposed by the pixel size (see Fig. 7.28).

Indeed, holograms have been recorded digitally on CCD chips [7.11]. Then, a digital reconstruction suggests itself, for instance, as a reverse of the hologram forming process. However, the reconstruction is even more demanding: more points in space have to be reconstructed to arrive at an image of the object because the object points are not known beforehand. In that case a reconstruction by using wave propagation instead of ray tracing is to be recommended because the fast Fourier transform algorithm that can be applied speeds up the calculation considerably. A real breakthrough, however, has to await larger pixel arrays and faster computers.

A second line of exploring the power of digitization is suggested via directly digitizing the hologram in a high-resolution film scanner [7.12]. As many as 8192×8192 pixels have been used for subsequent reconstruction of the holographic image, with scanning the hologram at 126 pixels/mm and even at 315 pixels/mm. The reconstructions are approaching the optical quality, albeit for small images only. Marine zooplankton has been investigated this way.

Problems

7.1 Specify the resolution (in lines per mm) that a holographic film must at least have to correctly record a fringe pattern when the angle of incidence is $\alpha = 30°$ for both the object and the reference wave (wavelength $\lambda = 488$ nm).

7.2 In a holographic exposure an emulsion is used whose transmission characteristic at the working point is approximately given by

$$T = a - bt_B I - ct_B^2 I^2 = a - b'I - c'^2 I^2.$$

Calculate the images upon reconstruction when the hologram is illuminated with the reference wave.

7.3 A reflection hologram is recorded with the reference wave R being incident at an angle $\alpha = 40°$, the object wave S having the angle $\gamma = 30°$ (see Fig.7.5). Upon reconstruction with R, at what angle is the conjugate image formed? Now, the angle of incidence of the reconstruction wave is gradually increased. At what angle α' does the conjugate image disappear?

7.4 The binary amplitude hologram of a coherent point source on the optical axis, recorded with a plane reference wave perpendicular to the holographic plate, is just a Fresnel zone plate (Fig. 7.29). Calculate the radii of the circles that form the boundaries between transparent and opaque regions of the plate. (Assume that during exposure the object and reference waves are in phase at the intersection of the hologram with the optical axis.)

7.5 A white-light hologram is recorded at $\lambda = 633$ nm, both the object and reference waves being perpendicular to the holographic plate. When the developed hologram is viewed, at what angle of incidence α does the image appear in green color ($\lambda = 500$ nm)?

7.6 Consider the imaging equations of holography, (7.32) to (7.37). Derive expressions for the longitudinal magnification, M_{lo}, and the lateral magnification, M_{lat}, of both the direct and the conjugate image. The recording wavelength λ_1 and the reconstruction wavelength λ_2 may be different. To start, analyse the image locations for a hologram of two point sources with lateral separation Δx and with longitudinal separation Δz, respectively, where $|\Delta x|, |\Delta z| \ll |z|$. Distinguish the following cases for the choice of reference wave: (a) divergent spherical wave upon recording and reconstruction; (b) divergent spherical wave upon recording and a plane wave upon reconstruction.
Show that for both the direct and conjugate image the following relation holds:

$$M_{lo} = \frac{\lambda_2}{\lambda_1}(M_{lat})^2.$$

7.7 Give a derivation of the conditions (7.54) and (7.55) for the minimum and maximum distance, respectively, an object point should have from a digital hologram.

7.8 Estimate the computing time required for the construction of a digital hologram having $4 \cdot 10^6$ raster points and containing 10^6 object points when all simplifications discussed in the text are used. Assume that one integer operation lasts 10 ns, and that the calculation of roots requires 200 ns.

7.9 Generate a digital hologram of a point source and of a square composed of point sources (located, for instance, at the corners of the square).

8. Interferometry

Interferometric methods allow the measurement of small changes of physical quantities (length, pressure, temperature, etc.) that have influence on the propagation properties of light. This is generally done by letting two waves interfere, one undisturbed and one changed by the object [8.1].

8.1 Mach–Zehnder Interferometer

We discuss the principle of classical interferometry by choosing the Mach–Zehnder interferometer (Fig. 8.1). The beam of a light source is divided at beam splitter 1. One of the beams is directed through the transparent object to be investigated, the other one through a reference object for comparison. Both beams are combined by beam splitter 2 and thus brought to interference. From the interference pattern, variations in the phase front

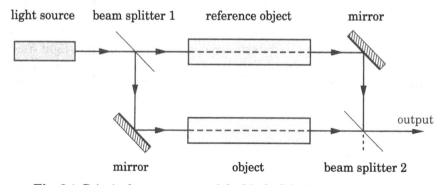

Fig. 8.1. Principal arrangement of the Mach–Zehnder interferometer.

of the test object as compared to the reference object can be determined. Often just a plane wave is taken as the reference wave. Only transparent objects can be investigated. The Mach–Zehnder interferometer is used for measuring slight changes in the refractive index of transparent objects. Areas of application are fluid flow, shock waves, heat transfer, and diffusion, to name the most important ones.

8.2 Sagnac Interferometer

Interferometry can also be used for rotation sensing. The instrument is called the Sagnac interferometer after the French physicist *Georges Sagnac* (1869–1928) who investigated it around 1910. The basic principle consists in splitting a light beam into two and letting the two beams interfere after they have traversed an optical loop in opposite directions [8.2]. Figure 8.2 shows an arrangement with one beam splitter and three mirrors. When the complete arrangement is rotated, for instance, by the

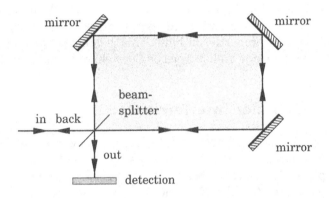

Fig. 8.2. A Sagnac interferometer.

rotation of the Earth, then a phase shift is observed between the two counterpropagating waves.

To understand the operation, we consider a modern form of the interferometer: the fiber Sagnac interferometer (Fig. 8.3). Coherent light from a light source (semiconductor laser) is fed by a lens into a single-mode, polarization preserving fiber. The beam splitter is realized by a coupling

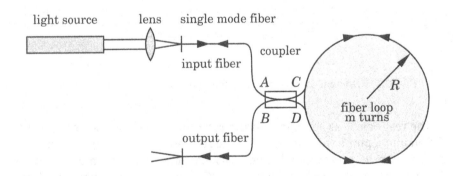

Fig. 8.3. Fiber Sagnac interferometer.

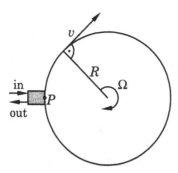

Fig. 8.4. Sketch of the one-loop interferometer.

element with input ports A and B and output ports C and D. The fiber is coiled up for increased sensitivity. The light traversing the multiturn loop in opposite directions is recombined by the same coupler and directed to the output fiber. Similarly to the Michelson interferometer, more or less light intensity is observed at the output according to the relative phase of the counterpropagating waves, depending on the angular velocity of the interferometer.

For simplicity, we consider a one-loop interferometer (Fig. 8.4), the loop radius being R, its angular velocity, Ω, and its tangential velocity, v. We know from mechanics that

$$v = \Omega R. \tag{8.1}$$

At $\Omega = 0$, the light traverses the loop in the time t_1, independently of the direction:

$$t_1 = \frac{2\pi R}{c}, \tag{8.2}$$

where $2\pi R$ is the circumference of the loop and c is the velocity of light in the fiber.

Let $\Omega \neq 0$. Then, in the time t_1, the point P of the fiber has been advanced by a distance s_1 by the rotation:

$$s_1 = vt_1 = v\frac{2\pi R}{c}. \tag{8.3}$$

To reach the starting point P again, the light, propagating in the direction of rotation, must additionally traverse this distance and for that it needs the additional time

$$t_2 = \frac{s_1}{c} = \frac{v2\pi R}{c^2}. \tag{8.4}$$

Thus, we get a round trip time of

$$t_1 + t_2 = \frac{2\pi R}{c}\left(1 + \frac{v}{c}\right). \tag{8.5}$$

The light propagating in the opposite direction needs less time because it saves the distance s_1. Now, the round trip time is

$$t_1 - t_2 = \frac{2\pi R}{c}\left(1 - \frac{v}{c}\right). \tag{8.6}$$

These times differ by

$$\Delta t = (t_1 + t_2) - (t_1 - t_2) = 2t_2 = 2\frac{v 2\pi R}{c^2} = \frac{4\pi \Omega R^2}{c^2} = \frac{4A}{c^2}\Omega, \tag{8.7}$$

where $A = \pi R^2$ is the area of the loop. When $v \ll c$ this approximation is sufficiently accurate. As physicists, however, we reach out for a more exact relation. We note that the interferometer has moved during the time t_2 in which the light traverses the distance s_1. We call this additional distance s_2:

$$s_2 = v t_2 = \frac{v^2 2\pi R}{c^2}. \tag{8.8}$$

Thus, the light also has to travel this distance s_2. This takes the time

$$t_3 = \frac{s_2}{c} = \frac{v^2 2\pi R}{c^3}. \tag{8.9}$$

Then, the total time for the light traveling around the loop in the direction of rotation is

$$t_1 + t_2 + t_3 = \frac{2\pi R}{c}\left(1 + \frac{v}{c} + \frac{v^2}{c^2}\right) \tag{8.10}$$

and for the light traveling in the opposite direction it is

$$t_1 - t_2 - t_3 = \frac{2\pi R}{c}\left(1 - \frac{v}{c} - \frac{v^2}{c^2}\right). \tag{8.11}$$

As is easily noticed, this operation has to be continued infinitely often. It is the famous problem that *Zeno of Elea* (490–430 B.C.) put forward of Achilles and the tortoise. Today, we are able to sum up the geometric series and do not detect any problem. In doing so, we obtain the following result for the total time the light traveling in the direction of rotation needs:

$$t_{\text{sum}}^+ = \frac{2\pi R}{c}\left(1 + \sum_{n=1}^{\infty}\left(\frac{v}{c}\right)^n\right) = \frac{2\pi R}{c}\sum_{n=0}^{\infty}\left(\frac{v}{c}\right)^n = \frac{2\pi R}{c}\frac{1}{1 - v/c}. \tag{8.12}$$

The light traveling in the opposite direction needs the time

$$t_{\text{sum}}^- = \frac{2\pi R}{c}\left(1 - \sum_{n=1}^{\infty}\left(\frac{v}{c}\right)^n\right) = \frac{2\pi R}{c}\left(1 - \left(\sum_{n=0}^{\infty}\left(\frac{v}{c}\right)^n - 1\right)\right)$$

$$= \frac{2\pi R}{c}\left(2 - \frac{1}{1 - v/c}\right). \tag{8.13}$$

These times differ by

$$\Delta t_{\text{sum}} = t_{\text{sum}}^+ - t_{\text{sum}}^- = \frac{2\pi R}{c}\frac{1}{1-v/c} - \frac{2\pi R}{c}\left(2 - \frac{1}{1-v/c}\right)$$

$$= \frac{4\pi R}{c}\left(\frac{1}{1-v/c} - 1\right) = \frac{4\pi Rv}{c^2(1-v/c)} \tag{8.14}$$

or

$$\Delta t_{\text{sum}} = \frac{4\pi Rv}{c^2(1-v/c)} = \frac{4A}{c^2(1-v/c)}\Omega . \tag{8.15}$$

For $v \ll c$ the approximate relation (8.7) is recovered.

How do we get the phase shift? If $T = \lambda/c$ is the period of the monofrequency light used, then $\Delta N = \Delta t_{\text{sum}}/T$ is the fringe shift and

$$\Delta\varphi = 2\pi\,\Delta N = 2\pi\frac{\Delta t_{\text{sum}}}{T} = \frac{8\pi A\Omega c}{c^2\lambda} = \frac{8\pi A}{c\lambda}\Omega \tag{8.16}$$

is the phase shift effected by the rotation ($v \ll c$). It is proportional to the angular velocity Ω.

To increase the sensitivity, the light is made to travel several times around the loop. An m-fold loop yields an m-fold phase shift

$$\Delta\varphi = m\frac{8\pi A}{c\lambda}\Omega . \tag{8.17}$$

Fiber technicians write this relation in the form

$$\Delta\varphi = \frac{4\pi LR}{\lambda c}\Omega , \tag{8.18}$$

where L is the length of the fiber wound up on the coil. With $L = m2\pi R$ and $A = \pi R^2$ the reader easily detects the equality of both formulations.

Finally, when the axis of rotation forms an angle Θ with the loop axis, only the component $\Omega\cos\Theta$ of the angular velocity Ω adds to the phase shift and $\Delta\varphi$ reads:

$$\Delta\varphi = \frac{4\pi LR\cos\Theta}{\lambda c}\Omega . \tag{8.19}$$

Ring-cavity lasers show the Sagnac effect, too. In this active device the frequencies adapt to the unequal optical paths, giving rise to a beat frequency of

$$\Delta\nu = \frac{4A}{\lambda P}\Omega , \tag{8.20}$$

P being the perimeter of the instrument. Using supercavity mirrors with a reflectance of $R \geq 99.9985\%$, one can achieve a finesse of $F \geq 30000$ and a quality factor of $Q \geq 4 \cdot 10^{11}$. The resulting frequency resolution of $1 \cdot 10^{-20}$ (according to $1\,\text{mHz}$ in $474\,\text{THz}$, the frequency of the He–Ne laser used) is high enough to measure deviations in the rotation of the earth [8.3] due to earthquakes, for example.

8.3 Holographic Interferometry

In classical interferometry, the objects to be compared have to be simultaneously present and must be transparent or highly reflecting. Holographic interferometry is free of all these restrictions [8.4]. Light waves can be stored with amplitude and phase for interferometric comparison at any later time! Therefore, the light waves to be compared may be derived from objects no longer existent. This makes autointerferometry possible, where the object is compared with itself at a later time. Arbitrarily formed, diffusely scattering objects that cannot be produced identically a second time for comparison are now available for interferometry. Tires of cars and of airplanes, for instance, can be checked for defects. The tire is compared with itself in an initial state and a more inflated state. A defect would lead to a larger local expansion visible in the density of the interference fringes.

Depending on the type of holographic recording essentially three types of holographic interferometric methods are distinguished.

8.3.1 Real-Time Method

A reference state of an object is taken holographically. The holographic plate is processed at the location of the recording without being moved. Then the object is changed in the desired way and illuminated in the same manner with coherent light as during recording. Simultaneously, by illuminating the hologram with the reference wave, the reference state of the object is reconstructed. Both waves interfere giving interference fringes that indicate the change, for instance, a deformation of the object. For documentation, the interference fringe system has to be photographed (see Fig. 8.5). This method has the advantage that the object may be changed continuously with the deformation being immediately visible in the fringe system. A disadvantage is that the relative location of the object and hologram may not be changed even slightly, except to produce a known amount of initial interference fringes. Consequently, the object may not be removed, since it is practically impossible to relocate it precisely enough with respect to the hologram.

8.3.2 Double-Exposure Method

In this method the holographic plate is exposed twice, once with the object in a reference state and a second time with it in a modified state. Thus, both wavefronts are recorded on the same holographic plate. When the hologram is illuminated with the reference wave, both waves are reconstructed and interfere. The interference pattern containing the information about the object modification can be said to be frozen in the hologram. It is put aside along with the hologram, so that upon reconstruction both waves are restored with their exact phase relationship (see Fig. 8.6).

Fig. 8.5. Two states of a stressed PMMA ring. Interference fringes made visible with real-time holographic interferometry.

Fig. 8.6. Reconstructed images of a hologram that has been exposed twice (double-exposure holographic interferometry). A water-filled wine glass played by a violin bow served as the object. The hologram was taken with two pulses of a ruby laser of about 20 ns duration separated by 300 µs.

8.3.3 Time-Average Method

The time-average method is best suited for periodically, preferably sinu-soidally, vibrating objects. Just one exposure of the oscillating object is taken with an exposure time usually large compared to the period of oscil-lation. A sinusoidally vibrating object stays at the turning points longer. Therefore, the light waves reflected from these positions lead to a stronger interference pattern in the holographic recording process than those from the intermediate positions do and thus are reconstructed more strongly. The exact theory is more involved. A drawback of this method is that the modulation of the interference fringes quickly decreases for higher elongations and thus the signal-to-noise ratio is reduced for higher order fringes (Fig. 8.7).

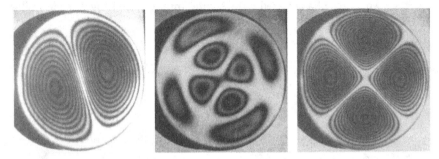

Fig. 8.7. Reconstructed images from holograms recorded with time-average holo-graphic interferometry. An oscillating tin cap served as the object. Three different oscillation modes are shown, belonging to different frequencies of forcing.

8.4 Theory of Holographic Interferometry

The theoretical description of the different methods of holographic inter-ferometry has much in common with that of the corresponding methods of speckle photography. However, the present methods do not work for in-plane displacements, but, instead, for displacement components normal to the surface.

8.4.1 Real-Time and Double-Exposure Method

The theoretical descriptions of the real-time and the double-exposure methods are quite simple and are similar for both methods. We choose the real-time method to formulate the theory.

First, a hologram of an object is recorded (see Fig. 7.15). The holographic plate is processed in place and the object stays untouched. Subsequently, the hologram is illuminated with the reference wave R. From the hologram a light wave E_1 proceeds that corresponds to the light wave from the original object up to a phase shift. Now the object is displaced a distance d towards the holographic plate and illuminated as when the hologram was taken. This gives rise to a light wave E_2 from the object that passes the hologram. The mutually coherent waves E_1 and E_2 interfere and cast a system of easily visible interference fringes over the object, when the displacements along the object are neither too large nor too small.

We consider a point P on the surface of the object. Assume the displacement d is only a few wavelengths. Then, the modulus of the complex amplitude stays unchanged, and we have only a phase difference in the viewing direction. Assume

$$E_1 = A_1 \exp(i\varphi_1) \quad \text{and} \quad E_2 = A_2 \exp(i\varphi_2) \tag{8.21}$$

at a point in the viewing direction, then, along with $A_1 = A_2$ and $\Delta\varphi = \varphi_2 - \varphi_1$, we have

$$E_2 = E_1 \exp(i\Delta\varphi). \tag{8.22}$$

The observer records the intensity I that depends on $\Delta\varphi$:

$$\begin{aligned}
I &= (E_1 + E_2)(E_1 + E_2)^* \\
&= E_1 E_1^* [1 + \exp(i\Delta\varphi)][1 + \exp(-i\Delta\varphi)] \\
&= I_1(2 + 2\cos\Delta\varphi) \\
&= 4I_1 \cos^2 \frac{\Delta\varphi}{2}.
\end{aligned} \tag{8.23}$$

This relation has already been obtained in a similar form for the Michelson interferometer (compare (4.9)). Figure 8.8 shows the intensity obtained upon superposition of two waves as a function of the phase difference $\Delta\varphi$ between them.

As with the Michelson interferometer, the phase shift $\Delta\varphi = kl$ is brought about by a path difference l between the interfering waves. How-

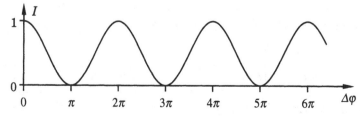

Fig. 8.8. Intensity upon superposition of two waves as a function of the phase difference $\Delta\varphi$.

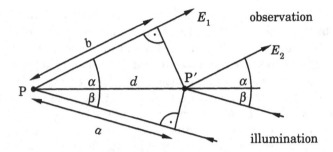

Fig. 8.9. Geometry for calculating the phase shift between E_1 and E_2, when the object point is displaced from P to P'.

ever, in this case one of the two waves is recalled from a hologram. With the notations given in Fig. 8.9 we have $l = a + b$ and

$$\Delta\varphi = k(a+b) = kd(\cos\alpha + \cos\beta). \tag{8.24}$$

This result, inserted into (8.23), yields

$$I = 4I_1 \cos^2\left(\frac{kd}{2}(\cos\alpha + \cos\beta)\right). \tag{8.25}$$

The angles α and β as well as k are known from the experimental arrangement. Figures 8.5 and 8.6 are examples of interference fringes obtained with this method. In these cases the phase difference $\Delta\varphi$ has been effected by variations in the optical path lengths in transparent objects due to changes in the refractive index by strains (Fig. 8.5) and due to changes in the thickness caused by vibration of the glass (Fig. 8.6).

8.4.2 Time-Average Method

For analysing oscillating objects, the time-average method is used. A hologram is recorded with an exposure time usually comprising many oscillation periods. Since the object is moving during the exposure, a time-dependent intensity (short-term intensity) $I(x,y,t)$ comes into play. We assume a sinusoidally oscillating, diffusely scattering object in the geometry of Fig. 8.9 with the illumination and observation directions perpendicular to the object. Then the phase of the object wave S is sinusoidally modulated. Using (8.24), we get, for $\alpha = \beta = 0$,

$$S(t) = A\exp(ik2d\sin\Omega t). \tag{8.26}$$

The quantities d and Ω are the amplitude and circular frequency of the oscillation, respectively. The factor of 2 appears because of front illumination and viewing. We then obtain for the short-term intensity

$$I(t) = (S(t) + R)(S(t) + R)^*. \tag{8.27}$$

The short-term intensity is integrated during the exposure time t_B:

$$B = \int_0^{t_B} I(t)\,dt.$$ (8.28)

According to (7.6) or (6.29), we obtain the following for the amplitude transmittance T in the vicinity of the working point of the emulsion:

$$T = a - b \int_0^{t_B} I(t)\,dt.$$ (8.29)

Inserting $I(t)$ from (8.27) yields the transmission distribution of the hologram. Upon reconstruction, that is, by forming the expression RT with the reference wave R, we obtain the usual four terms, one of them being the direct image

$$E_d \propto \int_0^{t_B} S(t)\,dt = \int_0^{t_B} A \exp(ik2d \sin \Omega t)\,dt.$$ (8.30)

We came across an integral of this type when discussing speckle interferometry and could relate it to the Bessel function J_0 defined in (6.32). Hence (8.30) reads:

$$E_d \propto J_0(2kd).$$ (8.31)

For the intensity we find

$$I_d = E_d E_d^* \propto J_0^2(2kd).$$ (8.32)

The graph of the function J_0^2 has already been given in Fig. 6.15. We recall that the roots of the function J_0^2 are not equally spaced and that the higher-order maxima get smaller with the phase shift. Thus, the signal-to-noise ratio decreases for higher fringe orders. This can be detected in Fig. 8.7.

8.4.3 Time-Average Method in Real Time

To get a quick overview of the oscillation modes of a plate, the time-average method may be used in combination with the real-time method. Unlike in the usual real-time method with stationary objects, a different intensity distribution with respect to the pure time-average hologram is obtained, because the reconstructed wave of the stationary object S_r and the oscillating object wave $S(t) = S_r \exp(i2kd \sin \Omega t)$ are simultaneously present.

For simplicity, we assume an illumination wave with a normal incidence and a normal viewing direction, and set, for ease of notation,

$$\varphi(t) = 2kd \sin \Omega t = \varphi_0 \sin \Omega t. \tag{8.33}$$

We know that the waves leaving the hologram interfere to form the sum

$$E_d = S_r + S_r \exp[i\varphi(t)] \tag{8.34}$$

for the direct image. Since we only perceive the intensity, we have in this case

$$I_d = \langle E_d E_d^* \rangle_{t_B} = \frac{1}{t_B} \int_0^{t_B} (S_r + S_r \exp[i\varphi(t)])(S_r + S_r \exp[i\varphi(t)])^* \, dt. \tag{8.35}$$

Here, t_B is the averaging time of the eye or the recording instrument (for instance, a CCD camera). We assume t_B to be an integer multiple of the oscillation period $T_S = 2\pi/\Omega$. Then we have

$$I_d = |S_r|^2 \frac{1}{t_B} \int_0^{t_B} (1 + \exp[i\varphi(t)])(1 + \exp[-i\varphi(t)]) \, dt$$

$$= |S_r|^2 \frac{1}{t_B} \int_0^{t_B} [2 + 2 \cos \varphi(t)] \, dt$$

$$= |S_r|^2 \, 2 \left(1 + \frac{1}{2\pi} \int_0^{2\pi} \cos(\varphi_0 \sin \Omega t) \, d(\Omega t) \right). \tag{8.36}$$

In this expression, again the Bessel function is hidden, this time as $J_0(\varphi_0) = J_0(2kd)$. Therefore, the intensity distribution of the fringe system reads:

$$I_d = 2|S_r|^2 (1 + J_0(2kd)). \tag{8.37}$$

In the general case, when neither the illumination nor the viewing is normal to the surface of the object (see Fig. 8.9), the expression $2kd$ has to be replaced by $kd(\cos \alpha + \cos \beta)$ as given by (8.24).

Comparing the interference pattern obtained by the usual time-average method with that of the time-average method in real time (Fig. 8.10), we notice that the usual time-average method yields twice the fringe density and a better contrast.

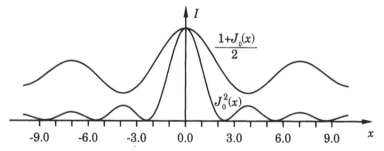

Fig. 8.10. Fringe density for the usual time-average method (J_0^2) and for the time-average method in real time ($1/2 + J_0/2$).

Problems

8.1 For the reconstruction of a double-exposure interferogram, a wavelength different from the recording wavelength is used. How does this affect the interference fringes visible in the reconstructed object image?

8.2 In Fig. 8.7, the reconstructed images of a vibrating tin lid from a time-average holographic recording are shown. Describe the mode of vibration of the cap in each of the pictures and determine the corresponding vibration amplitudes. A wavelength of $\lambda = 633$ nm has been used for recording and reconstruction. Assume that the angle between the vibration direction and the direction of illumination and viewing is $20°$. Roots of the function $J_0(z)$ are at $z_1 = 2.404$, $z_2 = 5.52$, ..., $z_9 = 27.49$, $z_{10} = 30.63$.

8.3 An acoustic transducer is excited by an asymmetric square-wave signal. A time-average interferogram is taken of the transducer's membrane. Give the intensity distribution of the fringe system that covers the object image upon reconstruction. How does this interference pattern depend on the duty cycle γ of the excitation signal?

9. Fourier Optics

The Fourier transform plays an important role in science and technology. The one-dimensional Fourier transform $\mathcal{F}$, for instance, connects a time signal $f(t)$ with its complex spectral function $A(\nu)$ which gives information about the frequency content of the signal:

$$A(\nu) = \mathcal{F}[f(t)](\nu) = \int\limits_{-\infty}^{+\infty} f(t) \exp(-2\pi i \nu t)\,\mathrm{d}t. \tag{9.1}$$

In optics, mainly the two-dimensional Fourier transform applies. For instance, the electric field in the back focal plane of a convex lens is the two-dimensional spatial Fourier transform of the electric field $E(x,y)$ in the front focal plane of the lens. Similarly to (9.1), the two-dimensional Fourier transform $\tilde{E}(\nu_x, \nu_y)$ is defined by

$$\tilde{E}(\nu_x, \nu_y) = \mathcal{F}[E(x,y)](\nu_x, \nu_y)$$

$$= \int\limits_{-\infty}^{+\infty}\int\limits_{-\infty}^{+\infty} E(x,y) \exp\left[-2\pi i(\nu_x x + \nu_y y)\right]\,\mathrm{d}x\,\mathrm{d}y. \tag{9.2}$$

Here, with a view to the frequency ν, the quantities ν_x and ν_y are called spatial frequencies. The basic definitions and relations pertaining to the Fourier transform are summarized in the Appendix. To understand why the Fourier transform occurs in many optical arrangements, we consider the diffraction of light at an aperture in the (x,y)-plane in the approximation of scalar diffraction theory.

9.1 Scalar Diffraction Theory

We consider the diffraction at a transparency having the transmittance distribution $\tau(x,y)$ in the plane $z = 0$. When illuminating it with a plane, linearly polarized monofrequency wave of wavelength λ, we obtain a wave field propagating into the half space behind, which we set out to describe (see Fig. 9.1).

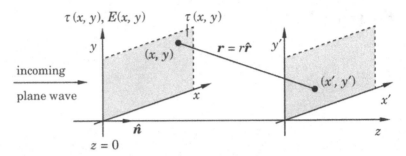

Fig. 9.1. Geometry for describing the diffraction of a plane wave at a transmittance distribution $\tau(x,y)$ in the plane where $z = 0$.

To begin, immediately behind the plane where $z = 0$ the field

$$E(x,y) = \tau(x,y)E_e(x,y) \tag{9.3}$$

is present, where $E_e(x,y)$ is the electric field strength of the incident wave in the $(x,y,0)$-plane. To a good approximation, further propagation can be described by the assumption that each point $(x,y,0)$ of the diffraction pattern is the source of a spherical wave (Huygens' principle).

Then, in a plane with coordinates (x',y',z) (see Fig. 9.1), a field $E(x',y',z)$ is obtained given by the Kirchhoff diffraction integral:

$$E(x',y',z) = \frac{1}{i\lambda} \int\limits_{-\infty}^{+\infty} \int\limits_{-\infty}^{+\infty} E(x,y)\frac{1}{r}\exp(ikr)\cos(\hat{n}\hat{r})\,\mathrm{d}x\,\mathrm{d}y. \tag{9.4}$$

Mainly, it corresponds to a sum of the spherical waves originating at all points $(x,y,0)$. Here, $1/i\lambda$ is a phase and amplitude factor and $\cos(\hat{n}\hat{r})$ is a directional factor. Both follow from Maxwell's equations, from which (9.4) is derived.

9.1.1 Fresnel Approximation

The Kirchhoff integral (9.4) is usually unwieldy and further approximations are needed for an analytical treatment. In the paraxial approximation, the coordinates of x and y and also x' and y' are restricted to values which are small compared to the distance z between the diffracting object and its diffraction pattern, that is, $|x|, |y| \ll z$ and $|x'|, |y'| \ll z$. This means that only rays which form a small angle with the optical axis, the z-axis, are considered. Then, to a good approximation, the cosine factor in the integrand can be neglected. Thus we can set

$$\cos(\hat{n}\hat{r}) = 1, \tag{9.5}$$

since all light rays travel almost parallel to the z-axis. Furthermore, the r-dependence of the amplitudes can be ignored. That is, since $r \approx z$, we can set $1/r = 1/z$. However, this approximation is too coarse for the exponential function, because small changes in r lead to large changes in the phase kr. Here, the approximation is carried one step further. The square root in

$$r = \sqrt{(x'-x)^2 + (y'-y)^2 + z^2} = z\sqrt{1 + \frac{(x'-x)^2}{z^2} + \frac{(y'-y)^2}{z^2}} \qquad (9.6)$$

is expanded into a power series and truncated after the second term:

$$r \approx z + \frac{(x'-x)^2}{2z} + \frac{(y'-y)^2}{2z}. \qquad (9.7)$$

With these approximations, the Fresnel approximation to the diffraction integral is obtained:

$$E(x',y',z) = \frac{\exp(ikz)}{i\lambda z} \int\limits_{-\infty}^{+\infty}\int\limits_{-\infty}^{+\infty} E(x,y)\exp\left(\frac{ik}{2z}\left[(x'-x)^2 + (y'-y)^2\right]\right) dx\,dy. \qquad (9.8)$$

This form of the diffraction integral lends itself to a different interpretation in the language of one-dimensional system theory. To this end, we define the function

$$h_z(x,y) = \frac{\exp(ikz)}{i\lambda z}\exp\left[i\frac{k}{2z}\left(x^2 + y^2\right)\right], \qquad (9.9)$$

the impulse response of free space. Then, the Kirchhoff diffraction integral can be written as a convolution of the complex electric field distribution in the diffracting plane with the impulse response of free space:

$$E(x',y',z) = E(x,y) * h_z(x,y). \qquad (9.10)$$

(Concerning the notion of convolution see the Appendix.) Indeed, from the definition of the convolution it is easily seen that the expression (9.10) is the Kirchhoff diffraction integral in the Fresnel approximation:

$$E(x',y',z) = E(x,y) * h_z(x,y)$$

$$= \int\limits_{-\infty}^{+\infty}\int\limits_{-\infty}^{+\infty} E(x,y)h_z(x'-x,y'-y)\,dx\,dy$$

$$= \frac{\exp(ikz)}{i\lambda z} \int\limits_{-\infty}^{+\infty}\int\limits_{-\infty}^{+\infty} E(x,y)\exp\left(\frac{ik}{2z}[(x'-x)^2 + (y'-y)^2]\right) dx\,dy. \qquad (9.11)$$

The Fresnel approximation can also be written differently with the help of the Fourier transform. To this end, the expressions $(x'-x)^2$ and $(y'-y)^2$ are expanded and suitably collected:

$$E(x',y',z) = \frac{\exp(ikz)}{i\lambda z} \exp\left[i\frac{\pi}{\lambda z}\left(x'^2 + y'^2\right)\right]$$

$$\cdot \int\limits_{-\infty}^{+\infty}\int\limits_{-\infty}^{+\infty} E(x,y)\exp\left[i\frac{\pi}{\lambda z}\left(x^2 + y^2\right)\right] \exp\left[-2\pi i\left(\frac{x'}{\lambda z}x + \frac{y'}{\lambda z}y\right)\right] dx\,dy.$$

$$(9.12)$$

With the abbreviation

$$A(x',y',z) = \frac{\exp(ikz)}{i\lambda z} \exp\left[i\frac{\pi}{\lambda z}\left(x'^2 + y'^2\right)\right] \tag{9.13}$$

and the definition (9.2) of the Fourier transform, we can write

$$E(x',y',z) = A(x',y',z)\mathcal{F}\left[E(x,y)\exp\left(i\frac{\pi}{\lambda z}\left(x^2 + y^2\right)\right)\right]\left(\frac{x'}{\lambda z}, \frac{y'}{\lambda z}\right). \tag{9.14}$$

Thus, the electric field in a plane where $z = $const is given by a Fourier transform of the electric field distribution in the diffracting plane after multiplication with a quadratic phase factor $\exp[(i\pi/\lambda z)(x^2 + y^2)]$.

The Fresnel approximation yields good results for quite short distances from the diffracting plane, in many cases down to ten wavelengths only.

9.1.2 Fraunhofer Approximation

For large distances from the diffracting plane (far field) and for a finite size of the diffracting pattern in x and y, the quadratic phase term becomes small and may be neglected. More precisely, it is required that

$$z \gg \frac{\pi}{\lambda}\left(x^2 + y^2\right). \tag{9.15}$$

Then the quadratic phase factor becomes

$$\exp\left[\frac{i\pi}{\lambda z}\left(x^2 + y^2\right)\right] \approx 1, \tag{9.16}$$

corresponding to negligible curvature of the wave fronts, and the Fraunhofer approximation is obtained:

$$E(x',y',z) = A(x',y',z)\int\limits_{-\infty}^{+\infty}\int\limits_{-\infty}^{+\infty} E(x,y)\exp\left[-2\pi i\left(\frac{x'}{\lambda z}x + \frac{y'}{\lambda z}y\right)\right] dx\,dy. \tag{9.17}$$

If we introduce new coordinates

$$\nu_x = \frac{x'}{\lambda z} \quad \text{and} \quad \nu_y = \frac{y'}{\lambda z}, \tag{9.18}$$

the expression for the electric field reads

$$E(x',y',z) = A(\lambda z\nu_x, \lambda z\nu_y, z)\mathcal{F}[E(x,y)](\nu_x, \nu_y) = \tilde{E}(\nu_x, \nu_y). \tag{9.19}$$

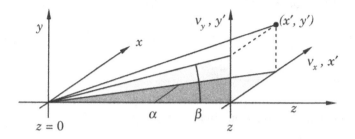

Fig. 9.2. Geometry for the definition of spatial frequencies.

The determination of the diffraction pattern in the far field has been reduced to a Fourier transform of the electric field distribution immediately behind the diffracting structure. The coordinates v_x and v_y introduced in (9.18) are called spatial frequencies. They are proportional to the corresponding diffraction angles α and β in the (x,z)- and the (y,z)-planes, respectively.

From Fig. 9.2 we find:

$$v_x = \frac{x'}{\lambda z} = \frac{\tan\alpha}{\lambda} \approx \frac{\alpha}{\lambda},$$
$$v_y = \frac{y'}{\lambda z} = \frac{\tan\beta}{\lambda} \approx \frac{\beta}{\lambda}. \tag{9.20}$$

The far-field or Fraunhofer approximation can be given a very intuitive interpretation. It is a decomposition of the light field $E(x,y)$ into plane waves propagating at angles α and β. As the approximation is valid for small α and β, we have $\tan\alpha \approx \sin\alpha \approx \alpha$ and $\tan\beta \approx \sin\beta \approx \beta$. Then the quantities x'/z and y'/z can be written as

$$\frac{x'}{z} \approx \sin\alpha \quad \text{and} \quad \frac{y'}{z} \approx \sin\beta. \tag{9.21}$$

This means that the exponential function in the integral of (9.17), along with the factor $\exp(ikz)$, represents a plane wave,

$$\exp(ikx\sin\alpha + iky\sin\beta + ikz), \tag{9.22}$$

that forms the angles α and β with the z-axis in the (x,z)- and (y,z)-planes, respectively. For given λ and z we have

$$v_x \propto \alpha \propto x', \qquad v_y \propto \beta \propto y'. \tag{9.23}$$

Thus, the rays arriving at the far-field point (x',y') leave the (x,y)-plane at the angles $\alpha \propto x'$ and $\beta \propto y'$. Large angles correspond to high spatial frequencies, small angles to low spatial frequencies.

The diffraction pattern at the point (x',y',z) on a screen is given by the intensity $I = |E(x',y',z)|^2$. Because the modulus has to be taken, the phase factor

$$A = \exp(ikz)\exp\left[\frac{i\pi}{\lambda z}\left(x'^2 + y'^2\right)\right] \tag{9.24}$$

drops out, and the diffraction pattern is given, up to a factor, by the spatial power spectrum of the input field:

$$I(\nu_x, \nu_y) = \frac{1}{\lambda^2 z^2}\left|\mathcal{F}[E(x,y)](\nu_x, \nu_y)\right|^2. \tag{9.25}$$

From this relation we see that the diffraction pattern gets fainter the farther it is taken from the diffracting object and the larger the wavelength. Both facts are intuitively comprehended. At large wavelengths the light is diffracted more strongly and thus, per given solid angle, less light is available. At a larger distance z, the light in a given solid angle is distributed over a larger area in the (x',y')-plane.

9.2 Fourier Transform by a Lens

A convex lens of focal length f_l focuses parallel rays (plane waves) into the back focal plane (Fig. 9.3). The coordinate v of the focal point in the back focal plane, the (u,v)-plane, corresponds to the angle β the rays form in the (y,z)-plane or (v,z)-plane with the optical axis:

$$\frac{v}{f_l} = \tan\beta \approx \beta. \tag{9.26}$$

A corresponding relation holds for the coordinate u and the angle α. This means that the lens projects the far-field diffraction pattern into its back

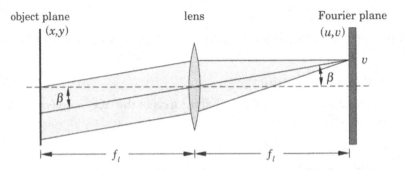

Fig. 9.3. Geometric configuration of rays for demonstrating the Fourier transform properties of a lens, $2f$ arrangement.

focal plane, since

$$v_x = \frac{x'}{\lambda z} = \frac{\alpha}{\lambda} = \frac{u}{\lambda f_l},$$

$$v_y = \frac{y'}{\lambda z} = \frac{\beta}{\lambda} = \frac{v}{\lambda f_l}.$$

$$(9.27)$$

Written in the coordinates $u = \lambda f_l v_x$ and $v = \lambda f_l v_y$ of the back focal plane, we have

$$\tilde{E}(u,v) = A(u,v,f_l) \int\limits_{-\infty}^{+\infty} \int\limits_{-\infty}^{+\infty} E(x,y) \exp\left[-2\pi i \left(\frac{u}{\lambda f_l}x + \frac{v}{\lambda f_l}y\right)\right] dx \, dy. \quad (9.28)$$

When the input field $E(x,y)$ is located in the front focal plane of the lens, the phase factor in A becomes independent of u and v. In this case the lens exactly performs a two-dimensional Fourier transform from the front to the back focal plane:

$$\tilde{E}(u,v) \propto \mathcal{F}[E(x,y)](u,v). \quad (9.29)$$

The function $\mathcal{F}[E(x,y)]$ is called the (complex) amplitude spectrum. Viewing or recording this spectrum again yields the power spectrum of $E(x,y)$:

$$I(u,v) = |\tilde{E}(u,v)|^2 \propto |\mathcal{F}[E(x,y)]|^2. \quad (9.30)$$

It is the diffraction pattern of the input and is simply called the spectrum. When the square of the modulus is taken, the phase factor drops out. Therefore the transparency can be placed anywhere in front of the lens for to obtain the diffraction pattern. To avoid vignetting effects the transparency is best placed directly in front of the lens (Fig. 9.4).

Often the diffraction pattern is very small. From the relationships $u = \lambda f_l v_x$ and $v = \lambda f_l v_y$ it follows that the diffraction pattern in the (u,v)-plane becomes larger the larger the focal length of the lens used. For the same reasons the ring system at the exit of a Fabry–Perot interferometer is projected with the help of a lens with a long focal length (e. g., $f_l = 1$ m).

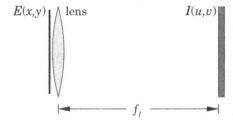

Fig. 9.4. Preferable geometry for obtaining the diffraction pattern.

9.3 Optical Fourier Spectra

In the following, we consider a few simple examples of the optical Fourier transform and illustrate them with a lens in the *2f* geometry. Once the correspondence between the object distribution (input transparency) and its Fourier transform is understood in the examples, the main structures of the diffraction patterns of more complex objects can often be guessed by intuition and experience.

9.3.1 Point Source

A point source located at (x_0, y_0) in the plane $z = 0$ can be described by a two-dimensional Dirac δ function. We write it formally as a product of two one-dimensional δ functions:

$$E(x,y) = E_0 \delta(x - x_0)\delta(y - y_0). \tag{9.31}$$

Then a formal integration yields:

$$\mathcal{F}[E(x,y)] = \int\limits_{-\infty}^{+\infty}\int\limits_{-\infty}^{+\infty} E_0 \delta(x - x_0)\delta(y - y_0)\exp\left[-2\pi i(v_x x + v_y y)\right]\,dx\,dy$$

$$= E_0 \exp\left[-2\pi i(v_x x_0 + v_y y_0)\right]. \tag{9.32}$$

Thus, the Fourier transform of a point source corresponds to the field of a plane wave. It appears in the back focal plane of a lens in a *2f* geometry (Fig. 9.5). When the point source is located on the optical axis, that is, when $(x_0, y_0) = (0,0)$, the plane wave propagates along the optical axis. For a point off the optical axis, $(x_0, y_0) \neq (0,0)$, the wave propagates at an angle to the optical axis with $\alpha \approx x_0/f_l$ and $\beta \approx y_0/f_l$. In both cases, the power spectrum is constant, $|\mathcal{F}[E(x,y)]|^2 = |E_0|^2$; the back focal plane is illuminated uniformly.

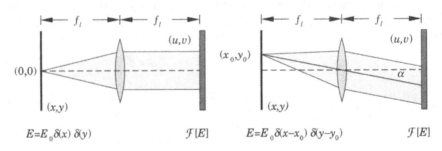

Fig. 9.5. Spectrum of a point source on the optical axis (*left*) and off axis (*right*).

9.3.2 Plane Wave

A plane wave propagating in the direction of the optical axis is of constant amplitude in the plane $z = 0$. Therefore the Fourier transform is given, up to a factor, by a two-dimensional δ function:

$$E(x,y) = E_0, \tag{9.33}$$

$$\mathcal{F}[E(x,y)] = \int\limits_{-\infty}^{+\infty}\int\limits_{-\infty}^{+\infty} E_0 \exp\left[-2\pi i(v_x x + v_y y)\right] dx\,dy$$

$$= E_0\,\delta(v_x)\delta(v_y). \tag{9.34}$$

Thus, a point appears at $(v_x, v_y) = (0,0)$ corresponding to $(u,v) = (0,0)$ in the back focal plane. The situation is depicted in Fig. 9.6.

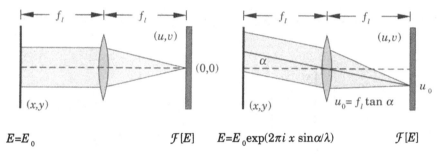

$$E = E_0 \qquad\qquad \mathcal{F}[E] \qquad\qquad E = E_0 \exp(2\pi i\, x \sin\alpha/\lambda) \qquad\qquad \mathcal{F}[E]$$

Fig. 9.6. Spectrum of a plane wave propagating along the optical axis (*left*) and at an angle α (*right*).

The electric field of a plane wave propagating obliquely to the optical axis at the angles α and β contains a space-dependent phase factor in the object plane $z = 0$. In the spectrum this leads to a shift of the focus point:

$$E(x,y) = E_0 \exp\left[\frac{2\pi i}{\lambda}(x\sin\alpha + y\sin\beta)\right], \tag{9.35}$$

$$\mathcal{F}[E(x,y)] = \int\limits_{-\infty}^{+\infty}\int\limits_{-\infty}^{+\infty} E_0 \exp\left[\frac{2\pi i}{\lambda}(x\sin\alpha + y\sin\beta)\right]\exp\left[-2\pi i(v_x x + v_y y)\right] dx\,dy$$

$$= E_0 \int\limits_{-\infty}^{+\infty} \exp\left[-2\pi i\left(v_x - \frac{\sin\alpha}{\lambda}\right)x\right] dx \int\limits_{-\infty}^{+\infty} \exp\left[-2\pi i\left(v_y - \frac{\sin\beta}{\lambda}\right)y\right] dy$$

$$= E_0\,\delta\left(v_x - \frac{\sin\alpha}{\lambda}\right)\delta\left(v_y - \frac{\sin\beta}{\lambda}\right). \tag{9.36}$$

Again, the resulting spectrum is a single point, but at the spatial frequency $(v_x, v_y) = (\sin\alpha/\lambda, \sin\beta/\lambda)$. This corresponds, in the back focal plane, to a focus point at $u_0 = f_l \sin\alpha \approx \alpha f_l$ and $v_0 = f_l \sin\beta = \beta f_l$ (Fig. 9.6).

9.3.3 Infinitely Long Slit

Slits are often found as aperture stops in optical devices such as in spectrographs. To begin, we consider an infinitely narrow, infinitely long slit. We assume an illumination of the slit by a plane monofrequency wave of normal incidence; that is, the slit acts as a coherent line source:

$$E(x,y) = E_0\,\delta(x), \tag{9.37}$$

$$\mathcal{F}[E(x,y)] = E_0 \int\limits_{-\infty}^{+\infty}\int\limits_{-\infty}^{+\infty} \delta(x)\exp\left[-2\pi i(v_x x + v_y y)\right]\,dx\,dy$$

$$= E_0 \int\limits_{-\infty}^{+\infty} \exp(-2\pi i v_y y)\,dy = E_0\,\delta(v_y). \tag{9.38}$$

The Fourier transform thus yields another line, rotated by 90° with respect to the slit (Fig. 9.7).

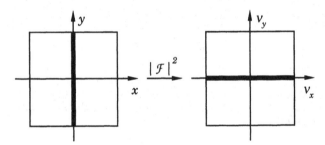

Fig. 9.7. Infinitely long, infinitely narrow slit and its spectrum.

An infinitely long slit of finite width a, illuminated as before, has an electric field given by the rect function:

$$E(x,y) = E_0\,\text{rect}\left(\frac{x}{a}\right) = E_0 \times \begin{cases} 1 \text{ for } & |x| < a/2, \\ 0 \text{ else,} \end{cases} \tag{9.39}$$

$$\mathcal{F}[E(x,y)] = E_0 \int\limits_{-\infty}^{+\infty}\int\limits_{-a/2}^{+a/2} \exp\left[-2\pi i(v_x x + v_y y)\right]\,dx\,dy$$

$$= E_0 \int\limits_{-\infty}^{+\infty} \exp(-2\pi i v_y y)\,dy \int\limits_{-a/2}^{+a/2} \exp(-2\pi i v_x x)\,dx$$

$$= E_0\,\delta(v_y)\frac{1}{-2\pi i v_x}\left[\exp(-2\pi i v_x a/2) - \exp(2\pi i v_x a/2)\right]$$

$$= E_0\,\delta(v_y)\frac{\sin(\pi v_x a)}{\pi v_x} = E_0 a\,\delta(v_y)\,\text{sinc}(a v_x). \tag{9.40}$$

The sinc function just introduced is defined by

$$\text{sinc}(x) = \frac{\sin \pi x}{\pi x}.$$ (9.41)

Again, the spectrum has (practically) no extension in the ν_y-direction. In the ν_x-direction, the intensity is modulated with the square of the sinc function, the slit function (Fig. 9.8).

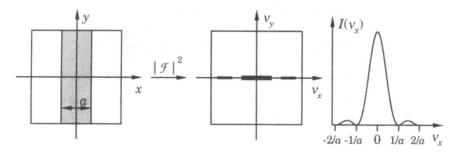

Fig. 9.8. Infinitely long slit of finite width and its power spectrum.

Figure 9.9 shows two experimentally obtained spectra of a long slit of different width a. When we use a variable slit whose width can be altered by turning a micrometer screw, the widening of the central maximum of the spectrum with decreasing slit width can be followed live. The broader the slit (that is, the larger a), the more closely spaced are the maxima in the spectrum. The roots of the sinc function are found at $\ldots -2/a, -1/a, 1/a, 2/a \ldots$.

Fig. 9.9. Two experimentally obtained spectra of a long adjustable slit. The slit width a is smaller in the lower spectrum.

It is easy to see that a modulation will also appear in the ν_y-direction, when the slit is of finite extent in the y-direction. To illustrate this case, Fig. 9.10 shows the diffraction pattern of a square aperture, obtained by illumination with a He–Ne laser. The modulation of the intensity in the ν_x- and ν_y-directions is obvious.

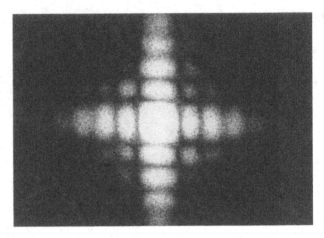

Fig. 9.10. Experimental diffraction pattern of a square aperture.

9.3.4 Two Point Sources

In Sect. 4.2 we considered the superposition of monofrequency spherical waves coming from two point sources. The far-field diffraction pattern given there can be calculated elegantly and simply with the formalism of the Fourier transform.

Let the two point sources be located on the x-axis, symmetrically to the origin, and have a separation of $2x_0$ (Fig. 9.11). This source arrangement leads to the electric field

$$E(x,y) = E_0 \, \delta(y) \left[\delta(x - x_0) + \delta(x + x_0) \right] . \tag{9.42}$$

The two-dimensional δ functions again are written as a product of one-dimensional δ functions. Then the far-field amplitude spectrum is given, along with (9.32), by

$$\mathcal{F}[E(x,y)] = \exp(-2\pi i v_x x_0) + \exp(2\pi i v_x x_0) = 2\cos(2\pi \, v_x x_0) . \tag{9.43}$$

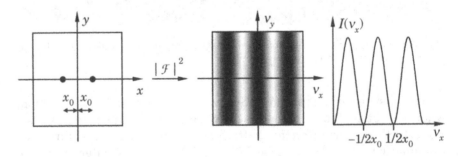

Fig. 9.11. Two point sources and their power spectrum.

The amplitude spectrum has a cosine modulation in the ν_x-direction with a spatial period of $1/x_0$, in the ν_y-direction the amplitude is constant. The intensity of the amplitude spectrum, the power spectrum of the input field, corresponds to a cosine grating:

$$|\mathcal{F}[E]|^2 = 4\cos^2(2\pi\,\nu_x x_0) = 2(1+\cos 4\pi\,\nu_x x_0). \tag{9.44}$$

Its fringe spacing is $\Delta\nu_x = 1/2x_0$, half the period of the amplitude spectrum.

9.3.5 Cosine Grating

In the previous example, we obtained a cosine grating as the intensity distribution in the Fourier spectrum of two point sources. This spectrum can be recorded on a photographic plate and used as the input pattern in a $2f$ arrangement. If we transilluminate the cosine grating with a plane wave and optically Fourier transform it, do we get the two point sources back?

The cosine grating, having the grating constant d, can be described by the transmittance (Fig. 9.12)

$$\tau(x,y) = \cos^2\left(\pi\frac{x}{d}\right) = \frac{1}{2} + \frac{1}{2}\cos\left(\frac{2\pi x}{d}\right)$$

$$= \frac{1}{2} + \frac{1}{4}\exp\left(i\frac{2\pi x}{d}\right) + \frac{1}{4}\exp\left(-i\frac{2\pi x}{d}\right). \tag{9.45}$$

When it is illuminated with a plane monofrequency wave, the electric field is modulated accordingly:

$$E(x,y) = E_0\,\tau(x,y) = E_0\left[\frac{1}{2} + \frac{1}{4}\exp\left(i\frac{2\pi x}{d}\right) + \frac{1}{4}\exp\left(-i\frac{2\pi x}{d}\right)\right]. \tag{9.46}$$

The spectrum is easily obtained when we have knowledge of the Fourier transform of plane waves [see (9.34) and (9.36)]:

$$\begin{aligned}\mathcal{F}[E(x,y)] = {} & \tfrac{1}{2}E_0\,\delta(\nu_y)\,\delta(\nu_x) && \text{0th order,}\\ & + \tfrac{1}{4}E_0\,\delta(\nu_y)\,\delta(\nu_x - 1/d) && \text{1st order,} \\ & + \tfrac{1}{4}E_0\,\delta(\nu_y)\,\delta(\nu_x + 1/d) && \text{-1st order.}\end{aligned} \tag{9.47}$$

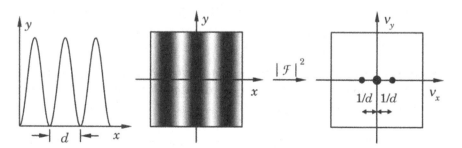

Fig. 9.12. Cosine grating and its power spectrum.

In the Fourier plane the spectrum is given by three points along the ν_x-axis which have a separation of $1/d$, the inverse of the grating constant. The central point at the origin, the zeroth order, has twice the amplitude of the two other points (Fig. 9.12).

The spectrum of two points yields a cosine grating, the spectrum of a cosine grating, however, yields three points. Why? The reason is the intensity formation in between, which destroys the phase information. A negative transmittance is not possible with a passive grating. Therefore, the cosine grating transmittance must have a direct part (a nonzero average). This part leads to undiffracted light, a plane wave along the optical axis that in the spectrum gives rise to a point at the origin.

We came across a similar phenomenon when discussing stellar speckle interferometry. There, the averaged autocorrelation function of a stellar speckle pattern is calculated corresponding to the Fourier transform of the spatial power spectrum of the (disturbed) object. In this case, too, no phase information of the object wave is available. This leads to ambiguities and artifacts upon image reconstruction. The same problem is encountered when processing specklegrams in flow diagnostics, for instance. It is known as the 'phase retrieval problem'.

9.3.6 Circular Aperture

A circular aperture of radius a is given by the transmittance

$$\tau(x,y) = \begin{cases} 1 & \text{for } x^2 + y^2 \leq a^2, \\ 0 & \text{otherwise,} \end{cases} \tag{9.48}$$

or, in polar coordinates (r, Θ), connected with the Cartesian coordinates by the transformation $x = r\cos\Theta$, $y = r\sin\Theta$, by

$$\tau(r,\Theta) = \tau(r) = \begin{cases} 1 & \text{for } r \leq a, \\ 0 & \text{otherwise.} \end{cases} \tag{9.49}$$

Thus, for the light field behind the aperture,

$$E(r,\theta) = \tau(r)E_0, \tag{9.50}$$

we obtain the Fourier transform:

$$\mathcal{F}[E](\nu_x, \nu_y) = E_0 \int_0^{2\pi} d\Theta \int_0^a \exp[-2\pi i(r\cos\Theta\,\nu_x + r\sin\Theta\,\nu_y)]r\,dr. \tag{9.51}$$

Because of the rotational symmetry it is convenient also to transform the spectral coordinates (ν_x, ν_y) to polar coordinates. With the transformation $\nu_x = \nu\cos\varphi$, $\nu_y = \nu\sin\varphi$ we get:

$$\mathcal{F}[E] = E_0 \int_0^a r\,dr \int_0^{2\pi} \exp[-2\pi i\nu r\cos(\Theta - \varphi)]d\Theta. \tag{9.52}$$

Here, the relation $\cos(\Theta - \varphi) = \cos\Theta \cos\varphi + \sin\Theta \sin\varphi$ has been used. The substitution $\xi = \Theta - \varphi$ leads to an integral over the Bessel function J_0 defined by

$$J_0(s) = \frac{1}{2\pi} \int_0^{2\pi} \exp(is\cos\xi)\,d\xi = \frac{1}{2\pi} \int_0^{2\pi} \exp(is\sin\xi)\,d\xi. \tag{9.53}$$

In a first step, we obtain:

$$\mathcal{F}[E](v,\varphi) = 2\pi E_0 \int_0^a rJ_0(-2\pi vr)\,dr. \tag{9.54}$$

The Bessel function J_0 is symmetric, that is, $J_0(-s) = J_0(s)$, and is connected with the Bessel function J_1 by

$$J_1(w) = \frac{1}{w} \int_0^w sJ_0(s)\,ds. \tag{9.55}$$

Thus, we finally obtain the following for the amplitude spectrum of a circular aperture of radius a:

$$\mathcal{F}[E](v,\varphi) = \frac{E_0 a}{v} J_1(2\pi va). \tag{9.56}$$

In Cartesian coordinates (v_x, v_y) this expression reads:

$$\mathcal{F}[E](v_x, v_y) = \frac{E_0 a}{\sqrt{v_x^2 + v_y^2}} J_1\left(2\pi a\sqrt{v_x^2 + v_y^2}\right). \tag{9.57}$$

The corresponding intensity distribution is given in Fig. 9.13. The central bright spot is called the Airy disk..

The first root in the radial direction occurs at

$$v_1 = \left(\sqrt{v_x^2 + v_y^2}\right)_1 = \frac{1.22}{2a}. \tag{9.58}$$

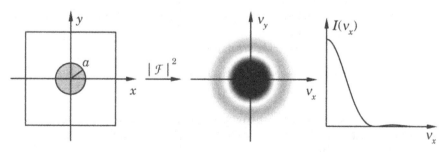

Fig. 9.13. Circular aperture and its power spectrum.

In optical instruments, circular apertures are usually applied as stops along the light path. Therefore, the Airy disk is often encountered in diffraction-limited imaging, for instance in microscopes or telescopes. The resolving power of these instruments is defined via the diameter of the Airy disk (see, for instance, [9.1]). Figure 9.14 shows two Airy disks and the adjacent diffraction rings obtained experimentally with two differently sized circular apertures. The actual intensity relation between the diffraction rings cannot be reproduced on paper because of the lack of grey scale dynamics.

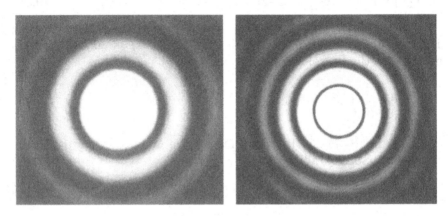

Fig. 9.14. The diffraction patterns of circular apertures of different radii.

9.3.7 Compound Diffracting Systems

In the examples considered so far, we observe recurrent relations between a diffracting pattern and its spectrum. For instance, a coarse structure as an input (a broad slit) yields a fine structure in the diffraction pattern (small distance between the maxima of the slit function). This is due to general relations valid for the Fourier transform.

In physics, the mathematical rules of the Fourier transform can be nicely illustrated optically.

The *linearity* of the Fourier transform,

$$\mathcal{F}[E_1(x,y) + E_2(x,y)](\nu_x, \nu_y) = \mathcal{F}[E_1(x,y)](\nu_x, \nu_y)$$
$$+ \mathcal{F}[E_2(x,y)](\nu_x, \nu_y), \qquad (9.59)$$

translates the superposition principle for light waves to the amplitude spectrum; that is, the Fourier transform of the superposition of two electric fields equals the superposition of the Fourier transform of the individual fields.

The *shift property* of the Fourier transform,

$$\mathcal{F}[E(x+\Delta x, y+\Delta y)](\nu_x, \nu_y) = \exp[2\pi i(\nu_x \Delta x + \nu_y \Delta y)]$$
$$\cdot \mathcal{F}[E(x,y)](\nu_x, \nu_y), \qquad (9.60)$$

states that only a linear phase shift is introduced when the diffracting pattern is shifted in x- and y-direction by Δx and Δy, respectively. The usual diffraction pattern, that is, the intensity of the amplitude spectrum, is not altered.

The *similarity property* of the Fourier transform,

$$\mathcal{F}[E(ax, by)](\nu_x, \nu_y) = \frac{1}{|a| \cdot |b|} \mathcal{F}[E(x,y)] \left(\frac{\nu_x}{a}, \frac{\nu_y}{b}\right), \qquad (9.61)$$

corresponds to the fact that an enlargement of the diffracting pattern (for instance, by the broadening of a slit) leads to a corresponding reduction in size of the diffraction pattern. Similarly, the diffraction pattern gets larger upon reduction of the diffracting structure. As light cannot get lost in this transformation, the brightness of the diffraction pattern changes according to its size.

The *convolution relations* of the Fourier transform are especially helpful for understanding the Fourier spectra of compound diffracting objects. When the electric field $E(x,y)$ can be written as a product, $E(x,y) = E_1(x,y)E_2(x,y)$, then its amplitude spectrum is given as a convolution of the Fourier transform of the individual fields:

$$\mathcal{F}[E_1(x,y) \cdot E_2(x,y)](\nu_x, \nu_y) = \mathcal{F}[E_1(x,y)] * \mathcal{F}[E_2(x,y)]. \qquad (9.62)$$

If, on the other hand, the input field can be written as a convolution of two fields,

$$E(x,y) = (E_1 * E_2)(x,y) = \int\limits_{-\infty}^{+\infty} \int\limits_{-\infty}^{+\infty} E_1(\xi, \eta) E_2(x-\xi, y-\eta) \, d\xi \, d\eta, \qquad (9.63)$$

the amplitude in the far-field diffraction pattern is given by the product of the individual Fourier transforms:

$$\mathcal{F}[(E_1 * E_2)(x,y)](\nu_x, \nu_y) = \mathcal{F}[E_1(x,y)](\nu_x, \nu_y) \cdot \mathcal{F}[E_2(x,y)](\nu_x, \nu_y). \qquad (9.64)$$

With these relations in mind, we can calculate the diffraction pattern of a grating consisting of M slits of width a and grating distance d ($>a$) in the x-direction (Fig. 9.15). The electric field behind the grating, when it is illuminated with a plane monofrequency wave at normal incidence, can be written as

$$E(x,y) = E_0 \sum_{m=1}^{M} \text{rect}\left(\frac{x}{a} - \frac{md}{a}\right) = E_0 \left[\sum_{m=1}^{M} \delta(x - md)\right] * \text{rect}\left(\frac{x}{a}\right). \qquad (9.65)$$

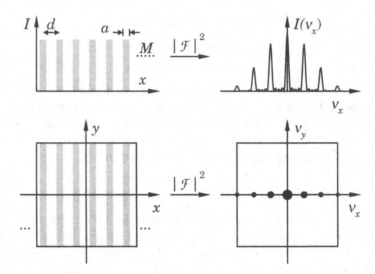

Fig. 9.15. Grating composed of m slits and its power spectrum.

Using linearity, the convolution relation, and the shifting property we get:

$$\mathcal{F}[E] = E_0\,\delta(\nu_y)\frac{\sin(\pi a\nu_x)}{\pi\,\nu_x}\sum_{m=1}^{M}\exp(-2\pi imd\nu_x)$$

$$= E_0\,\delta(\nu_y)\,a\,\text{sinc}(a\nu_x)\exp[-\pi id\nu_x(M+1)]\frac{\sin(\pi Md\nu_x)}{\sin(\pi\,d\nu_x)}\,. \tag{9.66}$$

When the intensity is calculated, the phase factor $\exp[-\pi id\nu_x(M+1)]$ drops out:

$$I(\nu_x,\nu_y) = |E_0|^2\,\delta(\nu_y)\,a^2\text{sinc}^2(a\nu_x)\frac{\sin^2(\pi Md\nu_x)}{\sin^2(\pi d\nu_x)}\,. \tag{9.67}$$

Looking at the diffraction pattern of this grating (see Fig. 9.15), we may discern the influence of the different components of the object. The coarse structure, proportional to $a^2\text{sinc}^2(a\nu_x)$, is determined by the individual slit, since the corresponding slit function forms the envelope of the diffraction pattern. The periodic repetition, on the other hand, becomes visible in a more detailed, periodic structuring of the spectrum [proportional to $\sin^2(\pi Md\nu_x)/\sin^2(\pi d\nu_x)$]. Thereby, the fine structure is determined by the grating size Md.

As an example for the spectrum of a compound diffracting system, Fig. 9.16 shows a honeycomb mesh and its spectrum. The hexagonal basic pattern being periodic, the spectrum is composed of a set of points showing the corresponding symmetries.

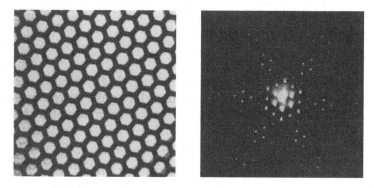

Fig. 9.16. Honeycomb mesh and its diffraction pattern.

9.4 Coherent Optical Filtering

As with the filtering of a time signal by the manipulation of its spectrum, a two-dimensional spatial pattern can be filtered by manipulating the optical spectrum in the spectral plane. Filter operations are employed for image enhancement, phase visualization and pattern recognition. For visualization of the filtered image, the modified spectrum has to be suitably transformed. The operation most suited would be the inverse Fourier transform. It cannot be realized by diffraction, however. But the Fourier transform, as given by a lens, yields the same result as we will see soon. This leads to the $4f$ arrangement, as it is called (Fig. 9.17).

The spectrum $F(u,v) = \mathcal{F}[f(x,y)]$ of the input pattern $f(x,y)$ is generated in the Fourier plane by lens 1. There, the spectrum may be modified, for instance, by letting only selected spatial frequency components pass. Thereby a field $F'(u,v)$ is generated. Lens 2 then produces a different image of the input pattern in the $(\bar{x},\bar{y})$-plane: $f'(\bar{x},\bar{y}) = \mathcal{F}[F'(u,v)]$. Without filtering, the input pattern is retained; but it is upside down. This can

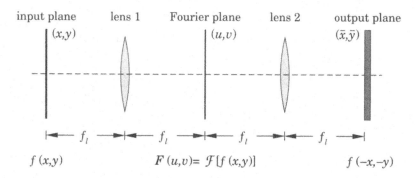

Fig. 9.17. The $4f$ arrangement.

be checked by following the imaging rays through the lens system or by applying the Fourier transform twice:

$$\mathcal{F}\left[\mathcal{F}[f(x,y)]\right] = f(-x,-y). \tag{9.68}$$

As with the filtering of one-dimensional signals, for instance, of electrical oscillations, the simplest way of manipulating the spectrum consists in rejecting certain frequency regions. This is done in the optical low-pass or high-pass filter.

9.4.1 Low-Pass Filter or Spatial Frequency Filter

Every optical system, because of its limited apertures, is an example of a low-pass filter: high spatial frequencies cannot pass. The extreme case is given by a pinhole with its idealized aperture function:

$$\tau(x,y) = \begin{cases} 1 & \text{for } (x,y) = (0,0), \\ 0 & \text{otherwise}. \end{cases} \tag{9.69}$$

Figure 9.18 shows a practical application: the spatial frequency filter, often simply called the spatial filter, we came across in holography. The incident laser beam is focused by a microscope objective at the pinhole. Then, in the focal plane (Fourier plane), we have the filtered amplitude spectrum

$$F'(\nu_x, \nu_y) = E_0\,\delta(\nu_x, \nu_y) \ \text{ or } \ F'(u,v) = E_0\,\delta(u,v). \tag{9.70}$$

The zeroth diffraction order only can pass. The second lens transforms the point source into a plane wave ($\mathcal{F}[\delta] = 1$), independently of the field in the input plane. Those spatial frequencies that may be generated by diffraction at dust spots on lenses are filtered out. Thus any disturbances of the wave front are removed.

The second lens (collimating lens) simultaneously acts for beam expansion. The beam diameter grows proportionally to the focal length of the collimating lens.

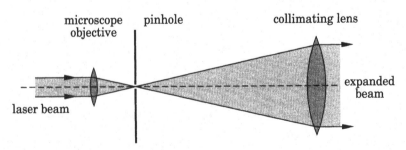

Fig. 9.18. Spatial frequency filter with a pinhole in the back focal plane of a microscope objective and a collimating lens. This unit acts as a beam expander with a homogeneous beam as output.

9.4.2 High-Pass Filter or Dark Field Method

With this filter, for rejecting all frequencies below a cutoff frequency, a disk is placed in the spectral plane. This way, edges are enhanced. By the loss of the 'dc part', however, the image usually becomes very dark. This type of filter is used in microscopy where it is called the dark field method. Practically little more than the zeroth order is rejected.

To begin, the effect of high-pass filtering is demonstrated with an amplitude object, the cosine grating discussed previously. The field of the grating in the object plane, along with (9.46) and $E_0 = 1$, can be written as

$$E(x,y) = \cos^2\left(\pi\frac{x}{d}\right) = \frac{1}{2} + \frac{1}{2}\cos\left(2\pi\frac{x}{d}\right). \tag{9.71}$$

In the spectral plane we have the field (see (9.47)):

$$\mathcal{F}[E(x,y)] = \delta(\nu_y)\left[\frac{1}{2}\delta(\nu_x) + \frac{1}{4}\delta\left(\nu_x - \frac{1}{d}\right) + \frac{1}{4}\delta\left(\nu_x + \frac{1}{d}\right)\right]. \tag{9.72}$$

When, as shown in Fig. 9.19, the zeroth order is eliminated from the spectrum by the high-pass filter, we have the modified spectrum:

$$\mathcal{F}_{\text{hp}}[E] = \delta(\nu_y)\frac{1}{4}\left[\delta\left(\nu_x - \frac{1}{d}\right) + \delta\left(\nu_x + \frac{1}{d}\right)\right]. \tag{9.73}$$

A second Fourier transform yields the field of the filtered image. We encountered this Fourier transform earlier when calculating the spectrum of two point sources [see (9.43)], so we are able to write down the result immediately. With $x_0 = 1/d$, we have:

$$\mathcal{F}[\mathcal{F}_{\text{hp}}[E]] = \frac{1}{4}2\cos\left(2\pi\frac{x}{d}\right) = \frac{1}{2}\cos\left(\pi\frac{x}{d/2}\right). \tag{9.74}$$

Notice that, after filtering, the image of the object has a structure twice as fine as it had originally, a substantial alteration (Fig. 9.19). The example shows that filter operations may have quite unexpected results.

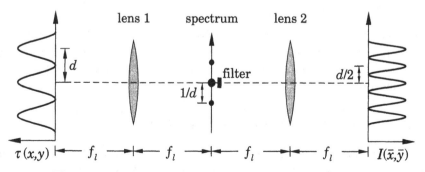

Fig. 9.19. Filtering a cosine grating with a high-pass filter.

The dark field method is most often used to visualize thin phase objects, for instance, thin sheets of organic specimens, air flows around bodies, vortices and shock waves in fluid flows, finger prints, strains in transparent materials, density alterations in heating, etc. In all these cases, only a spatial phase pattern is generated across the light beam, which has its origin in a spatially varying index of refraction. Without filtering, these phase patterns stay invisible, since phase modifications do not show up when forming the intensity. The field of a phase object is given by

$$E(x,y) \propto \tau(x,y) = a \exp[i\varphi(x,y)] \qquad (a = \text{const}). \qquad (9.75)$$

It is the phase pattern $\varphi(x,y)$ that contains the information on the object. The intensity of the field everywhere is the same:

$$I(x,y) = E(x,y)E^*(x,y) \propto \tau(x,y)\tau^*(x,y) = |a|^2 = \text{const}. \qquad (9.76)$$

For thin phase objects, $|\varphi| \ll 2\pi$, the exponential function can be expanded to yield approximately:

$$\tau(x,y) \approx a[1 + i\varphi(x,y)]. \qquad (9.77)$$

For simplicity, and since it does not alter the result, we assume that the phase distribution $\varphi(x,y)$ has zero mean. Due to the linearity of the Fourier transform we have

$$\mathcal{F}[\tau] = a\left(\delta(\nu_x)\delta(\nu_y) + i\mathcal{F}[\varphi(x,y)]\right). \qquad (9.78)$$

Rejecting the zeroth order yields the high-pass filtered spectrum:

$$\mathcal{F}_{\text{hp}}[E] = ia\mathcal{F}[\varphi(x,y)]. \qquad (9.79)$$

It no longer contains a dc term. The filtered image is formed by a further Fourier transform:

$$\mathcal{F}[\mathcal{F}_{\text{hp}}[E]] = ia\mathcal{F}[\mathcal{F}[\varphi(x,y)]] = ia\,\varphi(-x,-y). \qquad (9.80)$$

Now, the intensity is no longer uniform, but makes visible the phase pattern φ:

$$I = |\mathcal{F}[\mathcal{F}_{\text{hp}}[E]]|^2 = a^2\varphi^2(-x,-y). \qquad (9.81)$$

The intensity quadratically depends on the phase variation. For a constant phase the intensity vanishes. Therefore the visualized phase is given as a bright image on a dark background, thence the name dark field method.

9.4.3 Phase Filter or Phase Contrast Method

Up to now, we have considered amplitude filters only; that is, filters that stop some part of the spectrum and let the other pass. Pure phase objects can be made visible with filters that influence the phase of a suitable part of the spectrum. For thin phase objects we wrote in the previous section:

$$\tau(x,y) = a(1 + i\varphi(x,y)).$$ (9.82)

In this approximation we have a real dc term and a purely imaginary phase term. Because of the linearity of the Fourier transform, the dc term is shifted in phase by $\pi/2$ relative to the remaining part of the Fourier spectrum:

$$\mathcal{F}[\tau] = a(\delta(\nu_x)\delta(\nu_y) + i\mathcal{F}[\varphi(x,y)]).$$ (9.83)

If the zeroth order were shifted in phase by $\pi/2$, we would get an amplitude modulation in the image that is visible, in contrast to a pure phase modulation. A filter having this property can be fabricated quite easily. One only has to insert a quarter wave plate into the zeroth order. It shifts the phase of the field in the origin of the spatial frequency plane by $\pi/2$. Then, in (9.83) the dc term is multiplied by $\exp(i\pi/2) = i$, and the phase filtered spectrum is obtained:

$$\mathcal{F}_{pf}[\tau] = ai(\delta(\nu_x)\delta(\nu_y) + \mathcal{F}[\varphi(x,y)]).$$ (9.84)

A further Fourier transform yields the filtered image:

$$\mathcal{F}[\mathcal{F}_{pf}[\tau]] = ai(1 + \varphi(-x,-y)).$$ (9.85)

Now, the intensity reads:

$$I_{pf} = a^2\left[1 + 2\varphi(-x,-y) + \varphi^2(-x,-y)\right]$$
$$\approx a^2(1 + 2\varphi(-x,-y)).$$ (9.86)

In the last line, the term φ^2 is omitted, since we have assumed a weak phase modulation ($|\varphi| \ll 2\pi$).

Compared to the dark field method, now a linear relationship between image intensity and phase φ is obtained. This leads to an increased sensitivity for weak phase objects. The method is called the phase contrast method and was developed by *Frits Zernike* (1888–1966).

The method can be improved in a number of ways. The contrast in the image, for instance, is lowered by the dc term a^2. Therefore, the phase contrast method is combined with an attenuation of the amplitude of the zeroth order by a factor b ($0 < b < 1$):

$$I_{pf} = a^2\left[b^2 + 2b\varphi(-x,-y)\right].$$ (9.87)

Thus, although the intensity of the image decreases, the contrast is nevertheless increased.

9.4.4 Half-Plane Filter or Schlieren Method

The schlieren method also may be used to make phase objects visible [9.2]. In this case, again an amplitude filter is employed. One half of the spatial frequency plane is rejected (for instance, with a knife edge), including half of the zeroth order. This has the effect that in the filtered image the intensity is approximately proportional to the gradient of the phase (dependent on the orientation of the half plane):

$$I(-x, -y) \propto \left| \frac{\partial \varphi(x,y)}{\partial x} \right|. \tag{9.88}$$

This method achieves its task by disturbing the balance of the spectral components which leads to a uniform intensity of the image.

9.4.5 Raster Elimination

When an image is printed, it is difficult to generate shades of grey. Therefore, the image is sampled and the halftones are generated by inserting differently sized dots at the sample points. The same problem appears in digital image processing. The image must be sampled in the x-, the y- and even the z-direction and stored as a matrix of pixels (picture elements, two-dimensional) or voxels (volume picture elements, three-dimensional). How do we perceive grey scales in these images?

The sampling theorem we briefly mentioned in connection with digital holograms relates the sampled and the original image. It states that the original image can be restored exactly from the sampled image when it is bandlimited, that is, when it contains spatial frequencies up to a cutoff frequency only, and when the distance between adjacent sample points is not larger than a certain value determined by the cutoff frequency.

Mathematically, a sampled image (look closely at any halftone picture of this book) can be formulated in the following way:

$$g_{\mathrm{s}}(x,y) = \mathrm{comb}\left(\frac{x}{a}\right) \mathrm{comb}\left(\frac{y}{b}\right) g(x,y). \tag{9.89}$$

The comb function,

$$\mathrm{comb}\left(\frac{x}{a}\right) = \sum_{-\infty}^{+\infty} \delta(x - na), \tag{9.90}$$

describes the sampling; the function $g(x,y)$ describes the continuous (complex) amplitude distribution of the original image. The pixel distances in the x- and the y-directions are a and b, respectively.

The spectrum of $g_{\mathrm{s}}(x,y)$ is obtained from the spectrum of the continuous image, $G(\nu_x, \nu_y) = \mathcal{F}[g(x,y)](\nu_x, \nu_y)$, by means of the convolution relation

$$\mathcal{F}[g_{\mathrm{s}}(x,y)] = G_{\mathrm{s}}(\nu_x \nu_y) = a\, \mathrm{comb}(a\nu_x) b\, \mathrm{comb}(b\nu_y) * G(\nu_x, \nu_y). \tag{9.91}$$

Here, we have used the fact that the Fourier transform of a comb function again is a comb function.

Thus, the Fourier transform of the sampled image yields a grid that arises by repetition of the spectrum of the original image at each grid point. The vicinity of each grid point of the spectrum contains the complete information about the original image. Therefore, the grid points in the Fourier plane should have such a large distance that the individual, identical spectra do not overlap. When the vicinity of a single grid point in the spectrum of $g_s(x,y)$ is taken and the rest of the plane rejected, the original image is obtained by a Fourier transform. This type of filtering can be thought of as a multiplication of the spectrum $G_s(\nu_x, \nu_y)$ with an aperture function acting as a low-pass filter. A circular aperture of suitable diameter or a rectangular aperture of suitable size (edge lengths $1/a$ and $1/b$) may be taken as a low-pass filter. Mathematically formulated, we get

$$\mathcal{F}[G_{sf}(\nu_x,\nu_y)] = \mathcal{F}[G_s(\nu_x,\nu_y)A(\nu_x,\nu_y)] = g_s(-x,-y) * \mathcal{F}[A(\nu_x,\nu_y)], \qquad (9.92)$$

This is the convolution of the sampled image with the Fourier transform of the aperture function $A(\nu_x, \nu_y)$.

9.4.6 Demonstration Experiment

An experimental arrangement for demonstrating the filtering operations discussed above is given in Fig. 9.20. The $4f$ geometry of Fig. 9.17 has been augmented by a few components to simultaneously project the original image, its power spectrum, the filtered power spectrum, and the filtered image. In particular, large beam splitters are inserted into the main path to image the different planes onto a screen. In this way, the result of different optical filters can immediately be visualized.

Figure 9.21 shows an example for the influence of low-pass filtering on a halftone image. The upper row presents the halftone image (left) and its spectrum (right). In the lower row, the filtered spectrum (right) and the corresponding filtered image (left) are to be seen. The original raster points have disappeared and a grey-scale picture is obtained. However, the printing process has added a new grid.

9.4.7 Holographic Filters

Up to now, we have filtered images in the Fourier plane by rejecting certain spatial frequency regions. With these filters, except for in the phase contrast method, only the amplitude of the light wave has been influenced. The most general filter, a filter with complex transmittance, would modulate the amplitude and phase of the spectral field distribution as a function of (ν_x, ν_y) or (u,v). Filters of this kind are difficult to produce

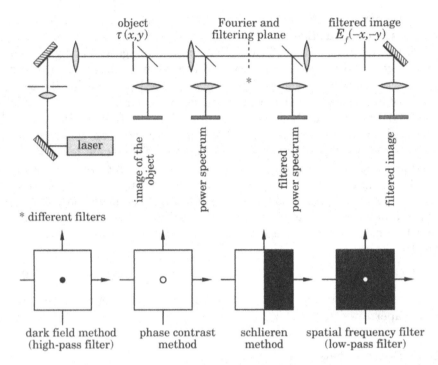

Fig. 9.20. Experimental arrangement for optical filtering.

directly. But in the form of holograms they can be realized. Thereby, the spectrum of an object is recorded holographically and used as a complex filter. The hologram is called the Fourier transform hologram or simply the Fourier hologram.

For recording the spectrum $F(u,v)$ of an object $f(x,y)$ holographically, a $2f$ arrangement may be used (Fig. 9.22). Let the reference wave R be a plane wave. It can be generated in a simple way by the Fourier transform of a point source with the same lens as used for transforming the object. The transmittance of the hologram then is given by [cf. (7.7)]

$$\tau(u,v)=a-bt_{\mathrm{B}}(R+F)(R+F)^*=a-bt_{\mathrm{B}}(RR^*+FF^*+R^*F+RF^*)\,.\,(9.93)$$

The Fourier transform of f is denoted by the capital letter F. When the hologram is illuminated with the plane reference wave R, the usual four terms appear:

$$\tau\cdot R=(a-bt_{\mathrm{B}}RR^*)R-bt_{\mathrm{B}}(FF^*R+R^*FR+RF^*R)\,.\tag{9.94}$$

The subsequent Fourier transform yields the image (see Fig. 9.23):

$$\mathcal{F}[\tau R]=(a-bt_{\mathrm{B}}RR^*)\mathcal{F}[R]-bt_{\mathrm{B}}\Big(\mathcal{F}[FF^*R]+\mathcal{F}[RR^*F]+\mathcal{F}[R^2F^*]\Big)\,.\,(9.95)$$

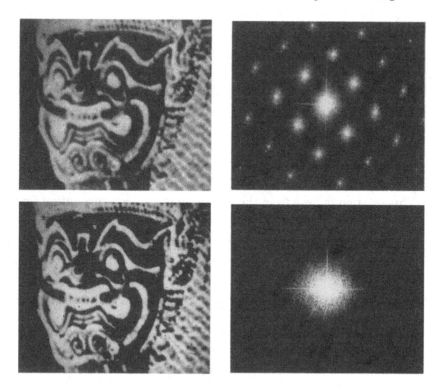

Fig. 9.21. Raster elimination of a halftone image by low-pass filtering.

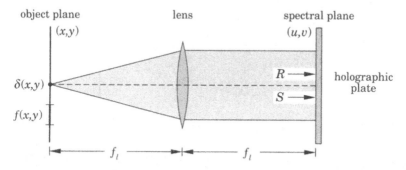

Fig. 9.22. 2f arrangement for recording a Fourier hologram. The object f produces the object wave S; the point source in the origin produces the reference wave R.

This result can be elaborated further with the following rules applying to the Fourier transform of two functions f and g. Let the amplitude spectra be termed as before with the corresponding capital letters F and G, that is $\mathcal{F}[f] = F$ and $\mathcal{F}[g] = G$. Then the following relations hold:

1. $\mathcal{F}[F] = f^{(-)}$.
   The minus sign is an abbreviation for $f(-x, -y)$. It means that the input pattern is reproduced upside down.

2. $\mathcal{F}[F^*] = f^*$,
   that is, taking the conjugate commutes with the operation Fourier transform.

3. $\mathcal{F}[F \cdot G] = \mathcal{F}[F] * \mathcal{F}[G] = (f * g)^{(-)}$,   convolution.

4. $\mathcal{F}[F \cdot G^*] = f^{(-)} * g^* = (f \otimes g)^{(-)}$,   crosscorrelation.

5. $\mathcal{F}[F \cdot F^*] = f^{(-)} * f^* = (f \otimes f)^{(-)}$,   autocorrelation.

6. $f \otimes g = f * (g^{(-)})^*$.
   This relation follows from the definition of the crosscorrelation function $f \otimes g$ and the convolution product $f * g$.

From these rules, the output from the hologram can be written as

$$\mathcal{F}[\tau \cdot R] = \text{const } \delta - b t_{\text{B}} \left((f \otimes f)^{(-)} + f^{(-)} + f^* \right). \tag{9.96}$$

The expressions const·δ and $-b t_{\text{B}}(f \otimes f)^{(-)}$ correspond to the zeroth order and broadened zeroth order, respectively; the expression $-b t_{\text{B}} f^{(-)}$ corresponds to the direct image and its counterpart $-b t_{\text{B}} f^*$ corresponds to the conjugate image. All images are real and can be observed on a ground glass plate (Fig. 9.23).

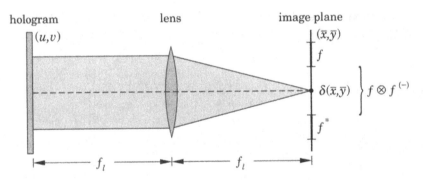

Fig. 9.23. Reconstruction and subsequent Fourier transform of a holographically recorded, complex spectrum.

9.4.8 Pattern Recognition

A Fourier transform hologram can be used as a holographic filter for pattern recognition. To achieve this, the hologram is illuminated with the Fourier transform G of an object g:

$$\tau \cdot G = (a - b t_{\text{B}} R R^*) G - b t_{\text{B}} (F F^* G + R^* F G + R F^* G). \tag{9.97}$$

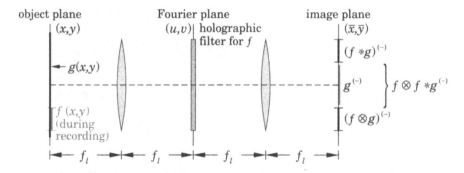

Fig. 9.24. Location of the images upon holographic filtering for pattern recognition.

A further Fourier transform yields the image

$$\mathcal{F}[\tau \cdot G] = (a - bt_\mathrm{B}RR^*)\mathcal{F}[G] - bt_\mathrm{B}(\mathcal{F}[FF^*G] + \mathcal{F}[R^*FG] + \mathcal{F}[RF^*G])$$
$$= \mathrm{const}\, g^{(-)} - bt_\mathrm{B}\Big((f \otimes f * g)^{(-)} + (f * g)^{(-)} + (f \otimes g)^{(-)}\Big). \qquad (9.98)$$

The first term, $\mathrm{const}\cdot g^{(-)}$ (zeroth order), is a reproduction of the input $g(x,y)$ in the input plane. The second term, $-bt_\mathrm{B}(f \otimes f * g)^{(-)}$, represents the broadened zeroth order. The third term, $-bt_\mathrm{B}(f * g)^{(-)}$, formerly the direct image, is the convolution of the reconstruction wave g and the object wave f, whereas the fourth term, formerly the conjugate image, is their crosscorrelation. Figure 9.24 shows the location of the individual images given by the above reconstruction terms in the holographic filtering process. The fourth term, the crosscorrelation between f and g, is a measure for the similarity of both patterns. Therefore it can be used for pattern recognition. The pattern f to be recognized in an input pattern g is stored in a Fourier hologram. When the image corresponding to the crosscorrelation term shows up bright upon reconstruction, then f is similar to a part of g. Figure 9.25 shows, to the left, the Fourier transform hologram of the letter **K**. It was taken with the arrangement of Fig. 9.22. Of course, it only displays the gross structure of the intensity spectrum of **K**, since the fine interference pattern introduced by the reference wave is not resolved. The reconstruction of the images of this hologram according to the arrangement of Fig. 9.23 yields two symmetric images of the letter **K**. They are to be seen in Fig. 9.25 to the right. When this filter for the letter **K** is inserted into the filter plane, the Fourier plane (see Fig. 9.24), then in an input pattern – a text placed in the object plane – this letter can be recognized from the output in the image plane.

In a demonstration experiment, the letters **HOK6** were used as an input filtered with **K**, and the image that appears in the image plane was photographed. In Fig. 9.26, three differently exposed, filtered images of the letters are reproduced to better cover the strongly different intensi-

Fig. 9.25. Fourier transform hologram of the letter **K** (*left*) and reconstruction of its images (*right*).

Fig. 9.26. Holographic filtering for pattern recognition. The letters **HOK6** are filtered with the letter **K**. Correlation to the right, convolution to the left. Three differently exposed photographs, to cover the strongly different intensities in the image, are shown.

ties in the image. In the center, both the input pattern **HOK6** and the broadened zeroth order are to be seen, superimposed. The crosscorrelation between **K** and **HOK6** appears to the right, the convolution to the left. In the upper photograph, a bright dot is to be seen at the location of the crosscorrelation of **K** with **HOK6** that locally is an autocorrelation. But at the location of **H** a high intensity is also found, and a point is detectable. This is due to the similarity of **H** with **K**.

Pattern recognition with holographic filters looks very promising as convolutions and correlations of two-dimensional functions can be calculated at the speed of light. Faster is impossible. However, for more than thirty years researchers have tried in vain to really make use of this possibility. There must be a reason behind it. The most important reason seems to be that holographic filtering is extremely sensitive. A shift of the filter by just a fraction of the wavelength leaves the filter useless.

Also, when the input pattern to be filtered, the function g, is only slightly rotated or tilted with respect to f, pattern recognition no longer works. Difficulties also arise because of the extremely high intensity of the zeroth order that overloads film materials and photodiodes. Moreover, the recording of the filters and the input and output of the pattern present difficulties. As pictures become more and more digital and ever faster processors are being built, the operations described here are today carried out by computer. Fourier optics, however, will never lose its place as an elegant and intuitive tool in diffraction theory.

Problems

9.1 A screen with two circular apertures of radius a, having a separation $b > 2a$, is illuminated by a plane monofrequency wave incident along the normal. Calculate the far-field diffraction pattern of the arrangement.

9.2 Consider a thin, annular aperture with inner radius R_1 and outer radius $R_2 = R_1 + \Delta R$, $\Delta R \ll R_1$. If the aperture is illuminated as in problem 9.1, determine its Fraunhofer diffraction pattern.
Hint: $dJ_1(z)/dz = J_0(z) - J_1(z)/z$.

9.3 A laser beam of diameter $d_e = 3$ mm falls onto the microscope objective of a pinhole acting as a spatial frequency filter (wavelength $\lambda = 514$ nm, focal length of the objective $f = 15$ mm). What diameter should the pinhole have? The laser beam is to be expanded to a diameter $d_a = 25$ mm. Calculate the focal length of the collimation lens.

9.4 A thin lens having a refractive index n and radii of curvature r_1 and r_2 is located in the plane $z = 0$, its optical axis being the z-axis. Give an expression for the transmission function $\tau(x,y)$ of the lens acting as a phase filter (no absorption). Using this result, show that the field distribution in the back focal plane, up to a position-dependent phase factor, is given by the Fourier transform of the field distribution immediately in front of the lens.

9.5 The far-field distribution of two orthogonal, infinitely thin slits, for example, is equal to its Fourier transform. Construct a general form for two-dimensional functions being invariant under the Fourier transform $\mathcal{F}$.

9.6 Show that a Fresnel zone plate (of radius a) acts like a lens with several foci upon illumination with a monofrequency plane wave incident along the perpendicular. The transmission function of the zone plate is given by

$$T(r) = \frac{1}{2}\left(1 + \text{sign}\left[\cos(\alpha r^2)\right]\right) \text{circ}_a(r) = f(\alpha r^2)\text{circ}_a(r) . \tag{9.99}$$

Hints: use the Fourier representation of a symmetric square wave with period 2π,

$$f(\xi) = \sum_{n=-\infty}^{n=+\infty} 2\,\text{sinc}\left(\frac{n}{2}\right) \exp(in\xi) , \tag{9.100}$$

and the Fresnel approximation of the diffraction integral.

9.7 A plane monofrequency wave, incident along the optical axis, illuminates an amplitude grating whose transmission function has the form of a square wave:

$$\tau(x,y) = \left[\text{comb}_a(\xi) * \text{rect}_b(\xi)\right](x) = \left[\sum_n \delta(\xi - na)\right] * \text{rect}\left(\frac{\xi}{b}\right). \tag{9.101}$$

Determine the images of the grating upon filtering by the dark field method and by the phase contrast method. Repeat these calculations for a corresponding phase grating with the transmission function ($\alpha \ll 1$):

$$\tau(x,y) = 1 + i\alpha\left(\left[\text{comb}(\xi/a) * \text{rect}(\xi/b)\right](x) - \frac{1}{2}\right). \tag{9.102}$$

9.8 A picture to be printed in halftone is usually discretized on a grid. Outline the Fourier spectra and the corresponding filters for smoothing of pictures rastered on a square grid and on a triangular grid, respectively (Fig. 9.27).

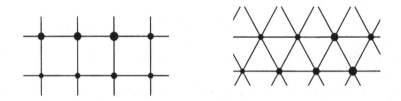

Fig. 9.27. Square grid and triangular grid.

9.9 A square aperture, centered on the optical axis and illuminated by a plane harmonic wave, serves as the object in a 4f arrangement upon the recording of a Fourier transform hologram. Let the reference wave also be a plane wave, incident along the optical axis. Calculate the transmission function of the hologram. Now, the square aperture is displaced laterally by ($\Delta x, \Delta y$) and again illuminated by the reference wave. Give an expression for the field distribution arising in the back focal plane of the arrangement. How does the distribution depend on the displacement vector ($\Delta x, \Delta y$) of the aperture?

10. The Laser

Today, the laser, with its ability to generate coherent light, has become an indispensable tool in optics and not only in optics. Lasers come in a variety of types, hardly to be counted. Coherent light can be generated from the far infrared to the near ultraviolet. It is hard to imagine optical diagnostics or communication techniques without the laser. In huge plants they serve for fusion research, and as pin-sized light sources they are found in CD players. They serve as a scalpel in the hands of surgeons for medical operations and for the cutting, drilling, and hardening of materials of all kinds. A vast literature exists where this variety is documented [10.1].

Nonlinear optics, allowing for a wide-range conversion of frequencies, did not become a significant field of optics before the advent of the laser. The laser itself is a basically nonlinear element and an object of research on its own. Its nonlinear dynamics shows a variety of forms ranging from regular to chaotic behavior [10.2]. The fundamentals of laser dynamics will be presented in this chapter.

10.1 The Laser Principle

The laser mainly consists of three parts: a (Fabry–Perot) resonator, an active medium in the resonator, and an energy source for activating the medium (Fig. 10.1). With these components it constitutes a self-excited oscillator: the active medium corresponds to the amplifying element, the resonator to the feedback element. Both the active medium and the resonator determine the light frequencies generated.

The phenomena in the laser may be divided into the propagation process of the light generated in the resonator and a light–matter interaction process in the active medium. The propagation in the resonator can be understood in the wave picture of light and is, in its simplest form, given by the theory of the Fabry–Perot interferometer (see the chapter on multiple-beam interference). To describe the light–matter interaction process, quantum mechanics or, more precisely, quantum electrodynamics is needed. However, the basic events can be understood in a model without explicit calculations, for instance, of matrix transition elements.

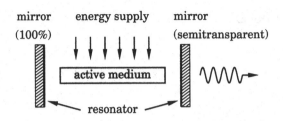

Fig. 10.1. The basic elements of the laser: resonator (two mirrors), active medium, and energy source.

The required numbers are taken as empirical constants from quantum theory or measurements.

Three basic interaction processes of light with matter are important for the laser: absorption, stimulated emission and spontaneous emission (Fig. 10.2). We assume that two states (levels), of energies E_1 and E_2, take part in the interaction. From the complete level scheme of an atom, we are interested at first only in these two energy levels. Upon absorption, a photon of energy $h\nu$ hits an atom of the laser medium in the state E_1 and disappears, exciting the atom to the higher state E_2. The photon, however, can only be absorbed, if the level E_2 'fits', that is, if $h\nu = E_2 - E_1$, so as not to violate energy conservation. When no suitable energy level is available, no absorption takes place, and the medium is transparent for photons of this energy. When the atomic system has absorbed the energy $h\nu$ and thus the upper level is occupied, a second photon of energy $h\nu$ may cause this energy to be emitted as a photon. This process is called stimulated emission. Then two photons having identical properties leave the atom. The quantum mechanical perturbation theory that allows us to calculate the processes of absorption and of stimulated emission, shows that both processes differ only in the initial conditions. Upon absorption, the atomic system starts from the state of lower energy, upon stimulated emission it starts from the state of higher energy. Hence, the transition probabilities are equal for both processes. The transition probability gives the number of transitions per atom and photon and per unit time. We call

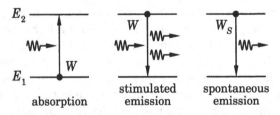

Fig. 10.2. Illustration of the processes of importance for the laser: absorption (*left*), stimulated emission (*middle*), spontaneous emission (*right*).

this quantity, having the unit s^{-1}, the normalized transition rate and denote it with the letter W.

A more difficult notion is that of spontaneous emission. In this process, an atomic system in the state of higher energy, E_2, decays into a state of lower energy, E_1, by emission of a photon. The word spontaneous indicates that the transition takes place with the randomness that is characteristic for quantum processes. To understand this kind of transition, quantum electrodynamics is required. For the subsequent, simple model of a laser it is, however, sufficient to assign an empirical value to the normalized transition rate W_s of spontaneous emission.

10.2 Laser Rate Equations

We consider a laser model with the following idealizations:

- The active medium possesses three energy levels that take part in the laser process (Fig. 10.3). The laser transition takes place between E_2 (upper laser level) and E_1 (lower laser level, ground state). The level E_3 (pump level) is assumed to have a very short life time, so that all electrons excited to this level instantaneously decay to the upper laser level E_2. The populations N_1, N_2, and N_3 give the number of atoms of the laser medium residing in the energy levels E_1, E_2, and E_3, respectively.
- Only one mode, that is, light of a single wavelength only, is present in the resonator. Then, the laser emits just one type of photons (Fig. 10.4).
- There is no spatial dependence of the populations N_1, N_2, and N_3 of the active medium in the resonator.

The basic equations we derive here for describing laser operation are called rate equations. They are balance equations for the total number of photons Q in the resonator and for the populations N_1, N_2, and N_3 of laser atoms in the levels E_1, E_2, and E_3, respectively. This yields four equations. Two of them prove to be insignificant. We have assumed a fast transition

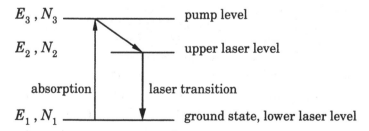

Fig. 10.3. Level diagram of a three-level laser. E_1, E_2, and E_3 are the three energy levels of the active medium, N_1, N_2, and N_3 are the number of atoms in the corresponding levels.

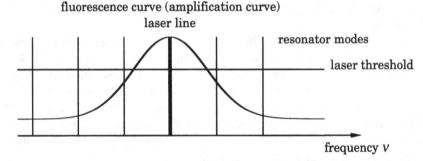

Fig. 10.4. Condition for the occurrence of a single laser mode only in the resonator.

from the pump level to the upper laser level. Thus, $N_3 = 0$, and we do not need an equation for N_3. In this case the laser is called an ideal three-level laser. With the total number N of laser atoms in the resonator being given we have $N = N_1 + N_2 + N_3$. With $N_3 = 0$ this expression simplifies to $N_1 + N_2 = N$. Moreover, just one equation for N_2 (or N_1) has to be set up. The other quantity then is known immediately. Hence, two equations are left. The first one describes the temporal change of the photon number Q and is called the intensity equation. The second equation describes the temporal change of one of the two populations of the laser levels and is called the matter equation.

The transition rates relevant for the transition between the individual levels are defined as follows:

W normalized transition rate for both absorption and stimulated emission (see Fig. 10.2).

W_s normalized transition rate for spontaneous emission (see Fig. 10.2). The spontaneous emission usually is negligible after laser action has set in. The upper laser level then responds mainly to stimulated emission, but spontaneous emission is indispensable for starting the laser action.

W_{sm} normalized transition rate for spontaneous emission into the resonator mode. The spontaneous emission takes place in all directions. A fraction of it only yields photons of the resonator mode.

W_{12} normalized transition rate from the ground state to the upper laser level via level three by the action of the pump.

W_{21} normalized transition rate for spontaneous emission and relaxation processes from level two to level one.

Besides these quantities, characterizing the interaction of light with matter, we introduce the decay constant γ for characterizing the resonator. It contains all the resonator losses: scattering, diffraction, and also output.

Now, all the quantities are defined that we need to formulate the rate equations. The rate of change of the number of photons in the resonator, dQ/dt, is given by the intensity equation:

$$\frac{dQ}{dt} = -\gamma Q - WN_1Q + WN_2Q + W_{sm}N_2. \tag{10.1}$$

The individual terms appearing in the equation have the following meaning. The term $-\gamma Q$ describes the damping during the propagation of the photons in the resonator. It is assumed to be proportional to the number of photons. Without any further processes in the resonator this term leads to an exponential decay of the photon number with time according to the decay constant γ. The second term, $-WN_1Q$, describes the absorption process from level 1 to level 2 in the active medium. Each absorption event diminishes the photon number by one. The transition rate connected with this process is proportional to the number N_1 of atoms in the ground state – the process cannot take place, when there are no atoms available in level one – and is also proportional to the number of photons. The term $+WN_2Q$ represents the stimulated emission. It is proportional to the population N_2 and to the photon number Q. The term $+W_{sm}N_2$ describes the increase of the photon number by spontaneous emission into the laser mode. It is independent of the photon number and is only proportional to the number N_2 of atoms in the upper laser level. This term generates photons even when Q vanishes, if $N_2 \neq 0$, and starts the laser light emission when inversion (see below) is reached.

The matter equation gives the rate of change in the number N_2 of atoms in the upper laser level:

$$\frac{dN_2}{dt} = -W_{21}N_2 + WN_1Q - WN_2Q + W_{12}N_1. \tag{10.2}$$

The two terms in the middle, describing absorption and stimulated emission , correspond to the two middle terms in (10.1), they are just of different signs. This is evident as each individual absorption event $dN_2 = +1$ extracts a photon from the light field; that is, it is necessarily accompanied by $dQ = -1$. Similarly, each stimulated emission event augments the number of photons in the resonator by one, that is, $dQ = +1$, while one atom leaves the state N_2, that is, $dN_2 = -1$. The term $+W_{12}N_1$ describes the pump process. It determines the rate at which atoms are transferred from the ground state via the pump level to the upper laser level by the energy supply, and is therefore proportional to N_1. The term $-W_{21}N_2$ determines the decay rate of the population of level 2 and thus is proportional to N_2. The losses mainly arise from the total spontaneous emission in all directions and from relaxation processes, that is, from radiationless energy transfer from coupling of the atom to the surroundings that takes up the energy upon transition of the atom to the ground state.

By changing the variables and subsequent normalization the two equations (10.1) and (10.2) can be brought into a quite simple form. Instead of using the population N_2, we introduce the population difference $N_d = N_2 - N_1$ as new variable. When $N_d > 0$, more atoms occupy the upper laser level than the lower one: inversion has been achieved. This is a necessary prerequisite for laser action. Written in terms of N_d, the intensity equation (10.1) reads:

$$\frac{dQ}{dt} = -\gamma Q + WQN_d + W_{sm}N_2.$$

(10.3)

As a further simplification we neglect the term $W_{sm}N_2$, the spontaneous emission into the laser mode. This process is needed for starting the laser, but usually does not play a big role afterwards, since then stimulated emission takes over. An exception is the semiconductor laser, which usually does not satisfy the condition $W_{sm} \ll WQ$. In this simplification the intensity equation is given by

$$\frac{dQ}{dt} = -\gamma Q + WQN_d.$$

(10.4)

From $N_1 + N_2 = N$ and $N_2 - N_1 = N_d$ it follows that $N_2 = (N + N_d)/2$ and $N_1 = (N - N_d)/2$. Substituting N_1 and N_2 into (10.2) we get for the population difference N_d:

$$\frac{dN_d}{dt} = \frac{d(N_2 - N_1)}{dt} = \frac{dN_2}{dt} - \frac{dN_1}{dt} = 2\frac{dN_2}{dt}$$
$$= 2\left(-W_{21}N_2 - WQN_d + W_{12}N_1\right)$$
$$= (W_{12} - W_{21})N - (W_{12} + W_{21})N_d - 2WQN_d.$$

To simplify the notation, two new parameters are introduced, $P = (W_{12} - W_{21})N$ and $\alpha = W_{12} + W_{21}$. The quantity P may be viewed as an effective pump term that, at $P > 0$ ($W_{12} > W_{21}$), begins to increase N_d. The quantity α describes the decay of the population difference N_d. Note that the transition rate W_{12} due to the pump process enters into the decay constant α for N_d. Thereby α is no longer a material constant. In terms of the new parameters we have:

$$\frac{dN_d}{dt} = P - \alpha N_d - 2WQN_d.$$

(10.5)

Our two new equations (10.4) and (10.5) may be simplified further by introducing the following normalized variables and the dot as the rescaled-time derivative:

$$q = \frac{W}{\gamma}Q, \quad n = \frac{W}{\gamma}N_d, \quad p = \frac{W}{\gamma^2}P, \quad b = \frac{\alpha}{\gamma}, \quad (\dot{\ }) = \frac{1}{\gamma}\frac{d()}{dt}.$$

Then the two laser rate equations are obtained in their simplest form:

$$\dot{q} = -q + nq,$$
$$\dot{n} = p - bn - 2nq.$$

(10.6)

The rate equations for the ideal three-level laser consist of a system of two ordinary differential equations of first order. The dynamics of the system (photon number, proportional to the intensity of the emitted laser light, and population difference as functions of time) can become utterly complicated when the pump p is time-dependent, $p = p(t)$. At constant p, on the other hand, a number of statements about the solutions of the system can be given.

10.3 Stationary Operation

Dynamical systems of type (10.6) with constant parameters are investigated by first looking for stationary solutions and their stability. For a stationary solution, q and n are constant by definition, that is, $\dot{q} = 0$ and $\dot{n} = 0$. This, from (10.6), renders the system of algebraic equations:

$$0 = (-1+n)q,$$
$$0 = p - bn - 2nq. \tag{10.7}$$

The first of the two equations is solved by $q = 0$ or $n = 1$. For both values a solution of the second equation exists, whereby physically acceptable solutions require $q \geq 0$. Thus there are two coexisting stationary solutions:

$$\text{solution one: } q = q_1 = 0, \qquad n = n_1 = p/b,$$
$$\text{solution two: } q = q_2 = (p-b)/2, \quad n = n_2 = 1. \tag{10.8}$$

The first solution is the less interesting one. It describes the laser below the laser threshold. Due to the condition $q \geq 0$, solution two is physically acceptable for $p \geq b$ only. It describes the light-emitting laser. Thus the laser threshold is given by $p = b$. Light is not emitted until $p > b$. This condition is called the Schawlow–Townes condition for lasing. Taking the spontaneous emission into account, a small contribution of light is also present below threshold. An investigation of the stability of the two solutions (see Sect. 10.4) yields that for $p < b$ solution one ($q = 0$) is stable, whereas solution two ($q < 0$) is unstable (anyway not realized physically). For $p > b$ the solutions exchange their stability. The graphs of the normalized photon number q and population difference n of the stable solutions are given in Fig. 10.5 in dependence on the pump parameter p.

The figures reveal an astonishing result: above the threshold $p = b$ the (normalized) population difference stays constant irrespective of the pump parameter p. The only explanation is that the additional pump energy above the threshold is completely converted into photons. The photon number then should increase linearly with the pump energy. It should be possible to check this result experimentally. This can easily be done with a laser whose pump power can be controlled, for instance with an argon ion laser or a semiconductor laser. Indeed the data sheets from manufacturers give the graph sketched in Fig. 10.5 for the output power of a

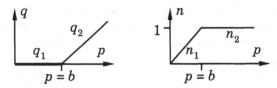

Fig. 10.5. Normalized photon number q and population difference n of the stable solutions of the laser rate equations (10.6) versus the pump parameter p.

semiconductor laser as a function of the pump current. Above threshold, the injected electrical energy is (almost) completely transformed into photon energy. To achieve a large overall efficiency of transforming electrical energy into photons, the threshold $p = b$ should be as small as possible. Indeed, the aim of current research is to design a "zero-threshold" laser. With semiconductor lasers of special construction, the threshold can be brought down to almost zero, and thus efficiencies approaching 100% are reached. In these lasers, the quantum mechanical properties of special semiconductor structures with small dimensions are used to advantage [10.3]. On the other hand, similar structures can be used for light detectors, yielding a quantum efficiency of nearly 100%. That is, almost every incident photon is detected.

10.4 Stability Analysis

In Sect. 10.2, we derived the rate equations for a three-level laser:

$$\dot{q} = -q + nq,$$
$$\dot{n} = p - bn - 2nq. \tag{10.9}$$

Here, q is the normalized photon number, n the normalized population difference between the upper and lower laser levels, p the normalized effective pump rate, and b the decay constant for the population difference. The system has two coexistent stationary solutions or points of equilibrium:

$$q_1 = 0, \qquad n_1 = p/b,$$
$$\text{and} \quad q_2 = (p-b)/2, \quad n_2 = 1. \tag{10.10}$$

A stability analysis yields which ones of the stationary solutions are stable and which are unstable, or, in the language of nonlinear dynamics, which ones are attractors and which are repellors, respectively. For two-dimensional systems, there exists a particularly simple scheme that is presented here. The methods discussed belong to the fundamentals of nonlinear dynamical systems theory. Nonlinear systems are of great general importance in physics, and knowledge of the basics of this theory is highly recommended.

To investigate the stability of an equilibrium point (q_i, n_i), $i = 1, 2$, a small perturbation and its time evolution near the point of equilibrium are considered. To this end the (usually nonlinear) system is linearized at the point of equilibrium. For short times and small distances from the point of equilibrium the linear system thus obtained shows the same qualitative behavior as the nonlinear system, except when the system, because of the nonlinear terms, remains no longer conservative but becomes dissipative (theorem of Poincaré).

Therefore, we first consider the linear system of two ordinary differential equations of first order with constant coefficients:

$$\begin{pmatrix} \dot{x} \\ \dot{y} \end{pmatrix} = \begin{pmatrix} ax + by \\ cx + dy \end{pmatrix} = \begin{pmatrix} a & b \\ c & d \end{pmatrix} \begin{pmatrix} x \\ y \end{pmatrix}. \tag{10.11}$$

The equilibrium point, or, in mathematical language, the singular point, is located at $(x_0, y_0) = (0, 0)$. There are essentially four different types of equilibrium, depending on the coefficient matrix

$$\mathbf{A} = \begin{pmatrix} a & b \\ c & d \end{pmatrix}. \tag{10.12}$$

They differ in the way trajectories (curves in state space) behave in their vicinity. Figure 10.6 shows portraits in state space in the neighborhood of equilibrium points of four different types. The trajectories are obtained as solutions of the differential equation via an exponential ansatz:

$$x = \xi \exp(\lambda t), \\ y = \eta \exp(\lambda t). \tag{10.13}$$

Inserting the ansatz into (10.11) yields the characteristic equation

$$\det(\mathbf{A} - \lambda \mathbf{1}) = 0, \tag{10.14}$$

where $\mathbf{1}$ is the unit matrix. Solving the characteristic equation supplies the eigenvalues of $\mathbf{A}$. Depending on the magnitude of the two eigenvalues λ_1 and λ_2 the four state-space portraits of Fig. 10.6 are obtained. Nodes may be stable (λ_1, λ_2 real, $\lambda_1 < 0$, $\lambda_2 < 0$) or unstable (λ_1, λ_2 real, $\lambda_1 > 0$, $\lambda_2 > 0$). Saddle points are always unstable (λ_1, λ_2 real, $\lambda_1 > 0$, $\lambda_2 < 0$).

node   saddle   focus   center

Fig. 10.6. State-space portraits in the vicinity of different types of equilibrium points of a linear two-dimensional dynamical system.

A focus may be stable (λ_1, λ_2 complex conjugate, $\text{Re}\{\lambda_1\} = \text{Re}\{\lambda_2\} < 0$) or unstable ($\lambda_1$, λ_2 complex conjugate, $\text{Re}\{\lambda_1\} = \text{Re}\{\lambda_2\} > 0$). A center is neutrally stable (λ_1, λ_2 purely imaginary, $\lambda_1 = -\lambda_2$). Stable means that the trajectories, the solution curves in the state space $\mathcal{M} = \{(x,y)\}$, ultimately head for the point of equilibrium. Unstable means that they ultimately recede from the point in question.

For two-dimensional systems as considered here, instead of the equilibrium points being characterized by the eigenvalues of the coefficient matrix $\mathbf{A}$, the classification can be done with the help of the invariants of $\mathbf{A}$: the trace $\text{tr}(\mathbf{A})$, the determinant $\det(\mathbf{A})$, and the discriminant $D(\mathbf{A})$. They are defined as follows:

$$
\begin{aligned}
\text{determinant:} \quad & \det(\mathbf{A}) = ad - bc, \\
\text{trace:} \quad & \text{tr}(\mathbf{A}) = a + d, \\
\text{discriminant:} \quad & D(\mathbf{A}) = \text{tr}^2(\mathbf{A}) - 4\det(\mathbf{A}).
\end{aligned}
\tag{10.15}
$$

In terms of these quantities we have:

$$
\det(\mathbf{A}) < 0: \text{ saddle}
$$
$$
\det(\mathbf{A}) = 0: \text{ indeterminate } (\mathbf{A} \text{ not invertible})
$$
$$
\det(\mathbf{A}) > 0: \left\{
\begin{array}{l}
D(\mathbf{A}) > 0: \left.\begin{cases} \text{tr}(\mathbf{A}) > 0 \text{ unstable} \\ \text{tr}(\mathbf{A}) < 0 \text{ stable} \end{cases}\right\} \text{node} \\[2ex]
D(\mathbf{A}) < 0: \left.\begin{cases} \text{tr}(\mathbf{A}) > 0 \text{ unstable} \\ \text{tr}(\mathbf{A}) < 0 \text{ stable} \end{cases}\right\} \text{focus} \\[2ex]
\phantom{D(\mathbf{A}) < 0:} \quad \text{tr}(\mathbf{A}) = 0 \qquad \text{center.}
\end{array}
\right.
\tag{10.16}
$$

These statements can be summarized pictorially in a diagram (Fig. 10.7) where $\det(\mathbf{A})$ is plotted versus $\text{tr}(\mathbf{A})$.

As mentioned above, the type of an equilibrium point of a general system of second order can be obtained by linearization at this point. Only in the case of center points is a further treatment necessary by consideration of the nonlinear terms, since they may lead to either a stable or unstable focus.

We now consider the general system

$$
\begin{aligned}
\dot{x} &= f_1(x,y), \\
\dot{y} &= f_2(x,y).
\end{aligned}
\tag{10.17}
$$

Let (x_0, y_0) be a point of equilibrium, that is, $f_1(x_0, y_0) = 0$ and $f_2(x_0, y_0) = 0$. To obtain the linearized equations, system (10.17) is expanded into a Taylor series at (x_0, y_0):

$$
\frac{d(x - x_0)}{dt} = (x - x_0)\frac{\partial f_1}{\partial x}(x_0, y_0) + (y - y_0)\frac{\partial f_1}{\partial y}(x_0, y_0) + R_1(x - x_0, y - y_0),
$$
$$
\frac{d(y - y_0)}{dt} = (x - x_0)\frac{\partial f_2}{\partial x}(x_0, y_0) + (y - y_0)\frac{\partial f_2}{\partial y}(x_0, y_0) + R_2(x - x_0, y - y_0).
\tag{10.18}
$$

Ignoring the nonlinear terms R_1 and R_2, we see that the Jacobian

$$J(x_0,y_0) = \begin{pmatrix} \frac{\partial f_1}{\partial x} & \frac{\partial f_1}{\partial y} \\ \frac{\partial f_2}{\partial x} & \frac{\partial f_2}{\partial y} \end{pmatrix} (x_0,y_0) \tag{10.19}$$

takes the place of the matrix $\mathbf{A}$ of (10.11). When R_1 and R_2 obey the condition

$$|R_1(x-x_0,y-y_0)| \le C_1((x-x_0)^2 + (y-y_0)^2),$$
$$|R_2(x-x_0,y-y_0)| \le C_2((x-x_0)^2 + (y-y_0)^2), \tag{10.20}$$

with C_1 and C_2 being constant, then, provided that $\mathrm{Det}(\mathbf{J}) \ne 0$, the Jacobian determines the type of the equilibrium points (x_0,y_0) of the system (10.18). This point can be shifted into the origin of the coordinate system by the transformation $\tilde{x} = x - x_0$, $\tilde{y} = y - y_0$. The linearization of (10.18) then attains the form of (10.11).

We now apply our knowledge to the laser rate equations (10.9). The Jacobian reads:

$$\mathbf{J}(q,n) = \begin{pmatrix} -1+n & q \\ -2n & -b-2q \end{pmatrix}. \tag{10.21}$$

We have two points of equilibrium and thus two matrices:

$$\mathbf{J}_1 = \mathbf{J}\left(0,\frac{p}{b}\right) = \begin{pmatrix} -1+p/b & 0 \\ -2p/b & -b \end{pmatrix} \tag{10.22}$$

and

$$\mathbf{J}_2 = \mathbf{J}\left(\frac{p-b}{2},1\right) = \begin{pmatrix} 0 & (p-b)/2 \\ -2 & -p \end{pmatrix}. \tag{10.23}$$

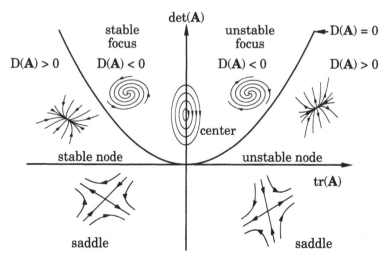

Fig. 10.7. Diagram giving the type of an equilibrium point after the determination of $\mathrm{tr}(\mathbf{A})$, $\det(\mathbf{A})$, and $D(\mathbf{A})$.

The invariants of $\mathbf{J}_1$ and $\mathbf{J}_2$ are

$$
\begin{aligned}
\det(\mathbf{J}_1) &= b - p, & \det(\mathbf{J}_2) &= p - b, \\
\operatorname{tr}(\mathbf{J}_1) &= p/b - b - 1, & \operatorname{tr}(\mathbf{J}_2) &= -p, \\
D(\mathbf{J}_1) &= (p/b + b - 1)^2, & D(\mathbf{J}_2) &= p^2 - 4(p - b).
\end{aligned}
\tag{10.24}
$$

The following cases can be distinguished:

1. $0 < p < b$

$$
\left.\begin{aligned}
\det(\mathbf{J}_1) &> 0 \\
\operatorname{tr}(\mathbf{J}_1) &< 0 \\
D(\mathbf{J}_1) &> 0
\end{aligned}\right\} \Rightarrow \begin{array}{c} (0, p/b) \\ \text{node, stable} \end{array}
\qquad
\begin{array}{l}
\det(\mathbf{J}_2) < 0 \\
\Rightarrow ((p - b)/2, 1) \\
\text{saddle}
\end{array}
$$

2. $b = p$
   $\mathbf{J}_1 = \mathbf{J}_2$, $\det(\mathbf{J}_1) = 0$.
   Both points of equilibrium coincide. The system is degenerate. A detailed discussion reveals a stable node.

3. $0 < b < p$

$$
\begin{array}{l}
\det(\mathbf{J}_1) < 0 \\
\Rightarrow (0, p/b), \text{saddle}
\end{array}
\qquad
\begin{array}{l}
\det(\mathbf{J}_2) > 0 \\
\operatorname{tr}(\mathbf{J}_2) < 0 \\
\text{Two cases exist:} \\
\text{1. } D(\mathbf{J}_2) > 0 \text{ node, stable,} \\
\text{2. } D(\mathbf{J}_2) < 0 \text{ focus, stable.}
\end{array}
$$

The result is summarized graphically in the (p,b)-parameter plane (see Fig. 10.8). The analysis shows that for $p < b$ the point of equilibrium $(q_1, n_1) = (0, p/b)$ is stable, whereas the point of equilibrium $(q_2, n_2) = ((p - b)/2, 1)$ is unstable. At $p = b$ the two points of equilibrium exchange their stability and since then, for $p > b$, the point of equilibrium (n_1, q_1) is unstable, the point of equilibrium (n_2, q_2) is stable. The system is said to undergo a transcritical bifurcation.

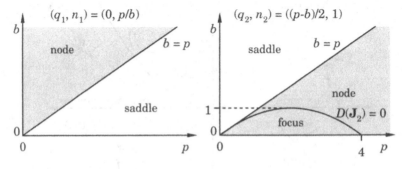

Fig. 10.8. Type of the equilibrium points of the laser rate equations (10.9). The stable regions are shaded.

The laser rate equations represent a quite simple system as long as p and b stay constant. When p or other parameters, for instance the decay time in the cavity, are made time dependent, interesting dynamical effects occur. They can be used for the production of short pulses and, in the case of periodic modulation, may lead to chaotic solutions.

10.5 Transient dynamics

In the previous section we have discussed solutions of the laser rate equations that correspond to stationary operating conditions. The light output of the laser working above threshold will then be constant (continuous wave or cw emission). Our analysis has shown that these states are stable and it can be shown that they will be approached from all physically reasonable initial conditions. The stable equilibrium is globally stable.

10.5.1 Relaxation Oscillations

When the laser is switched on, the light intensity usually approaches its stationary value via relaxation oscillations. This behavior is demonstrated in Fig. 10.9 for an ideal four-level laser (the population of the pump level and of the lower laser level are always equal to zero):

$$\dot{q} = -q + nq,$$
$$\dot{n} = p - bn - nq,$$

(10.25)

with the parameters given in the figure caption. We observe that with changing parameters, demonstrated here by switching the pump term p from zero to a constant value $p > 0$, the stable stationary solutions give way to a more complex, time-dependent dynamics. The initial transient is characterized by a significant overshoot of light intensity above the stationary value, caused by an initial excess inversion and associated high gain of the laser medium. The inversion n is diminished sharply by the first pulse and has to be replenished by the pump, whereupon a second but weaker emission peak is generated. The system finally settles on a new stationary value.

This interplay between light amplification and depletion of population inversion may lead to the phenomenon of spiking. It occurs with certain types of lasers that are pumped by pulsed sources or operated not far above threshold. For example, Fig. 10.10 shows the oscilloscope traces of light intensity emitted by a flash-lamp pumped ruby laser. The signal consists of a train of irregularly spaced pulses of different peak intensity. It is observed that there is a correlation between pulse height and interpulse distance: the larger the temporal separation of one pulse to

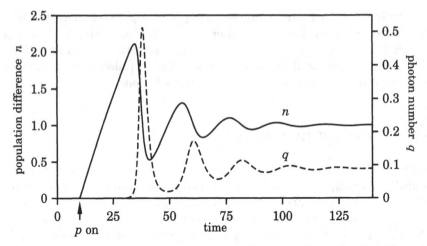

Fig. 10.9. Photon number q and population difference n when a four-level laser is switched on. At the time indicated by the arrow, the pump term is switched from $p = 0$ to $p = p_0 = 0.1$ (decay constant $b = 0.01$).

its predecessor, the larger is its peak intensity. Furthermore, the taller the pulses the shorter they are (Fig. 10.10 (*right*)). This behavior can be understood quite naturally as relaxation oscillations of the laser, irregularly triggered by fluctuations of spontaneously emitted light after the population inversion has been sufficiently replenished by the pump.

10.5.2 Q-Switching

In general the irregular output of a spiking laser is not very useful for practical applications. The available pump energy is split into many short

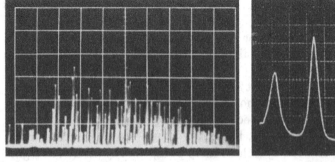

Fig. 10.10. Spiking of a ruby laser. Attenuated laser light is detected by a photodiode and its signal displayed on an oscilloscope. *Left:* train of irregular spikes, horizontal scale 50 μs/div., vertical scale 0.5 V/div.; *right:* enlarged view at higher temporal resolution (2 μs/div.), vertical scale 50 mV/div.

and relatively weak pulses that are not sufficiently reproducible. However, the very phenomenon of spiking points out a way to generate very powerful single laser pulses: try to avoid the formation of a large optical field in the cavity until the population inversion has built up to a strong level. This can be achieved by artificially increasing the cavity losses (or equivalently, decreasing the cavity's quality factor (Q-factor)). Thus, the active medium acts as an energy store for the giant pulse to be produced until it is able to deliver the energy into the light field. This, of course, is only possible if the lifetime of the upper laser level is sufficiently long to be filled up by the pump.

The buildup of a large population inversion can be understood by reference to the non-normalized rate equations (10.4) and (10.5). Let us denote by γ_0 the damping parameter of the good cavity. The population inversion N_{d0} at the onset of laser emission is given by

$$N_{d0} = \frac{\gamma_0}{W}. \tag{10.26}$$

It is proportional to the damping parameter of the cavity. If we thus increase the cavity losses such that $\gamma \gg \gamma_0$ we obtain a population inversion

$$N_d = \frac{\gamma}{\gamma_0} N_{d0} \tag{10.27}$$

that is larger by a factor $\tilde{\gamma} = \gamma/\gamma_0$. When the cavity losses are suddenly decreased to their undisturbed value γ_0, the energy stored in the medium is released and gives a powerful laser pulse, called giant pulse. Figure 10.11 shows a numerical calculation based on the rate equations for a four-level system

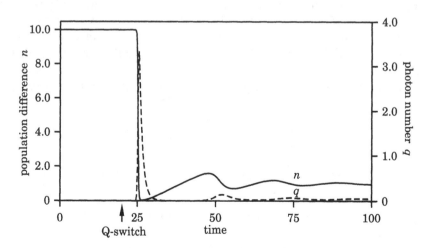

Fig. 10.11. Photon number q and population difference n of a four-level laser with Q-switching. At the time $t = 20$ the Q-factor of the cavity is increased by a factor of 10 (pump parameter $p = 0.1$, decay constant $b = 0.01$).

$$\dot{n} = -\tilde{\gamma}q + nq,$$
$$\dot{q} = p - bn - nq,$$

where the variables n and q are normalized with respect to the good-cavity damping constant γ_0. In the figure, for $t < 20$ the cavity losses are 10 times larger than for the undisturbed cavity, that is, $\tilde{\gamma} = 10$. The initial conditions were calculated by letting the system relax to the corresponding equilibrium state, with the photon number q being nearly zero and the population inversion n being nearly ten times as large as its normal value. At the time $t = 20$ the cavity damping is suddenly restored to γ_0, i.e. $\tilde{\gamma} = 1$. The energy deposited in the active material is discharged into the light field, leading to a short giant pulse with a peak intensity of more than the 70-fold of the equilibrium (cw) output intensity. The calculation also shows that there is a certain delay between the Q-transition and the pulse emission depending on the initial conditions and, in a real laser system, on the level of spontaneous emission. The calculation of Fig. 10.11 also tells us that there will be further secondary pulses of lower intensity following the initial pulse. It is thus desirable to switch again to high cavity damping after the giant pulse was emitted.

There are several methods to implement Q-switching in real laser systems, which can be divided into active and passive methods. Active Q-switching is achieved by using mechanical means such as shutters, rotating apertures, prisms or mirrors. These devices are, however, rather slow and thus electro-optical switches such as Pockels or Kerr cells are preferred. Their principle of operation relies on the fact that certain materials can be made birefringent by application of a strong electric field. By placing such a medium behind a polarizer in the laser cavity it is possible to control its optical transmission by the applied voltage.

Passive Q-switching can be realized very elegantly by insertion of a saturable absorber, for example, a dye solution, into the laser cavity, see Fig. 10.12. The absorber acts as an intensity-dependent switch; the absorption is caused by an electronic transition between two levels at resonance with the laser frequency. The absorption coefficient depends on

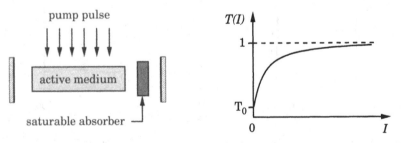

Fig. 10.12. *Left:* Laser system containing a saturable absorber as a passive Q-switch; *right:* Intensity dependence of light transmission by a saturable absorber.

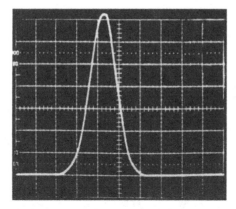

Fig. 10.13. Giant pulse emitted by a flash-lamp pumped ruby laser with a dye-filled cuvette the dye solution, cryptocyanine in methanol, acting as saturable absorber (time scale 50 ns/div.)

the population difference of the two levels and saturates when the levels, at high intensities, have nearly the same occupation number. The saturable absorber can be imagined to act as a soft switch which becomes transparent in the course of buildup of the laser pulse.

Figure 10.13 gives, as an example, the giant pulse emitted by a ruby laser with an intracavity saturable absorber (cryptocyanine in methanol). The pulse width is about 70 ns, the pulse energy is of the order of 1 J, the peak power of the order of 10 MW. With Q-switching, pulse widths between 10 and 100 ns are readily available, but no less because the light has to pass the cavity several times to be sufficiently amplified. These pulses are quite useful for a number of applications, but by far they do not represent the limit achievable in terms of pulse duration, peak power and intensity. Other methods such as mode-locking have to be applied to generate optical pulses in the ps- or even fs-range (see Section 11.2.2).

10.5.3 Cavity Dumping

Q-switching for getting short pulses is only advantageous when the upper laser level has a sufficiently long life time to build up a high population difference to the lower laser level as in the ruby laser with its life time of about 3 ms. When the life time of the upper laser level is short as in the argon ion laser (about 1 μs) the energy can be stored in the electromagnetic field of the cavity instead and released as a short pulse by opening the cavity for a short time (usually less than one cavity round-trip time for a photon). A realization of this idea, called cavity dumping, is given in Fig. 10.14 for an argon-ion laser. The output mirror normally present is replaced by a cavity dumping unit consisting of a "Bragg cell", two spherical mirrors of 100 % reflectivity and an output prism to intercept

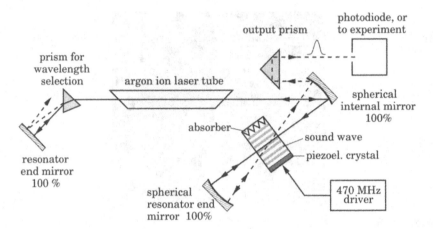

Fig. 10.14. Cavity dumping of an argon ion laser with Bragg deflection of the light at a high-frequency sound wave.

the deflected photons for redirection as output light pulse. The Bragg cell serves as a fast shutter to open the cavity by deflecting the light beam out of the cavity. To this end a short acoustic pulse of some 50 ns duration at a carrier frequency of several 100 MHz (e. g. 470 MHz) is generated and propagated through the transparent material placed in the path of the laser beam (inside the cavity). For this short time a running sound pulse is set up in the material with its corresponding modulation of the refractive index. A (running) phase grating is present leading to diffraction of the light wave. The cell is inclined to the laser light storage beam to achieve Bragg deflection for optimal output. Figure 10.15 (*left*) shows the output light pulse obtained in this way. Its duration is about 10 ns. By repeatedly pulsing the Bragg cell with acoustic pulses a corresponding train of light pulses can be tapped out of the laser as shown in Fig. 10.15

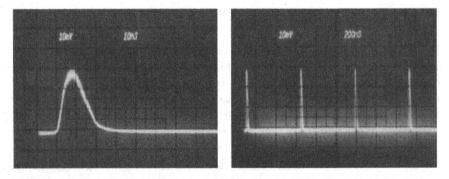

Fig. 10.15. Light output of a cavity-dumped argon ion laser: pulse shape (*left*) and train of pulses with 550 ns separation (*right*).

(*right*). The pulses have a coherence length of about 20 mm and can thus be used for holographic cinematography of small objects [10.4] (see also Section 7.4 on holographic cinematography).

10.6 Chaotic Dynamics

In this section, we want to discuss a mainly scientific aspect of the laser: chaotic laser dynamics. We have seen that while the one-mode laser described by the rate equations (10.6) with constant parameters will settle down to a stable stationary state the introduction of time- or state-dependent parameters introduces more complicated dynamics.

A particularly important and common form of external perturbation of a physical system is periodic driving. In the laser it can be accomplished by modulation of the pump, $p = p(t)$, for instance in laser sources for optical communication devices. However, the periodic driving may yield peculiar effects. Even with purely sinusoidal modulation, $p(t) = p_0(1 + p_m \sin \omega t)$, the light intensity may not follow the modulation, but oscillates at a fraction of the modulation frequency or even completely erratically, never repeating itself. This behavior, observed for a variety of simple deterministic equations, has become known as deterministic chaos or simply chaos.

We discuss this behavior for the ideal four-level laser with periodic pump modulation:

$$\dot{q} = -q + nq + sn,$$
$$\dot{n} = p_0(1 + p_m \sin \omega t) - bn - nq. \tag{10.28}$$

Here, p_m is the degree of modulation ($0 \leq p_m \leq 1$), p_0 is the constant pump term, and s is the spontaneous emission rate into the laser mode. The other quantities are the same as before. Figure 10.16 shows some solutions from a period-doubling cascade that may occur upon alteration of the modulation frequency ω. In the figure, trajectories in state space are plotted, that is, in the plane spanned by the system variables n and q. Each trajectory represents the temporal evolution of the system by a plot of its state $(n(t), q(t))$ at consecutive times t. For example, at $\omega = 0.0145$, an oscillation with the modulation frequency ω (period 1) is obtained. Lowering the frequency ω, an oscillation of half the modulation frequency, $\omega/2$ (period 2), then of $\omega/4$ (period 4) is observed. These period doublings repeat infinitely often, whereby they occur ever faster as ω is decreased. The frequency points of period doubling accumulate at a finite limiting value of ω, where the solution becomes of infinite period, that is, aperiodic. Beyond this limiting value, various periodic and aperiodic solutions are possible. An example of an aperiodic, erratic solution is given in the fourth image of Fig. 10.16. A number of loops and rings is to be seen, but however long the system may oscillate, it never returns to settle down on

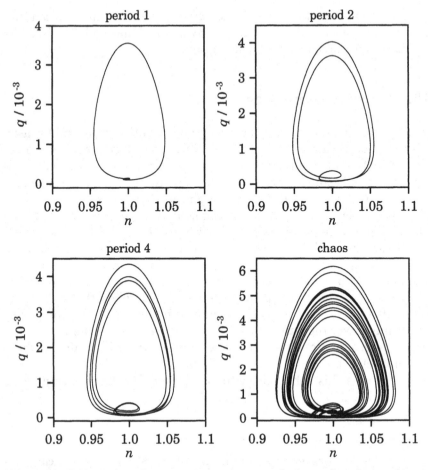

Fig. 10.16. Trajectories of attractors (stationary solutions after decay of transients) of a four-level laser system out of a period-doubling cascade. The individual graphs have been obtained at the following modulation frequencies: $\omega = 0.0145$ (period 1), $\omega = 0.01385$ (period 2), $\omega = 0.01375$ (period 4), and $\omega = 0.0136$ (chaotic attractor). The other parameters were set to $s = 10^{-7}$ (normalized spontaneous emission), $p_0 = 6 \cdot 10^{-4}$ (constant pump term), $p_m = 1$ (degree of modulation), and $b = 0$ (decay constant of the population difference).

a periodic solution. We have a chaotic attractor. Its characteristic property is a sensitive dependence of the dynamics on initial conditions. This means that two typical, close system states separate exponentially fast on an infinitesimal scale. Macroscopically, their distance gets comparable with the attractor dimensions in a finite time. This property makes impossible a long-term forecast of the dynamics, in our case, of the light emission of a chaotic laser.

The period-doubling cascade given in Fig. 10.16 represents just an arbitrarily selected example of the rich dynamics of a periodically pump modulated laser in the rate equation approximation. Such cascades exist in large number also in other parameter regions. To get a survey of the dynamics of a system, bifurcation diagrams may be calculated. A bifurcation, for instance a period doubling, is defined as a change in the qualitative behavior of the system upon alteration of a parameter. In Fig. 10.16, for example, a period-doubling bifurcation must occur at a frequency ω_{12} between $\omega_1 = 0.0145$ (period 1) and $\omega_2 = 0.01385$ (period 2).

In a bifurcation diagram one of the dependent variables in the steady state, that is, one of the variables of the attractor, is plotted versus one of the system parameters. In our case, the variables are q and n, which, of course, are functions of time, $q(t)$ and $n(t)$. Therefore, to obtain graphics in a plane, a reduction of the curves $q(t)$ or $n(t)$ must be envisaged. Here, we want to plot a bifurcation diagram with the modulation frequency $\omega = 2\pi/T$ of the pump. It suggests itself to plot one of the variables – we choose n – only at a fixed phase of the modulation, that is, to plot $n(t_0)$, $n(t_0+T), n(t_0+2T), \ldots$. This type of sampling is known among physicists as stroboscopic sampling. It is often used to stop rotating motions. The same is the case here. For instance, when we have a solution $n(t)$ that is periodic with the modulation frequency – an orbit of period 1, as it is called, supposing a normalization with the period T – then all values $n(t_0), n(t_0+T), \ldots$ are the same. Thus, only a single point has to be plotted at the corresponding parameter value. If we vary the parameter, here ω, a different value $n_\omega(t_0) = n(t_0,\omega)$ is obtained. But not for all values of ω is the laser light modulated with the frequency of the pump. Figure 10.16 reveals that periods 2, 4, … may also occur. How do these periods show up in the bifurcation diagram? We consider the period 2, alias $2T$. As the oscillation does not repeat before two modulation periods, we have $n(t_0) = n(t_0+2T)$, $n(t_0+T) = n(t_0+3T), \ldots$. To be sure, we have $n(t_0) \neq n(t_0+T)$, since otherwise the oscillation would be of period 1. We thus get two different points for n to be plotted at the corresponding ω value. Similarly, a period-4 solution renders four points, and a chaotic attractor renders infinitely many. From a chaotic attractor, usually about 50 to 100 points are used for visualization.

Figure 10.17 shows two bifurcation diagrams for a four-level laser, one with a coarse sampling in the parameter ω and one blown up to demonstrate the intrinsic richness of the solutions. The points of period doubling stand out clearly. There, the 'curve' splits into two branches, it bifurcates. This process obviously repeats infinitely often, up to 'period 2^∞', leading to an aperiodic attractor. In this case, despite periodic driving, the oscillation never repeats; a quite peculiar but common behavior of nonlinear systems. The laser is but one of numerous examples nature keeps at hand.

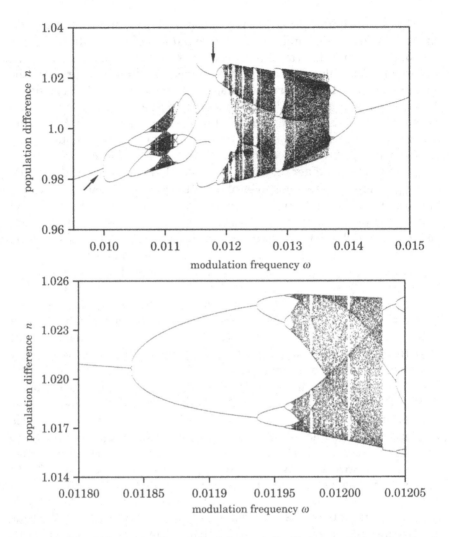

Fig. 10.17. Two bifurcation diagrams of the periodically pumped ideal four-level laser. The lower diagram is an enlarged section from the upper diagram, the part indicated by the vertical arrow. The parameters s, p_0, p_m, and b are the same as in Fig. 10.16.

In the bifurcation diagram, parameter values also exist where three points appear. The solutions are then obviously of period 3. They appear suddenly after a chaotic solution. The scenario at this parameter value again is connected with a bifurcation, called tangent bifurcation or saddle-node bifurcation (see, for example, [10.5]). Moreover, jumps are to be seen between two period-2 solutions that overlap. There, two period-2 solutions coexist. This phenomenon is known from the resonance curves of nonlinear oscillators which may lean over. This leads to hysteresis phenomena upon slow (adiabatic) variation of the parameter.

The bifurcation diagrams shown only represent part of the behavior of the laser system. Besides the modulation frequency ω, the system has further parameters: the degree of modulation p_m, the pump strength p_0, the decay constant b of the population difference, and the rate s of spontaneous emission. The parameter space, the set of possible parameter values, is thus high-dimensional. The complete investigation and presentation of the dynamics of lasers therefore poses a difficult problem.

The parameter space for the normalized laser rate equations (10.6) is two-dimensional and is spanned by the two parameters b and p. Figure 10.8 shows, in the language of nonlinear dynamical systems theory, a phase diagram or parameter space diagram of the simple laser rate equations. For each point of the parameter space (here only the upper right quadrant of the (p,q)-plane), the type of the equilibrium point of the system is indicated. More complicated attractors do not occur. The dynamics of the system is entirely known.

The modulated system, despite also looking quite innocent, requires substantially greater effort to arrive at a complete description of its dynamics [10.6]. To this end, the points of bifurcation are determined in dependence on two parameters of the system, since the bifurcations represent the qualitative, and therefore essential, changes of its dynamics. The two parameters span a plane in parameter space. Bifurcation points of the same type form curves in this plane that form the boundaries between different types of dynamic behavior. With thermodynamics in mind (compare p–T diagrams) we call the resulting diagrams also phase diagrams. Figure 10.18 gives an example of a phase diagram for the modulated four-level laser (10.28). Only the period-doubling bifurcation curves from period 1 to period 2 (dashed lines) and the saddle-node bifurcation lines belonging to period 1 (solid lines) are shown. To relate the phase diagram and the bifurcation diagram of Fig. 10.17 a horizontal bar is inserted indicating the parameter interval for which the bifurcation diagram has been calculated. One particular (period-doubling) bifurcation point is indicated in both diagrams by an oblique arrow. A remarkable feature of the phase diagram is the periodic arrangement of the bifurcation lines. A similar bifurcation structure has been found in other nonlinear oscillator systems that have been investigated in depth [10.7]. This points to underlying, may be universal, ordering principles for classes of nonlinear systems.

Modern chaos theory has shown that chaotic dynamics and therefore long-term unpredictability are ubiquitous in nature. We should not be astonished to encounter chaos also in the laser, a necessarily nonlinear system. The laser thus has become, as did hydrodynamics in the first place, an object of chaos research [10.2]. Because of the controllability of

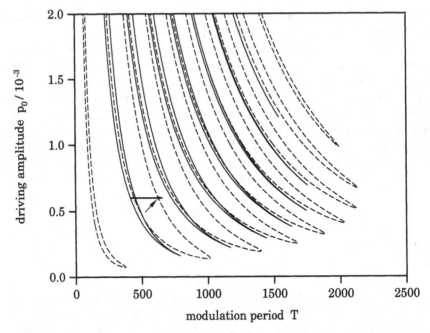

Fig. 10.18. Phase diagram of the periodically pumped four-level laser with parameters $p_m = 1$, $b = 0$, and $s = 10^{-7}$, showing part of the bifurcation set in the (T, p_0) parameter plane, $T = 2\pi/\omega$ being the modulation period. The dashed lines correspond to period-doubling bifurcations of period-1 oscillations, the solid lines give the locations of period-1 saddle-node bifurcations. The horizontal bar indicates the range of parameter variation for which the bifurcation diagram in Fig. 10.17 has been calculated. The oblique arrow pointing to the intersection of the bar with the dashed curve marks the same period-doubling bifurcation point as in Fig. 10.17.

parameters, laser dynamics and therewith coherent optics will contribute substantially to the investigation of nonlinear and complex dynamical systems.

10.7 Synchronization

It is a principle of nature to couple many (often identical) systems to form new ones with new properties. When we look at just two coupled systems an important feature is whether they synchronize or not. The earliest reported example is that of Christiaan Huygens who observed that two pendulum clocks hooked at the same bar were going to oscillate at the same frequency [10.8] obviously by coupling of the oscillations through the bar. This example with regular dynamics can easily be described theoretically with simple pendula introducing a coupling between them. Not

Fig. 10.19. Semiconductor laser (laser diode) with optical feedback via an external resonator.

as obvious is that chaotic systems (for instance lasers) are also susceptible to synchronization, that experiments can be conducted to show the effect and that even applications are envisaged for secure communication [10.9].

For optic communication semiconductor lasers are used [10.10]. They are small and compact and pumped by current injection. However, they are of low resonator quality and therefore amenable to disturbation by external light reflected back into the resonator. Figure 10.19 shows a standard arrangement to study this effect by optical feedback [10.11]. This configuration already leads to chaotic light fluctuations [10.12]. In the case of low feedback ($\lesssim 1\%$) a special phenomenon occurs called low frequency fluctuations (LFF). An experimental example of the appearance and time scale of low frequency fluctuations is given in Fig. 10.20.

When two lasers with external mirrors are coupled synchronization phenomena may occur. Figure 10.21 shows an arrangement for unidirectional coupling. When both lasers are operated alone without coupling, a fluctuating light output as in Fig. 10.22 is observed. When the two lasers are coupled unidirectionally as shown in Fig. 10.21, the second laser synchronizes its fluctuations to the first one (Fig. 10.23). This locking of a second laser to a chaotic one is of interest for secure communication schemes [10.13].

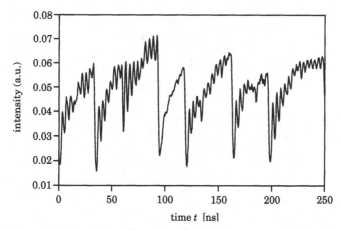

Fig. 10.20. Low frequency fluctuations of a semiconductor laser taken with a bandwidth of 200 MHz.

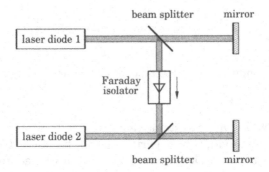

Fig. 10.21. Uni-directional coupling of two semiconductor lasers with two beam splitters and a Faraday isolator letting light pass in one direction only.

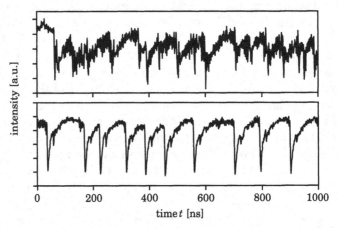

Fig. 10.22. Independent low frequency fluctuations of two uncoupled semiconductor lasers with external cavity.

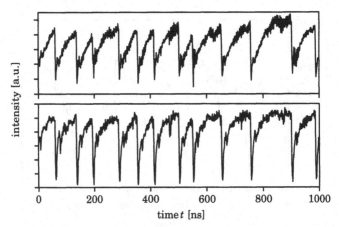

Fig. 10.23. Two lasers synchronize their fluctuations when coupled.

Problems

10.1 Derive the rate equations for an ideal four-level laser system:

$$\dot{q} = -q + nq,$$
$$\dot{n} = p - bn - nq. \qquad (10.29)$$

For an ideal four-level system it is assumed that the populations of both the pump level and the lower laser level are zero.

10.2 A well-known model of nonlinear dynamics, also describing the laser in certain limiting cases, is the *Lorenz* equations:

$$\dot{x} = \sigma(y - x),$$
$$\dot{y} = \rho x - y - xz, \qquad (10.30)$$
$$\dot{z} = -\beta z + xy,$$

where σ, ρ, and β denote the parameters of the model ($\beta > 0$). Determine the stationary solutions and their stability.

10.3 Write a computer code to integrate the laser rate equations (10.28) for arbitrary parameter values and initial conditions. Plot the computed trajectories on the screen or on a printer. By checking the bifurcation diagram of Fig. 10.17 for suitable parameter values, try to find a chaotic attractor and verify numerically its sensitive dependence on initial conditions.
Hint: methods for integration of ordinary differential equations are described, for instance, in [10.14].

11. Ultrafast Optics

The fastest events in nature occur in optics. That has to do with the speed of light that cannot be surmounted. But are we also able to handle this speed in measurements on fast events? Not quite, but we are approaching the limits more and more. To measure ever shorter events is like approaching absolute zero in temperature. You are very close, but you may come closer. To stop a motion, everybody knows that a short exposure time is needed. This can be done with a fast shutter or by illuminating the moving object by a short enough light pulse. Both, to build ever faster shutters and to produce ever shorter light pulses thus persists as an important topic in optics. New developments in optical technology have brought substantial progress in this area. Even light in flight can now be photographed.

11.1 Properties of Ultrashort Pulses

Ultrashort light pulses represent a unique and most useful form of light, with widths ranging from a few nanoseconds (ns, 10^{-9} s), down to picoseconds (ps, 10^{-12} s), and further down to a few femtoseconds (fs, 10^{-15} s). Even the production of pulses in the attosecond (as, 10^{-18} s) region, then lying in the XUV/X-ray region of the electromagnetic spectrum, is now feasible [11.1]. No other light source but the laser is capable of producing such short packets of light in a controllable and reproducible way.

Ultrashort laser pulses concentrate the available pulse energy onto a very small time interval and are thus ideal probes for scientific investigation and a versatile tool in technological applications. They open up new areas of research in ultrafast phenomena and highly energetic light-matter interactions [11.2]. The short pulse duration is put to advantage in picosecond and femtosecond spectroscopy to unravel e. g. ultrafast molecular dynamics in chemical reactions or biological processes. Powerful and very short light pulses are also required for optical long-range measurements and surveillance such as LIDAR (LIght Detection And Ranging). The tremendous peak powers of up to 10^{20} W/cm^2 achievable by amplification and focusing of fs-pulses now make it possible to study highly nonlinear interactions of light fields with matter, including

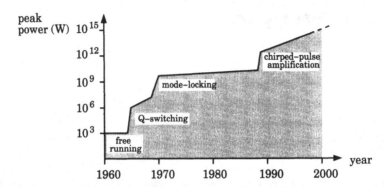

Fig. 11.1. Rise in pulsed-laser peak powers during the last decades with the corresponding advances in laser technology (adapted from Perry and Morou, Ref. [11.2]).

multi-photon ionization and absorption, high-harmonic generation and relativistic laser-plasma interactions, with the prospect to realize table-top particle accelerators, for example. Ultrashort pulses hold also a great promise for all-optical signal transmission and information processing, from ultra-high bit rate solitonic fiber transmission systems to ultrafast optical switching and all-optical computing. Many still unsolved questions of ultrashort pulse propagation and interaction along with their plethora of applications should give us a strong motivation to enter the discussion of their properties and methods of generation.

The minimum pulse durations achieved in the generation of ultra-short pulses have decreased remarkably during the past three decades; from several ns with, e.g., giant pulses of arc lamp excited ruby lasers or Q-switched Nd:YAG lasers to the ps range and finally to the fs and as region by employment of sophisticated methods of mode locking and pulse compression, to be discussed in the following sections [11.3]. Accordingly, the available peak powers or intensities have risen by nearly a dozen orders of magnitude (Fig. 11.1). Nowadays, laser systems producing stable pulses with durations of the order of 50 fs (and peak powers of 10^{10} W and more) are commercially available, and in research labs optical pulses as short as two optical cycles, corresponding to a pulse length of about 4 fs, have been produced in the visible and infrared spectral range [11.4]. This already approaches the fundamental limits of ul-trashort pulse generation in the visible spectral region, demonstrated by the fact that the difficulties of generating and transmitting ever shorter pulses are increasing substantially with the pulse shortness, as demonstrated by Fig. 11.2. Nevertheless it is to be expected that some day the shortest pulses will be measured in attoseconds (10^{-18} s) and that we will have light sources at our disposal capable of emitting pulses as short as a single light oscillation!

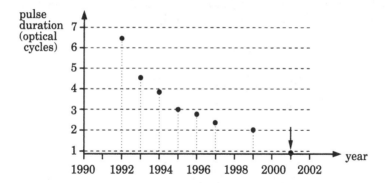

Fig. 11.2. Reduction of laser pulse durations during the past decade (adapted from Reid, Ref. [11.5]). Pulse widths level off as further reduction turns out to be increasingly difficult. The arrow indicates the generation of subfemtosecond pulses in high-intensity laser-matter interactions. No laser is available to directly produce pulses that short.

11.1.1 Time-Bandwidth Product

We know from the discussion of Fourier optics that spatially small features of a light field such as narrow apertures entail a broad spatial frequency spectrum. This is a general property of Fourier transform pairs (see the Appendix) which also has to be taken into account when considering light pulses of very short duration. The shorter the pulse, the larger is the frequency range involved and the more important become the dispersive properties of the medium carrying the pulse. Let us therefore first consider the spectral features of a one-dimensional Gaussian wave packet with center frequency ω_0, the optical field taken at $z = 0$,

$$E(z = 0, t) = E(t) = A(t)\exp(-i\omega_0 t) = E_0 \exp(-i\omega_0 t)\exp(-at^2). \quad (11.1)$$

With this ansatz it is implicitly assumed that the envelope function $A(t)$ is a slowly varying function on the time scale of the optical period $T = 2\pi/\omega_0$. This is quite a good approximation for pulse widths down to a few tens of fs. For example, a 100 fs pulse delivered by a Ti:sapphire laser oscillator at 780 nm will still contain more than 40 oscillations of the light field.

In (11.1) a Gaussian envelope function was chosen because its Fourier transform is also a Gaussian function. Furthermore, real laser pulses often can be approximated quite well by a Gaussian shape. The width Δt_G of the Gaussian pulse (FWHM) is defined as twice the time τ_G at which the intensity $I(t) = |E_0|^2 \exp(-2at^2)$ attains half of its maximum value:

$$\Delta t_G = 2\tau_G = \sqrt{\frac{2\ln 2}{a}}. \quad (11.2)$$

The Fourier transform of the pulse (11.1) is obtained by reference to the Appendix (Fig. A.5, convolution and scaling theorem):

$$
\begin{aligned}
\tilde{E}(v) &= \mathcal{F}\left[E_0 \exp(-i\omega_0 t)\exp(-at^2)\right](v) \\
&= E_0 \mathcal{F}[\exp(-i\omega_0 t)](v) * \mathcal{F}[\exp(-at^2)](v) \\
&= E_0\delta(v+v_0) * \sqrt{\frac{\pi}{a}}\,\mathcal{F}[\exp(-\pi t^2)]\left(v\sqrt{\frac{\pi}{a}}\right) \\
&= E_0\delta(v+v_0) * \sqrt{\frac{\pi}{a}}\exp(-\frac{\pi^2}{a}v^2) \\
&= E_0\sqrt{\frac{\pi}{a}}\exp\left[-\frac{\pi^2}{a}(v+v_0)^2\right]
\end{aligned}
\tag{11.3}
$$

or

$$
\tilde{E}(\omega) = E_0\sqrt{\frac{\pi}{a}}\exp\left[-\frac{(\omega+\omega_0)^2}{4a}\right].
\tag{11.4}
$$

As expected, the spectrum of the pulse has a Gaussian shape and is centered around $-\omega_0$, where $v_0 = \omega_0/(2\pi)$ is the carrier frequency. The bandwidth (FWHM) Δv_G of the pulse is obtained, analogously to (11.2), by requiring the power spectrum to drop to half its maximum value:

$$
\Delta v_G = \frac{1}{\pi}\sqrt{2a\ln 2}.
\tag{11.5}
$$

We recognize that while the pulse width (11.2) is inversely proportional to $\sqrt{a}$, $\Delta t_G \sim 1/\sqrt{a}$, the bandwidth is directly proportional to it, $\Delta v_G \sim \sqrt{a}$. Thus, the time-bandwidth product is a constant, independent of a,

$$
\Delta t_G \Delta v_G = \frac{2\ln 2}{\pi} \approx 0.44.
\tag{11.6}
$$

This result, derived for the Gaussian pulse, can be generalized and it can be shown that for arbitrary localized pulses the time-bandwidth product is bounded from below by a constant of the order of 1,

$$
\Delta v \Delta t \geq \text{const.}
\tag{11.7}
$$

The exact numerical value depends on the pulse shape and the definition of temporal width and bandwidth.

The relation (11.7) has an immediate consequence for the design of lasers producing very short pulses: the smaller the desired minimum pulse width is, the larger has to be the gain bandwidth of the active medium, i. e., the frequency or wavelength range in which the amplification factor exceeds unity (see Fig. 5.2). In fact, the gain bandwidth has to be larger than the pulse bandwidth required by (11.7) due to the inevitable losses in the laser system which have to be compensated for by the amplifying medium.

Therefore, lasers for ultrashort pulses are broad-bandwidth systems. For example, taking (11.6), a minimum pulse width of $\Delta t = 10\,\text{fs}$ calls for an amplification bandwidth of at least $\Delta \nu = 44\,\text{THz}$. At a center wavelength of $\lambda_0 = 780\,\text{nm}$ of a Ti:sapphire laser this bandwidth corresponds to a wavelength range of $\Delta \lambda \approx 90\,\text{nm}$. Then, dispersive effects such as the dispersion of group velocity cannot be ignored and have to be taken into consideration in the system design.

11.1.2 Chirped Pulses

The gain curve as outlined in Fig. 5.2 gives us information on the action of the medium on the power spectrum of the amplified light. However, the medium also modulates the phase spectrum of the light by linear and nonlinear dispersion. This phase modulation affects the pulse shape and duration and must be controlled to achieve the desired optimum pulse quality.

In (11.1) is implied a very special phase distribution in the pulse spectrum, namely, a constant phase (see (11.4)). To investigate the effect of phase modulation on pulse shape we analyze two special cases: linear and quadratic frequency dependence of the phase (chirped pulses).

In the first case, the pulse spectrum is modified in the following way:

$$\tilde{E}_{\text{lp}}(\omega) = E_0 \exp[-i\tau(\omega + \omega_0)] \sqrt{\frac{\pi}{a}} \exp\left[-\frac{(\omega + \omega_0)^2}{4a}\right], \tag{11.8}$$

where the parameter τ measures the rate of phase change around the carrier frequency. The inverse Fourier transform, whose elaboration is left as an exercise to the reader, yields the corresponding temporal shape of the pulse,

$$E(t) = \mathcal{F}^{-1}[\tilde{E}_{\text{lp}}(\omega)](t) = E_0 \exp(-i\omega_0 t) \exp[-a(t - \tau)^2]. \tag{11.9}$$

which turns out to be just the original Gaussian pulse, displaced in time by the amount τ (Fig. 11.3). This time is called the group delay. Therefore, linear phase modulation has the effect of temporally shifting the pulse envelope without change of its shape.

In the case of a quadratic phase factor, we start from the following spectrum,

$$\tilde{E}_{\text{qp}}(\omega) = E_0 \exp\left[-i\beta(\omega + \omega_0)^2\right] \sqrt{\frac{\pi}{\alpha}} \exp\left[-\frac{(\omega + \omega_0)^2}{4\alpha}\right], \tag{11.10}$$

where $\alpha > 0$ measures the spectral width of the pulse and β the degree of phase modulation. This equation can be written as

$$\begin{aligned}
\tilde{E}_{\text{qp}}(\omega) &= E_0 \sqrt{\frac{\pi}{\alpha}} \exp\left[-\left(\frac{1}{4\alpha} + i\beta\right)(\omega + \omega_0)^2\right] \\
&= E_0 \sqrt{\frac{\pi}{\alpha}} \exp\left[-\frac{(\omega + \omega_0)^2}{4\Gamma}\right]
\end{aligned} \tag{11.11}$$

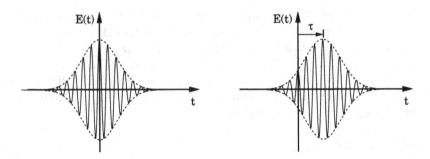

Fig. 11.3. A Gaussian pulse (*left*) is shifted in time by the amount τ when its phase is modulated linearly according to (11.8).

or

$$\tilde{E}_{qp}(\nu) = E_0 \sqrt{\frac{\pi}{\alpha}} \delta(\nu + \nu_0) * \exp\left(-\frac{\pi^2}{\Gamma}\nu^2\right), \qquad (11.12)$$

where we have introduced the complex parameter

$$\Gamma = \frac{\alpha}{1 + i4\alpha\beta}. \qquad (11.13)$$

Applying the inverse Fourier transform operation we obtain

$$\mathcal{F}^{-1}[\tilde{E}_{qp}(\nu)](t) = E_0 \sqrt{\frac{\pi}{\alpha}} \mathcal{F}^{-1}[\delta(\nu + \nu_0)](t) \mathcal{F}^{-1}[\exp(-\frac{\pi^2}{\Gamma}\nu^2)](t)$$

$$= E_0 \sqrt{\frac{\pi}{\alpha}} \exp(-i2\pi\nu_0 t) \sqrt{\frac{\Gamma}{\pi}} \exp(-\Gamma t^2)$$

$$E_{qp}(t) = E_0 \sqrt{\frac{\Gamma}{\alpha}} \exp(-i\omega_0 t) \exp(-\Gamma t^2), \qquad (11.14)$$

where we have used that the Fourier transform relations for a Gaussian function also hold for a complex 'width' parameter Γ.

To proceed let us split the parameter Γ into real and imaginary part, $\Gamma = a + ib$, which gives

$$a = \text{Re}\{\Gamma\} = \frac{\alpha}{1 + 16\alpha^2\beta^2}, \qquad (11.15)$$

$$b = \text{Im}\{\Gamma\} = -\frac{4\alpha^2\beta}{1 + 16\alpha^2\beta^2}. \qquad (11.16)$$

The complex amplitude of the pulse (11.14) is denoted by

$$E_0' = = E_0 \sqrt{\frac{\Gamma}{\alpha}}. \qquad (11.17)$$

Now we can write (11.14) in a form analogous to (11.9) or (11.1) with real

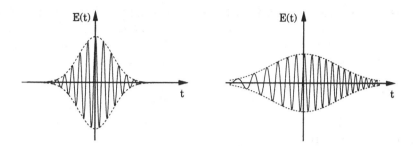

Fig. 11.4. Result of quadratic phase modulation in the spectrum of a Gaussian pulse. Pulse without phase modulation (*left*) and with quadratic phase modulation, a chirped pulse (*right*).

parameters a, b:

$$E(t) = E_0' \exp(-i\omega_0 t) \exp(-\Gamma t^2)$$
$$= E_0' \exp\left[-i(\omega_0 t + bt^2)\right] \exp(-at^2). \tag{11.18}$$

This describes the temporal pulse shape belonging to (11.10), having a Gaussian envelope and a quadratic phase modulation. Conversely, since

$$\frac{b}{a} = -4\alpha\beta, \qquad a^2 + b^2 = \frac{\alpha^2}{1 + 16\alpha^2\beta^2} = \alpha a, \tag{11.19}$$

if we are given the parameters $a > 0$ and b of the special pulse described by (11.18), its spectral parameters α and β in (11.10) are determined by

$$\alpha = \frac{a^2 + b^2}{a}, \qquad \beta = -\frac{b}{4(a^2 + b^2)}. \tag{11.20}$$

The pulse described by (11.18) has two interesting features that are demonstrated in Fig. 11.4. First, the envelope of the pulse is a Gaussian function whose width is larger than the width Δt_G of the non-modulated pulse (11.2) (where a is replaced by α):

$$\Delta t_G' = \sqrt{\frac{2\ln 2}{a}} = \Delta t_G \sqrt{\frac{\alpha}{a}} = \Delta t_G \sqrt{1 + 16\alpha^2\beta^2}, \tag{11.21}$$

where (11.15) has been used. It is seen that for $|\beta| \gg 1/\alpha$ the width increases proportionally to $|\beta|$ while the pulse amplitude (11.17) decreases with $|\beta|$ since the total energy is constant.

This property implies that, in order to achieve a small pulse width in a laser system, not only the amplitude gain has to be flat over a large frequency interval but also the phase modulation has to be kept small in this spectral region. Then the width of the generated pulses is as small as possible and approximately given by the time-bandwidth product, (11.6) or (11.7). Such pulses are said to be transform limited.

As a second result we note in Fig. 11.4 that the pulse shows a marked frequency variation of the carrier wave over the duration of the pulse. The instantaneous frequency $\omega(t)$ of a quasimonochromatic signal $E(t) = A\exp(i\Phi(t))$ is defined by the time derivative of its phase,

$$\omega(t) = \frac{d\Phi(t)}{dt}. \tag{11.22}$$

In the case of (11.18) we have $\Phi(t) = -(\omega_0 t + bt^2)$ and thus, with the positive value of frequency taken,

$$|\omega(t)| = \omega_0 + 2bt. \tag{11.23}$$

The instantaneous frequency of the pulse considered depends linearly on time; the quadratic phase modulation turns out to correspond to a linear frequency modulation. This pulse is said to be linearly chirped. For $b > 0$ the frequency $|\omega(t)|$ increases with time – the pulse is positively chirped. Correspondingly, when $b < 0$ and the frequency decreases with time, the chirp is negative. As we shall soon see, chirped pulses are important for pulse compression and in the formation of optical solitons.

Up to now we have not considered the propagation of ultrashort pulses but only their field at a fixed point in space. It is easy to transfer the obtained results to the case of one-dimensional linear wave propagation. Upon transmission in a linear medium the monofrequent wave components that constitute the propagating pulse do not interact with each other but travel with their phase velocities c given by the dispersion relation

$$c(\omega) = \frac{\omega}{k(\omega)} = \frac{c_0}{n(\omega)}. \tag{11.24}$$

After traveling a distance z each wave component has gained a phase increment $\Delta\Phi = kz$, which induces a phase modulation in the signal $E(z,t)$ when the wave number k is frequency dependent. When the pulse spectrum is confined to a narrow spectral region around ω_0 we can expand the wave number k around ω_0,

$$k(\omega) = k(\omega_0) + \frac{dk}{d\omega}\bigg|_{\omega_0}(\omega - \omega_0) + \frac{1}{2}\frac{d^2k}{d\omega^2}\bigg|_{\omega_0}(\omega - \omega_0)^2 + \ldots$$

$$= k_0 + k_0'(\omega - \omega_0) + \frac{1}{2}k_0''(\omega - \omega_0)^2 + \ldots. \tag{11.25}$$

Thus we see that the phase modulation of the pulse at z is given by

$$\Delta\Phi(\omega) = k_0 z + k_0' z(\omega - \omega_0) + \frac{1}{2}k_0'' z(\omega - \omega_0)^2 + \ldots. \tag{11.26}$$

As we have seen before, the term linear in ω gives a shift of the pulse by the group delay $\tau = k_0' z$. The corresponding velocity v_g is given by

$$v_g = \frac{z}{\tau} = \frac{z}{k_0' z} = \frac{1}{dk/d\omega|_{\omega_0}} = \frac{d\omega}{dk}\bigg|_{k_0} \tag{11.27}$$

and is called the group velocity. The quadratic phase modulation, which is responsible for the dispersive widening of the pulse, is described by the parameter k_0'', the *group velocity dispersion* of the medium, abbreviated as GVD. We have

$$k_0'' = \frac{\mathrm{d}k'}{\mathrm{d}\omega}\bigg|_{\omega_0} = \frac{\mathrm{d}}{\mathrm{d}\omega}\left(\frac{1}{v_g(\omega)}\right)\bigg|_{\omega_0} = -\frac{1}{v_g^2}\frac{\mathrm{d}v_g}{\mathrm{d}\omega}\bigg|_{\omega_0}. \tag{11.28}$$

The group velocity dispersion and even higher dispersive terms are of utmost importance for ultrashort pulse generation and propagation [11.6]. We will come across it again in the chapter on fiber optics in connection with optical solitons in fibers. Solitons additionally need intensity dependent propagation to come into existence.

11.2 Generation of Ultrashort Pulses

In the chapter on the laser we discussed the method of Q-switching which makes feasible the generation of giant pulses with durations of the order of ten nanoseconds. Q-switching could be described in the framework of the single-mode rate equations as being essentially a transient relaxation from an initial condition with largely overpopulated upper laser level.

As we have further seen, still shorter pulses than those obtainable by Q-switching, in the range of 1 ps or less, require a relatively wide gain bandwidth of the active medium. A large number of longitudinal modes will participate in the emission of such a broadband laser system, the number being determined by the mode spectrum of the resonator, the gain characteristics and the losses. To generate a periodic sequence of well-defined ultrashort, intense pulses a further requirement has to be met: the laser modes have to be properly phased to all superimpose constructively. This can be achieved by a technique called mode locking.

11.2.1 Principle of Mode Locking

To understand the basic idea behind mode locking, let us consider the superposition of a number $2M + 1$ of resonator modes with frequencies

$$\omega_n = \omega_0 + n\Delta\omega, \quad n = -M, -M+1, \ldots, 0, \ldots M \tag{11.29}$$

where the center frequency ω_0 corresponds to the resonator mode at the peak of the gain curve and $\Delta\omega = \pi c/L$ is the frequency separation of adjacent modes of the laser, see Fig. 11.5. The resonator modes are standing waves, satisfying the boundary conditions of the resonator, and can each be thought of as being composed of two counterpropagating waves of equal amplitude. Since we are interested in the light leaving the resonator at the exit mirror it is sufficient to consider the superposition of the partial waves traveling into the positive z-direction, say:

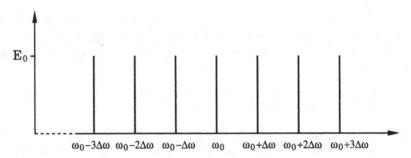

Fig. 11.5. Resonator modes used in the calculation of the output field.

$$E(z,t) = \sum_{n=-M}^{+M} E_n \exp[i(k_n z - \omega_n t)]. \tag{11.30}$$

Let us assume that the exit mirror is located at $z = 0$. Since we only want to know about the temporal shape of the light output we consider the field immediately behind the exit mirror, at $z = 0$:

$$E_{\mathrm{ml}}(t) = \sum_{n=-M}^{+M} E_n \exp(-i\omega_n t) = \exp(-i\omega_0 t) \sum_{n=-M}^{+M} E_n \exp(-in\,\Delta\omega t). \tag{11.31}$$

As a further simplifation we suppose that all partial waves have equal amplitudes $E_n = E_0$ =const, that is, they have equal intensities and phases (a more general case is discussed in one of the problems at the end of the chapter). Then it is easy to derive a simple expression for the superposition of modes:

$$\begin{aligned}
E_{\mathrm{ml}}(t) &= E_0 \exp(-i\omega_0 t) \sum_{n=-M}^{+M} \exp(-in\,\Delta\omega t) \\
&= E_0 \exp(-i\omega_0 t) \exp(-iM\,\Delta\omega t) \sum_{n=0}^{2M} \exp(in\,\Delta\omega t). \tag{11.32}
\end{aligned}$$

The sum constitutes a finite geometric series, and thus

$$\begin{aligned}
E_{\mathrm{ml}}(t) &= E_0\exp(-i\omega_0 t)\exp(-iM\,\Delta\omega t)\frac{1 - \exp[i(2M+1)\Delta\omega t]}{1 - \exp(i\,\Delta\omega t)} \\
&= E_0\exp(-i\omega_0 t)\exp(i\frac{\Delta\omega}{2}t)\frac{\exp[-i(M+\frac{1}{2})\Delta\omega t] - \exp[i(M+\frac{1}{2})\Delta\omega t]}{1 - \exp(i\,\Delta\omega t)} \\
&= E_0\exp(-i\omega_0 t)\frac{\exp[-i(M+\frac{1}{2})\Delta\omega t] - \exp[i(M+\frac{1}{2})\Delta\omega t]}{\exp(-i\frac{\Delta\omega}{2}t) - \exp(i\frac{\Delta\omega}{2}t)}. \tag{11.33}
\end{aligned}$$

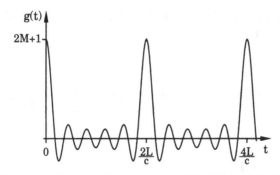

Fig. 11.6. Envelope function $g(t)$ of (11.35) for $M = 5$.

We finally obtain

$$E_{\mathrm{ml}}(t) = -E_0 \exp(-i\omega_0 t) \frac{\sin[(2M+1)\frac{\Delta\omega}{2}t]}{\sin(\frac{\Delta\omega}{2}t)} . \tag{11.34}$$

The term modulating the carrier wave is also encountered in the description of diffraction by a grating (and, accordingly, called a grating function):

$$g(t) = \frac{\sin((2M+1)\frac{\Delta\omega}{2}t)}{\sin(\frac{\Delta\omega}{2}t)} . \tag{11.35}$$

Its graph is depicted in Fig. 11.6 for the case $M = 5$. The function is periodic in t; the period T is determined by the roots of the denominator:

$$T = \frac{2\pi}{\Delta\omega} = \frac{1}{\Delta\nu} = \frac{2L}{c} . \tag{11.36}$$

Thus the period T is given by the roundtrip time of the resonator, and the superposition of the modes results in a pulse traveling back and forth in the resonator at the speed of light. The amplitude of the pulses is given by

$$E_0 g_0 = E_0 \lim_{t \to 0} g(t) = (2M+1)E_0 , \tag{11.37}$$

that is, the output field at maximum is by a factor of $N = 2M + 1$, the number of modes, larger than the single-mode field and, accordingly, the peak intensity is larger than the single-mode intensity by a factor of N^2. The width 2τ of the pulse (FWHM of the intensity) can be estimated in the following way. The half width τ is determined by the condition

$$\frac{\sin(N\frac{\Delta\omega}{2}\tau)}{\sin(\frac{\Delta\omega}{2}\tau)} = \frac{N}{\sqrt{2}}, \tag{11.38}$$

or, setting $\alpha = N\Delta\omega\tau/2$, by

$$\sin(\alpha) = \frac{N}{\sqrt{2}} \sin\left(\frac{\alpha}{N}\right) . \tag{11.39}$$

Assuming the number of modes N to be large the right-hand side can be expanded to yield

$$\sin(\alpha) = \frac{\alpha}{\sqrt{2}}. \tag{11.40}$$

The first non-zero root of this nonlinear equation can be numerically computed; it is given by $\alpha \approx 1.39$. Thus, the pulse width of the mode locked pulses is

$$\tau_{ml} = 2\tau = 2 \cdot \frac{2.78}{N\Delta\omega} = 0.88\frac{2L}{Nc}. \tag{11.41}$$

The important fact to note is that the pulse width decreases inversely proportional to the number of modes; in fact it is approximately equal to the fraction $1/N$ of the resonator round-trip time.

Our analysis shows that to generate ultrashort pulses by mode locking it is necessary that a large number of modes of the laser be excited. For example, to have 1 ps pulses at a repetition rate of 80 MHz ($T = 12.5$ ns) approximately 11000 longitudinal modes have to participate in the emission.

The calculation leading to (11.34) rests on the assumption that all the modes have equal amplitudes and phases. It is in particular the phase relationship between the modes that allows the constructive interference of the waves in a periodic fashion. If this relationship is not maintained, e.g. disturbed by spontaneous emission noise, the mode locking principle will not work. Figure 11.7 illustrates this point with a numerical example. On the left part of the figure, a superposition of 500 equidistant modes with zero phase shifts is plotted. The resulting intensity gives a well defined, very short pulse. On the right part, the signal formed by the same number of modes with random phase shifts is virtually indistinguishable from noise.

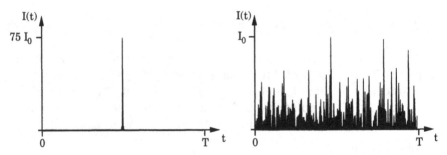

Fig. 11.7. Superposistion of 500 equidistant modes with zero phases (*left*) and with random phases (*right*). Note the two plots have different intensity scales. The reference intensity I_0 is arbitrarily chosen to represent the maximum value of the random signal.

11.2.2 Methods of Mode Locking

In chapter 10 on the laser we had already described two ways of getting short pulses: Q-switching and cavity dumping. Pulse durations in the ns range can be obtained. In these methods the modes of the laser cavity play no role. As we have seen in the preceding section, a superposition of laser modes with equal phases at some instant in time will give a train of short pulses the shortness depending on the number of modes coupled and locked together.

Methods of mode locking are either active or passive. Active here means that a device is inserted into the laser cavity that modulates the light beam via an external action. Passive means that the light itself generates the necessary modulation in a nonlinear medium placed in the laser cavity. This type is also called self mode locking.

Active Mode Locking. A common device for active mode locking is an acousto-optic modulator (AOM) placed inside the laser cavity near the output mirror (Fig. 11.8). This device was first introduced for active mode locking already in 1964 [11.7]. It consists of a transducer that converts an electric signal into a corresponding sound wave by the piezoelectric

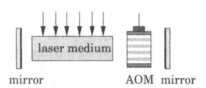

mirror   AOM mirror

Fig. 11.8. Active mode locking with an acousto-optic modulator (AOM).

effect. This sound wave is sent down a block of glass to intercept the light for diffracting it. Usually standing acoustic waves are used. The diffraction of the light leads to a loss modulation whose frequency ν_{mod} is two times the frequency of the sound wave ν_s because the acoustic field in a standing sound wave undergoes two zeroes with no acoustic field at all in one period of the sound field. Coupling of the laser modes is achieved when (c = light velocity, L = mirror spacing of the laser cavity)

$$\nu_{\mathrm{mod}} = 2\nu_s = \frac{c}{2L} = \Delta\nu, \tag{11.42}$$

the frequency mode spacing of the laser modes (see also chapter 5), because an amplitude (loss) modulation of anyone laser mode $\nu_m = mc/(2L)$ leads to the generation of side bands $\nu_{m\pm1} = \nu_m \pm \Delta\nu$ and thus to a coupling of the modes. Pulse widths attained this way reach down to the ps scale, a factor of thousand down from Q-switching. The output then is a pulse train as described in the previous section.

Passive Mode Locking. Like passive Q-switching (Fig. 10.12) passive mode locking relies on a saturable absorber inside the laser cavity; the life time of the upper transition level, however, must be short, much shorter than a cavity round trip time (Fig. 11.9). Passive mode locking was first

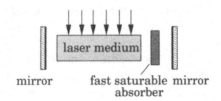

mirror   fast saturable mirror
                    absorber

Fig. 11.9. Passive mode locking with a fast saturable absorber (medium with intensity dependent loss and fast recovery time in the ps range).

observed in 1965 [11.8]. A saturable absorber introduces an intensity dependent loss that saturates at high intensity with virtually no loss. The operation can best be described in the time domain. The intensity peaks of the fluctuating laser emission are amplified more than the pedestals of the peaks sharpening them. The peaks run back and forth through the laser cavity experiencing amplification in the laser medium and narrowing in the absorber. This eventually leads to a pulse repeating every round trip time τ and being as short as the absorption recovery time of the absorber, if the bandwidth of the amplifying laser medium is large enough. As in the case of active mode locking the result is a pulse train with pulse width down to a few ps and pulse separation of $\tau = 2L/c$, but without the need of precise external driving at $\nu_{mod} = 1/\tau$.

Kerr Lens Mode Locking. Because of the absorption recovery time of fast saturable absorbers (mostly dyes) lying in the range of a few ps the pulse duration cannot be made shorter than these few ps. But there exists a faster effect, the optical Kerr effect, that can be exploited to simulate an almost instantaneous absorber recovery time. It is estimated to lie in the fs range, a factor of thousand shorter than for fast dyes. The optical Kerr effect relies on the nonlinearity of bound electron motion in strong electric fields. It leads to a nonlinear intensity dependent refractive index in an optical material

$$n = n_0 + n_2 I \tag{11.43}$$

and thereby to self focusing ($n_2 > 0$) of a wave with, for instance, a Gaussian transverse profile. This is because the high intensity along the optical axis propagates slower than the laser intensities away from the axis leading to a phase distortion of the wave front similar to that of a focusing lens. When this "Kerr lens" is applied properly in a laser cavity together

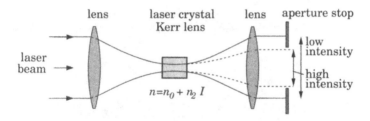

Fig. 11.10. Principle of Kerr lens mode locking. At low intensity the self-induced lens action of the nonlinear crystal is weak and a large fraction of the laser beam is blocked by the aperture stop. At high intensity the self-focusing is strong enough to let the beam pass through the stop without significant absorption.

with an aperture stop (Fig. 11.10) what is called **Kerr lens mode locking** is achieved, at present giving the shortest pulses directly out of a laser (in conjunction with dispersion compensation as dispersive effects of pulse broadening then take over and also must be minimized). This type of mode locking was first observed in 1991 [11.9] and today is routinely used in commercially available Ti:sapphire lasers to produce pulse trains of ~ 100 fs broad pulses and less.

11.2.3 Sonoluminescence

The laser, one of the ingenious scientific inventions of the twentieth century, certainly is the most versatile and powerful device to produce ultrashort light pulses in the laboratory. Nature, however, provides us with other, quite unexpected sources of pulsed, incoherent light, for example, with the phenomenon of sonoluminescence. It denotes the light emission from small oscillating bubbles in a liquid, driven by an intense ultrasonic field. Such bubbles are highly nonlinear oscillators, undergoing a strong collapse during each cycle of oscillation. During the final stages of collapse the bubble interior gets strongly compressed and highly excited. Under appropriate conditions a short, tiny flash of light is emitted. It is possible to conduct experiments with only a single light-emitting bubble, which is captured at the pressure anti-node of a standing acoustic field (Figs. 11.11 and 11.12).

The trapped bubble emits light pulses in synchrony with the ultrasonic field; the phenomenon is then called single-bubble sonoluminescence (SBSL). The SBSL light pulses have remarkable properties: they are very short (their duration being of the order of 35...350 ps) [11.10] with a spectrum indicating temperatures of the bubble interior in excess of 10^4 K. The experimental setup to achieve this kind of light pulse emission is fascinatingly simple; it consists of a water-filled flask, ultrasonic

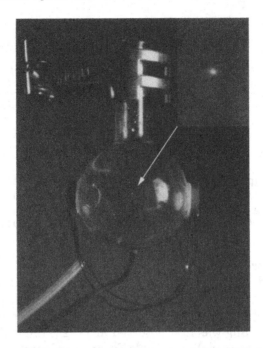

Fig. 11.11. Tiny, acoustically excited bubble emitting light (single-bubble sonoluminescence). An enlarged view of the flask center is given in the inset.

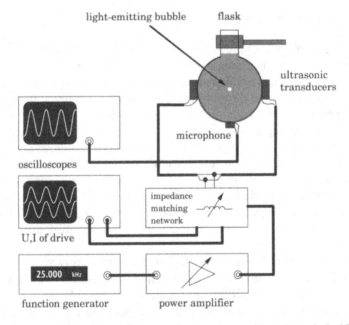

Fig. 11.12. Experimental arrangement for the observation of single-bubble sonoluminescence.

transducers glued to the flask, a frequency generator and amplifier to drive the transducers, a small transducer acting as a microphone to pick up the acoustic signal and a supply of purified and partially degassed water (Fig. 11.12). Nevertheless, at the time of this writing (2002) the physics and chemistry of bubble luminescence is still the subject of active and ongoing research [11.11]. And, maybe, experiments conducted with ultrashort laser pulses will aid us to finally arrive at a satisfactory explanation of this intriguing phenomenon.

11.2.4 Chirped Pulse Amplification

Short laser pulses with their tremendous peak intensities pose great demands on the quality of optical elements such as mirrors, lenses or crystals of the laser. But even carefully designed optics can only tolerate a certain incident light flux which, without further measures, limits the pulse energies deliverable by the system. Remedy is available by the method of chirped pulse amplification (Fig. 11.13) first realized in 1985 [11.12]. The basic idea behind it is very simple: an ultrashort laser pulse produced by the laser oscillator is phase modulated (chirped) in a configuration of dispersive elements, the stretcher. We know from the discussion in Sect. 11.1.2 (see also Fig. 11.4) that thereby the pulse width is increased and its peak amplitude diminished, allowing for amplification of the pulse without exceeding the damage threshold of the device. Upon amplification, the pulse gains energy while keeping its phase modulation, giving much higher intensity at the large pulse width. Finally, the amplified pulse is unchirped in a compressor. It reverses the phase modulation introduced by the stretcher and gives an output pulse with about the same width as the incoming pulse, but with much larger amplitude.

Pulse stretching and compression can be achieved by various dispersive elements such as gratings, prisms, multi-layer (chirped) mirrors or fibers. Figs. 11.14 and 11.15 show just two examples of dispersive devices, a Gires-Tournois interferometer and a pair of parallel, blazed gratings,

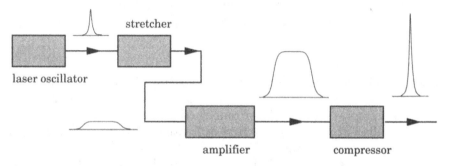

Fig. 11.13. Principle of chirped-pulse amplification.

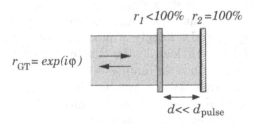

Fig. 11.14. A Gires-Tournois interferometer.

respectively. The Gires-Tournois interferometer is a Fabry-Perot type etalon whose back surface is 100 % reflecting and whose front surface is partially reflecting. The thickness of the etalon has to be small compared to the spatial extent of the incoming laser pulse; that is, the free spectral range of the etalon has to be large enough to accomodate the whole pulse spectrum. This requirement usually limits the application of the Gires-Tournois interferometer to picosecond laser systems. The complex reflectivity of the device as a function of wavelength varies periodically; it gives, at certain frequencies, the desired dispersive behavior (see Problem 11.5).

The dispersion of the grating pair outlined in Fig. 11.15 can be understood easily by calculating the rays' path lengths at different frequencies. Thus, consider two plane, collinear waves of frequencies ω_1 and $\omega_2 > \omega_1$ incident on the first grating under the angle α, measured w.r.t. the perpendicular to the gratings. Because the gratings are parallel the exit angle from the second mirror has to be α again, and the rays leave the arrangement parallel to the incoming light. The angular dispersion by the first grating is cancelled by the second grating but is transformed into a frequency dependence of the optical path length and path offset. Thus, the rays with frequencies ω_1 and ω_2 need slightly different times to traverse the arrangement.

The exit angle of the rays, β, from the first grating is their angle of incidence onto the second grating and depends on the frequency $\omega = 2\pi c/\lambda$ by

$$\frac{2\pi c}{\omega d} - \sin\alpha = \sin\beta, \tag{11.44}$$

where d is the grating constant, i. e. the spacing of the rulings, and the angles are counted positive along the arrows shown in the Figure. This formula expresses the condition of constructive interference of waves scattered at adjacent rulings in the first diffraction order. Consider the reference line r shown in Fig. 11.15 (*right*) that is perpendicular to the exit beams and passes through A. The optical path length $l(\omega)$ of a ray with

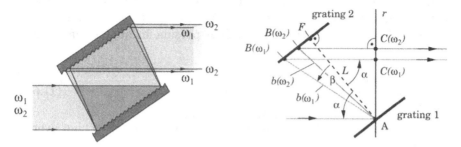

Fig. 11.15. *Left*: A pair of blazed gratings with negative group velocity dispersion for pulse chirping; *right*: ray paths for two rays of different frequencies for calculating the dispersion formula.

frequency ω, measured from the point of incidence A to the exit point $C(\omega)$ via $B(\omega)$, is given by

$$l(\omega) = b(\omega)(1 + \cos[\alpha - \beta(\omega)]),\tag{11.45}$$

where $b(\omega)$ is the distance AB that depends on the grating separation L as follows (compare Fig. 11.15):

$$b(\omega) = \frac{L}{\cos\beta(\omega)}.\tag{11.46}$$

To calculate the phase shift $\varphi(\omega)$ of a wave on the way from A to C as a function of frequency, phase matching of partial waves at the grating has to be taken into account. In the first diffraction order, waves scattered at adjacent rulings differ in phase by 2π. When by a small change of frequency from ω_1 to ω_2 the point $B(\omega_1)$ goes into $B(\omega_2)$ it moves a distance $\overline{B(\omega_1)B(\omega_2)}$ which corresonds to $\overline{B(\omega_1)B(\omega_2)}/d$ rulings, each giving a phase jump of -2π because $\overline{AB}(\omega_2) < \overline{AB}(\omega_1)$. With F, the base point of the perpendicular through A, chosen as the reference point, this phase term can therefore be written as

$$-2\pi\frac{L}{d}\tan\beta.\tag{11.47}$$

It adds to the phase shift $kl = \omega l/c$ due to the propagation over the path length l to give the phase with the correct dependence on frequency,

$$\varphi(\omega) = \frac{\omega}{c}l(\omega) - 2\pi\frac{L}{d}\tan\beta(\omega).\tag{11.48}$$

Collecting the results, (11.44), (11.45) and (11.48) and differentiating $\varphi(\omega)$ with respect to ω, we obtain after a few intermediate steps not reproduced here:

$$\frac{\partial\varphi}{\partial\omega} = \frac{l(\omega)}{c},\tag{11.49}$$

that is, the group delay $\tau = \partial\varphi/\partial\omega$ of the arrangement is just given by the time the light needs to traverse the optical path of length l. Differentiating (11.49) once more w.r.t. ω gives the dispersion formula of the grating pair

$$\frac{\partial^2\varphi}{\partial\omega^2} = -\frac{4\pi^2 cL}{\omega^3 d^2}\left\{1-\left(\frac{2\pi c}{\omega d}-\sin\alpha\right)^2\right\}^{-3/2}. \tag{11.50}$$

It is a nonlinear function of the frequency and of negative sign everywhere.

It is clear from the previous discussion that a single grating pair will cause a transverse spatial chirp, and thus an elongated profile of the exit beam. Different frequencies leave the device at different offsets to the beam axis. This distorting effect can be avoided by using a second grating pair that is mirror-symmetric with respect to the normal of the exit beam. Compared to the first pair, the light path in the second pair is reversed. It thus cancels the spatial chirp and restores the original beam profile while increasing the frequency chirp.

With the grating arrangements just discussed it is possible to imprint on an pulse a negative chirp only. By adding a telescopic lens system as shown in Fig. 11.16 the sign of the group velocity dispersion is reversed. The two arragements have completely symmetrical dispersion properties: the pulse modified by one unit is restored exactly to its original shape by the complementary unit (up to amplitude scaling, because of amplification or losses). In practical chirped pulse amplifiers usually the telescopic arrangement is used as the pulse stretcher on the low energy side, because of the presence of lenses, and the lensless arrangement of Fig. 11.15 is employed as the compressor for the amplified pulses.

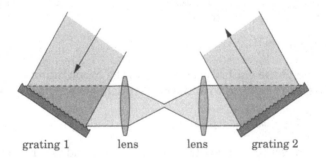

grating 1   lens   lens   grating 2

Fig. 11.16. A grating pair with telescopic inverter having positive group velocity dispersion. The lens system transforms plane waves into plane waves with the negative spatial frequency.

Instead of gratings, prisms can be used for stretcher or compressor units. To assess their dispersion properties, also the material (chromatic) dispersion has to be taken into account. Prisms are less dispersive than gratings but also have smaller losses. They are used, for example, in femtosecond laser oscillators to compensate for the group velocity dispersion introduced by other optical elements (mirrors, laser crystal).

11.3 Measurement of Ultrashort Pulses

Even the fastest photodetectors available today are not able to faithfully record the shape, less the phase, of ultrashort laser pulses. The fastest semiconductor photodiodes have rise times of the order of $\gtrsim 10\,\text{ps}$ that require a bandwidth of the electronic equipment of $\gtrsim 30\,\text{GHz}$, touching the limits of current electronic technology. Direct electro-optic sampling, e. g. by the electron tube of a streak camera, achieves a time resolution of $\gtrsim 1\,\text{ps}$, still falling short of the femtosecond scale needed to characterize ultrashort pulses.

Optics itself must be used to probe these pulses. The easiest way is to split the pulse under consideration and to let its parts collide head on in a nonlinear medium yielding a signal that is sensitive to whether only one pulse at a time or both pulses are present [11.13, 11.14]. Figure 11.17 sketches the arrangement that has become known as the two-photon fluorescence method [11.15]. The physical basis is two-photon absorption, a process where two photons are absorbed simultaneously to arrive at a high energy level of the medium. As this process depends quadratically on the intensity, the overlap region absorbs quadratically more photons than the other regions in the path of the beams. The two absorbed photons are reradiated from the molecule as one photon of double frequency (the

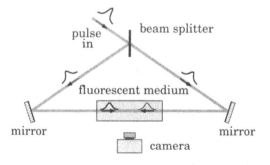

Fig. 11.17. Two-photon fluorescence method for measuring the autocorrelation of a short pulse.

fluorescent light). Thus, by photographing the medium with the colliding pulses, the overlap region stands out and its intensity trace essentially gives the intensity autocorrelation of the pulse. The pulse shape itself cannot be retrieved, unless assumptions on the pulse shape are made, for instance, a Gaussian shape. Moreover, despite its simplicity, the method encounters quantitative difficulties in that the contrast between overlap and nonoverlap parts is sensitive to many experimental parameters.

Therefore, other methods of ultrafast pulse characterization have been conceived. All nonlinear methods have in common that the incoming light pulse is used to probe itself by a nonlinear interaction, the time calibration being given by an adjustable optical delay. As nonlinear mechanism of interaction, besides two-photon absorption, two-photon fluorescence, second-harmonic or third-harmonic generation, self-diffraction or optical gating (see the next Section) have been successfully employed.

A method routinely used in the laboratory, interferometric autocorrelation, is based on a Michelson interferometer arrangement (compare Fig. 4.1). On the detector side the two pulses split by the beam splitter are sent through a nonlinear crystal where, e. g., second harmonic (SH) generation takes place, that is the more effective, the stronger the pulses overlap. The intensity of the second harmonic is recorded as a function of pulse delay (mirror displacement) to give the autocorrelation trace.

As will be seen in Chapter 12, the amplitude $E^{(2\omega)}$ of the SH wave for collinear (phase-matched) beams is given by

$$E^{(2\omega)} \propto (E^{(\omega)})^2, \tag{11.51}$$

$E^{(\omega)}$ being the field of the incoming wave. Ignoring the factor of proportionality, we can write for the average intensity of the second harmonic, $I_{12}(\tau)$, calculated over a time interval $[-T, T]$ (e. g. the pulse repetition period) large enough to contain the support of both pulses, $E(t)$ and $E(t-\tau)$,

$$I_s(\tau) = \frac{1}{2T} \int_{-T}^{T} I^{(2\omega)}(t, \tau) \, dt, \tag{11.52}$$

where

$$I^{(2\omega)}(t, \tau) = |E^{(2\omega)}(t)|^2 = |(E(t) + E(t-\tau))^2|^2. \tag{11.53}$$

Thus, we have for the average intensity

$$I_s(\tau) = \frac{1}{2T} \int_{-T}^{T} |(E(t) + E(t-\tau))^2|^2 \, dt. \tag{11.54}$$

For a single, transform limited pulse

$$E(t) = E_0(t) \, \text{Re}\{\exp(-i\omega_0 t)\}, \tag{11.55}$$

where the slowly varying envelope function $E_0(t)$ can assumed to be real,

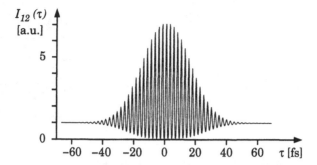

Fig. 11.18. Numerically computed interferometric autocorrelation trace of a transform limited Gaussian pulse (11.56) of width $\tau_G = 23\,\text{fs}$ and center wavelength $\lambda = 732\,\text{nm}$.

one can approximate (11.54) to arrive at (see Problem 11.6):

$$I_s(\tau) = \frac{3}{4}\left\{\frac{1}{2T}\int_{-T}^{T}E_0^4(t)\,dt\right\} + \frac{3}{2}\left\{\frac{1}{2T}\int_{-T}^{T}E_0^2(t)E_0^2(t-\tau)\,dt\right\}$$

$$+\frac{3}{2}\cos(\omega\tau)\left\{\frac{1}{2T}\int_{-T}^{T}[E_0^3(t)E_0(t-\tau)+E_0(t)E_0^3(t-\tau)]\,dt\right\}$$

$$+\frac{3}{4}\cos(2\omega\tau)\left\{\frac{1}{2T}\int_{-T}^{T}E_0^2(t)E_0^2(t-\tau)\,dt\right\}. \tag{11.56}$$

The first term on the right-hand side of this equation gives a constant background, the second term represents the autocorrelation function of the envelope's intensity. The last two terms give a modulation of the autocorrelation trace that, for pulses with a more complicated phase structure, also contains information on this structure (e. g., on the chirp of the pulse). Figure 11.18 shows a sketch of an interferometric autocorrelation trace, numerically computed for a transform-limited Gaussian pulse.

From (11.56) it is easily seen that the ratio of peak value (at $\tau = 0$) to background value (at $\tau \to \infty$) of $I_{12}(\tau)$ is 8:1. For pure intensity autocorrelation, i. e., when the interference fringes are averaged over and the oscillating terms vanish, this ratio is reduced to the value 3:1, implying a smaller signal-to-noise ratio. The interferometric method, on the other hand, requires a high stability of the laser and the optomechanical setup, at least on the timescale of the data acquisition that may take many laser shots.

As even non-overlapping, collinear pulses produce some background light at the second harmonic that is detected, the autocorrelation trace levels off at a constant value. This can be avoided, e. g., by letting the two beams enter the nonlinear crystal obliquely at equal, but opposite

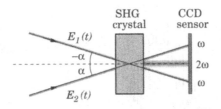
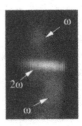

Fig. 11.19. *Left:* Noncollinear beam geometry of a background-free autocorrelator based on second-harmonic generation; *right:* photograph of the light reaching the detector, the middle spot belonging to the second harmonic beam.

angles with respect to the axis of observation (Fig. 11.19, *left*). Then, only two photons from different beams will generate a new photon at 2ω having the right propagation direction to reach the detector. With this background-free method a better contrast of the autocorrelation trace can be achieved. For illustration, the right part of Fig. 11.19 shows a photograph of the light reaching the detector of an autocorrelator of this type. The circular spots at the top and the bottom of the image are caused by the incoming beams (red light), the elongated spot in the middle belongs to the frequency-doubled light, shining in deep blue. The intensity profile of the middle spot is proportional to the intensity autocorrelation and is recorded by a CCD sensor array.

Sophisticated experimental techniques for ultrashort pulse characterization have been invented in recent years, culminating in the frequency-resolved optical gating (FROG) method [11.16]. In principle this method is able to yield both the intensity and the phase of a pulse without significant ambiguity. This is achieved by measuring spectra of the autocorrelation signal to obtain time-frequency diagrams (spectrograms) similar to visible speech in acoustics. A possible experimental arrangement is depicted in Fig. 11.20. It is quite similar to the background-free autocorrelator discussed before, but augmented by a spectrometer to take the optical spectra of the signal coming from the nonlinear element. The signal is given by Eq. (11.51) in the case of second-harmonic generation, by

$$E_{\mathrm{sd}}(t,\tau) \propto E^2(t)E^*(t-\tau),$$
(11.57)

for a self-diffraction gate, or by

$$E_{\mathrm{pg}}(t,\tau) \propto |E(t-\tau)|^2 E(t),$$
(11.58)

for a polarization gate [11.16, 11.17]. The result of the measurement is a two-dimensional spectrogram as shown in Fig. 11.21 (*left*) where the measured power spectrum is plotted as a function of frequency (or wavelength) and of the time delay τ. The key idea of the FROG method is

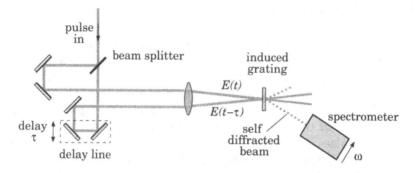

Fig. 11.20. Experimental arrangement for frequency-resolved optical gating using self-diffraction by a nonlinear target to determine pulse shape and phase of the incoming pulse.

that with the additional information obtained in the frequency domain the inverse problem of finding the phase and amplitude of the generating pulse from this diagram is solvable without essential, or no ambiguity (depending on the gate type), contrary to the case of the one-dimensional autocorrelation trace (or the one-dimensional power spectrum). The pulse reconstruction is done on a computer with a 'phase-retrieval' algorithm that is essentially an iterative search algorithm for the unknown function $E(t)$. Fig. 11.21 (*right*) sketches the pulse (intensity and phase) reconstructed from the spectrogram on the left. It is a positively chirped pulse – a feature already suggested by the oblique form of the spectrogram.

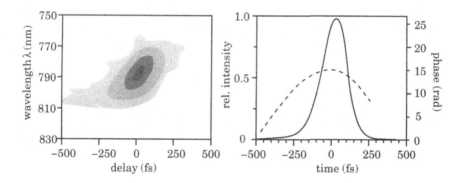

Fig. 11.21. *Left:* Sketch of a FROG spectrogram (obtained with a polarization gate); *right:* corresponding pulse shape (—) and phase (- - -) of a positively chirped pulse (after Trebino *et al.*, ref. [11.17]).

11.4 Optical Gating

To have a look at fast events their motion must be stopped by a short exposure time. Mechanical systems do not reach far, electronic systems come down to about nanoseconds, but for picosecond and femtosecond shutter times only light is available. Thus the idea arises of how light can be used for building a shutter or optical gate to let other light pass, in particular for a short time only. The first realisation succeeded in 1969 with picosecond pulses [11.18] and light in flight could be visualised [11.19].

Central to optical gating is the Kerr effect, first reported by *John Kerr* (1824–1907) in 1875 for static electric fields. A transparent medium was made birefringent by application of a strong electric field and thus a light beam traversing the medium could be influenced. The physical reason for the alteration of a medium by an electric field is twofold, comprising the alignment of polarized molecules in a liquid and electric deformation of the molecules or atoms of the medium. As a light wave comes along with an (alternating) electric field the same effect should be present in principle. It is then called the optical Kerr effect. The Kerr effect is a non-linear optical effect, quadratic in the electric field, so that also the oscillating optical field leads to a resulting net birefringence. In a birefringent medium there are two different polarization directions of a propagating light wave, where linearly polarized beams of the respective polarization have different propagation speeds, a fast and a slow speed. Placing a polarizer and an analyzer (= polarizer turned at 90° with respect to the first polarizer) in front and behind a medium will stop all light, because the polarized light cannot pass the analyzer. In a Kerr cell shutter the polarizer–analyzer pair is rotated by 45° with respect to one of the (fast or slow) polarization direction (see Fig. 11.22). Then the polarized beam is rotated in the birefringent medium upon propagation and can, at least partially, pass the analyzer. That way, by switching the birefringence on

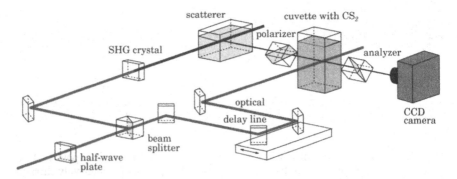

Fig. 11.22. Experimental arrangement for taking light in flight with a transversally gated optical Kerr shutter.

Fig. 11.23. Light in flight: scattered ps laser pulse observed through a longitudinally activated Kerr gate utilizing a 5 mm thick cuvette filled with CS_2. The time delay of the gating pulse with respect to the probe pulse was increased from one exposure to the next, giving a movie-like sequence running from left to right and from top to bottom.

and off, a light beam can be made pass or not. When the switching is done optically by light pulses, we have an optical gate.

Figure 11.22 shows an arrangement for gating with a light pulse transverse to the gating direction for photographing light in flight. For this demonstration experiment a 100 fs pulse (extending 30 µm in the propagation direction) was split at a beam splitter for both switching the birefringence in a cuvette filled with CS_2 and, after being frequency doubled, to be photographed when propagating in a slightly scattering medium. The optical delay line in the path of the switching pulse shifts the shutter opening time with respect to the scattered pulse so that the (latter) pulse can be caught at different positions in the scattering medium. Figure 11.23 shows the result of a light in flight experiment. In this case a slightly different arrangement was used where the switching pulse was applied not transversally but longitudinally in the CS_2-cell. (This arrangement is more difficult to sketch graphically, therefore the transverse case is given in Fig. 11.22). The light pulse indeed can be photographed at different positions during its flight.

Optical gating has found a number of applications. A prominent one is looking through diffusing media by catching the first-arrival light or "ballistic photons" that are not scattered in the medium [11.20]. As the ballistic photons have the shortest path compared to the scattered or diffused ones, they arrive first and can form an image without the diffuse background normally obscuring an image viewed through a diffusing medium [11.21]. As the time differences are small a fast gate must be employed for catching just the few first photons of a fast pulse to get rid of the later arriving photons scattered around in the medium.

11.5 Optical Coherence Tomography

An alternative but physically similar way to look into scattering media is by using low-coherence light and interferometry instead of ultra short laser pulses and gating. This method has got the name of optical coherence tomography [11.22] because of its ability to give a full 3-D view of an object. The basic idea is sketched in Fig. 11.24. It is essentially a Michelson interferometer with a low coherence light source for illuminating the object. The detector gets a modulated signal (fringes) only for the coherent part of the light, i. e. for light coming from a small depth interval of the object, its location corresponding to the distance of the reference mirror from the beam splitter. Thus the gating is done by interference. The analogy to short pulse gating is complete when we recollect that a short pulse has a broad spectrum and therefore is of low coherence. Thus also short pulses can be used in this interferometric arrangement. Interference fringes only occur when the pulses from the reference mirror and from the respective object depth overlap. To obtain a 3-D view, the object is translated in the x- and y-directions for a slice at fixed-depth z and the reference mirror is translated for scanning the depth z of the object. The depth resolution is the better the lower the coherence of the source or the shorter the pulse.

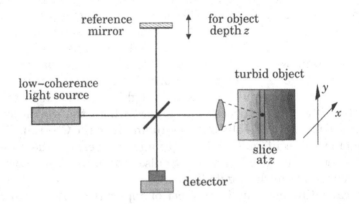

Fig. 11.24. Basic arrangement for imaging through a turbid medium in a Michelson design (compare Fig. 4.1) for optical coherence tomography.

Improvements and extensions of optical coherence tomography include full field optical coherence microscopy [11.23] to dispense with the x-y scanning of the object, introduction of differential phase contrast to measure phase changes inside an object not accessible with the normal reflection type optical coherence tomography [11.24] and other interferometric arrangements instead of a Michelson design to improve the low photon yield [11.25]. Work is in progress in this area to arrive at instruments for routine in vivo in-depth imaging of biological tissue at high resolution that otherwise is difficult to obtain, for instance of the cardiovascular system and the gastrointestinal tract [11.26].

Problems

11.1 Calculate the peak power of a laser pulse having the following properties: the pulse shape be Gaussian of width $\Delta t_G = 30\,\mathrm{fs}$, the pulse energy be $W_P = 12\,\mathrm{nJ}$. Determine the peak intensity of the corresponding unfocused laser beam (diameter 2 mm) and of the beam focused down to a spot of diameter $2r_s = 10\,\mu\mathrm{m}$, assuming a Gaussian transverse beam profile in both cases: $E(r) \propto \exp(-r^2/(2r_s^2))$, r being the radial coordinate perpendicular to the beam axis.

11.2 Determine the time-bandwidth product of the following pulse shapes, taking the FWHM definition of temporal and spectral width:
(i) a rectangular pulse of duration τ,

$$E_{\mathrm{rp}}(t) = \begin{cases} E_0 \exp(-i\omega_0 t) & \text{for} \quad |t| < \tau/2, \\ 0 & \text{elsewhere}; \end{cases} \tag{11.59}$$

(ii) an exponentially decaying pulse, beginning at $t = 0$:

$$E_{\mathrm{ep}}(t) = \begin{cases} E_0 \exp(-i\omega_0 t)\exp(-\gamma t) & \text{for } t \geq 0, \\ 0 & \text{for } t < 0. \end{cases} \tag{11.60}$$

11.3 Verify Eq. (11.9) by applying the inverse Fourier transform to the linearly phase-modulated spectrum (11.8) of a Gaussian wave packet.

11.4 What value of the group velocity dispersion $|k_0''|$ (in $\mathrm{ps}^2/\mathrm{km}$) must be realized to stretch the unchirped pulse of problem **11.1** to a length of 3 ps, after it has travelled a distance of 10 m?

11.5 Derive the complex amplitude reflectance r_{GT} of a lossless Gires-Tornois interferometer of thickness d (Fig. 11.14), analogously to problem 5.2. The amplitude reflectance of the second mirror be $r_2 = -1$, the angle of incidence be $= 0$. Verify that $|r_{\mathrm{GT}}| = 1$. Give an expression for the phase φ_{GT} of r_{GT} as a function of frequency, and show that the dispersion of the interferometer is described by

$$\frac{\mathrm{d}^2\varphi}{\mathrm{d}\omega^2} = -\frac{2t_0^2(1-r_1^2)r_1\sin\omega t_0}{(1+r^2-2r\cos\omega t_0)^2}, \tag{11.61}$$

where $t_0 = 2nd/c$ is the transit (or round-trip) time of the interferometer.

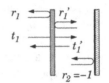

Hint: The Stokes relations should be used to relate amplitude reflectances r_1, r_1' and transmittances t_1, t_1' of the entrance mirror: $r_1 = -r_1'$, $t_1 t_1' = 1 - r_1^2$ (unprimed quantities for light entering the device, see the Figure).

11.6 (a) Derive (11.56) as an approximation for (11.54) by exploiting the assumption that $E_0(t)$ is slowly varying compared to the period of the light oscillation. (b) Using (11.56), calculate the interferogram $I_s(\tau)$ of an unchirped Gaussian pulse $E(t) = \exp(-i\omega_0 t) \exp(-at^2)$.

12. Nonlinear Optics

Light waves in a vacuum as given by the Maxwell equations propagate as linear waves. Accordingly, light waves can be superimposed and cross each other without mutual influence. To be sure, quantum electrodynamics predicts an interaction between photons. It can, however, be completely neglected at all practically attainable light intensities.

The situation looks different when light waves propagate in matter. There, an incident wave generates polarization charges and currents that themselves are sources of waves altering the overall wave field. At high field strengths, matter goes nonlinear: polarization (and magnetization) is no longer proportional to the inner fields. Thus, nonlinearity occurring in the interaction of light with matter is the cause of a series of interesting and technically important phenomena that include the generation of harmonics, mixing, optical bistability, and switching. These phenomena, amongst others, form the subject of nonlinear optics.

Classically, nonlinear optical effects are described by nonlinear susceptibilities having their origin in the nonlinear binding forces. Viewed from the photon aspect, which is more appropriate than the classical description due to the intrinsic quantum nature of the interaction between light and matter, many nonlinear effects can be explained simply and quite naturally. In nonlinear optics, at least two photons are involved in an interaction. If both photons give rise to a direct transformation into a different photon, that is, three photons are involved in the interaction process, the process is called three-wave interaction. It has found its most important technical application in second-harmonic generation. In this chapter the basic nonlinear optical phenomena are discussed.

12.1 Two-Wave Interaction

From low-intensity optics it is known that a photon passing through matter may be absorbed. It may also kick off an electron from its bound state, leaving the atom ionized. The condition for these phenomena to happen is that the photon carries the appropriate amount of energy. At high intensities, this condition can be relaxed in the sense that not one, but two photons together must carry the appropriate amount of energy. Then the above phenomena are called two-photon absorption and ionization.

12.1.1 Two-Photon Absorption

When a harmonic wave of circular frequency ω_1 passes through a suitable optical medium that at normal intensities shows no absorption at this frequency, it nevertheless may show strong absorption at high intensities. In the photon picture this can be understood in the following way. The medium does not possess an energy level at $\Delta E = \hbar\omega_1$ above the ground level, but does have one at $\Delta E = 2\hbar\omega_1$. The medium then may take two photons from the light field to attain this excited state. At higher intensity, an appreciable number of these processes takes place since its rate is proportional to the square of the intensity, or photon flux. The medium then shows absorption at the frequency ω_1 despite no transition being present corresponding to this frequency (Fig. 12.1). Two-photon absorption was predicted in 1931 [12.1], but was not experimentally observed before 1961 [12.2].

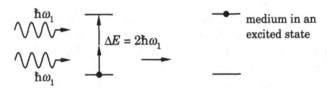

Fig. 12.1. Illustration of two-photon absorption with an incident wave of circular frequency ω_1.

When two harmonic waves of different frequency ω_1 and ω_2 propagate through a medium, each may contribute a photon for absorption. This process, experimentally realized in 1963 [12.3], is a generalization of the usual two-photon absorption with $\Delta E = 2\hbar\omega_1$. A photon of frequency ω_1 and a photon of frequency ω_2 are absorbed simultaneously and leave the medium in an excited state with an acquired energy of $\Delta E = \hbar(\omega_1 + \omega_2)$ (Fig. 12.2).

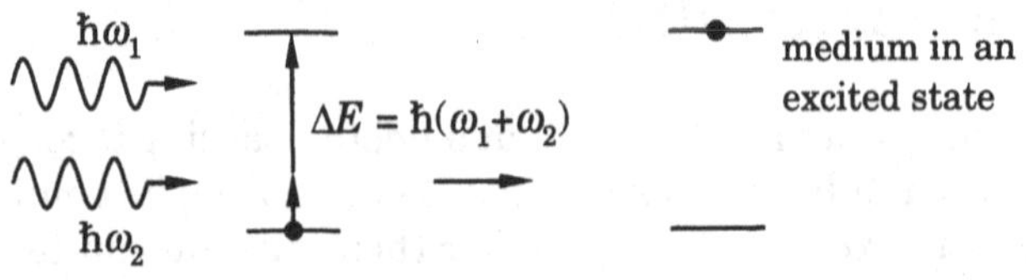

Fig. 12.2. Illustration of two-photon absorption with two incident waves of circular frequency ω_1 and ω_2.

Two-photon absorption has found application in two-photon absorption spectroscopy [12.4]. It is mainly used to explore the energy level structure of matter. Some energy levels cannot be reached by single-photon absorption due to 'forbidden' transitions, while two-photon absorption is possible because different transition rules apply. In a typical experiment of two-photon absorption spectroscopy, the change of transmittance of a sample for a light wave of tunable frequency ω_1 in the presence of a second strong light wave of fixed frequency ω_2 is measured (Fig. 12.3). The change in transmittance leads to a change of the intensity

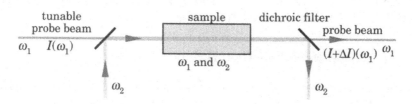

Fig. 12.3. Principle of two-photon absorption spectroscopy.

$I(\omega_1)$ to $I(\omega_1) + \Delta I(\omega_1)$ in the presence of the light wave of frequency ω_2. With special filters (dichroic filters) the two waves of frequency ω_1 and ω_2 can be separated and thus the difference $\Delta I(\omega_1)$ can be determined.

12.1.2 Two-Photon Ionization

When the sum of the energy of two photons in a radiation field is larger than the binding energy of the electron to the atom or molecule, it may be ionized in a two-photon process as illustrated in Fig.12.4. That this process may indeed occur has been shown in 1964 by two-photon ionization of hydrogen [12.5].

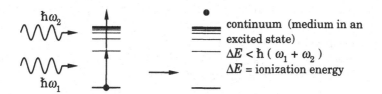

Fig. 12.4. Illustration of two-photon ionization.

12.2 Three-Wave Interaction

When two light waves are sent through a crystal with nonlinear properties, combination frequencies may arise. Their efficient generation relies
on special properties of the medium. Usually only three waves take part
in the interaction process, since phase matching (see Sect. 12.7.1) cannot be ensured for the remaining combination frequencies. Depending
on the conversion process two basic types of three-wave interaction may
be distinguished: sum-frequency generation and the optical parametric
amplifier. Special cases are second-harmonic generation and optical rectification. Three-wave interaction processes are introduced in this section
and discussed theoretically in Sect. 12.7.

12.2.1 Second-Harmonic Generation

In the process of second-harmonic generation, often abbreviated as SHG,
two photons of frequency ω_1 are converted into a photon of frequency ω_2
in a suitable nonlinear medium, whereby $\omega_2 = 2\omega_1$ (Fig. 12.5). Second-

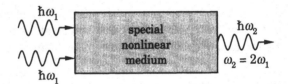

Fig. 12.5. Illustration of second-harmonic generation.

harmonic generation has been realized in 1961 [12.6]. In acoustics and
electrical engineering, conversion processes of this kind have long been
known. Efficient conversion is only possible when the waves of frequency
ω_1 and of twice the frequency, ω_2, both have the same propagation velocity.
In optics, we usually have large chromatic dispersion in a medium, and
thus second-harmonic generation needs special geometries to be achieved.
Today, second-harmonic generation is routinely used to convert, for instance, the infrared emission of a Nd:YAG laser at $\lambda = 1064$ nm into the
visible, that is, to $\lambda = 532$ nm. Lasers of this kind, being pumped with a
semiconductor laser, are commercially available.

12.2.2 Sum-Frequency Generation

In the process of sum-frequency generation a photon of frequency ω_1 and
a photon of frequency ω_2 are converted into a photon of frequency $\omega_3 =
\omega_1 + \omega_2$ (Fig. 12.6). This process is a generalization of second-harmonic
generation. Sum-frequency generation was first realized experimentally

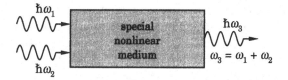

Fig. 12.6. Illustration of sum-frequency generation.

in 1962 [12.7]. It is called up-conversion when $\omega_2 \gg \omega_1$ and the intensity of the wave of frequency ω_2 is large compared to the intensity of the wave of frequency ω_1. This case is used in the detection of infrared photons. With the help of sum-frequency generation they are shifted into a higher frequency region, for instance, the visible. There, the detection is easier since suitable photo detectors are available [12.8].

12.2.3 Difference-Frequency Generation

In the process of difference-frequency generation a photon of frequency ω_1 and a photon of frequency ω_2 ($\omega_1 > \omega_2$) are converted into a photon of frequency $\omega_3 = \omega_1 - \omega_2$ (Fig. 12.7). A special form of difference-

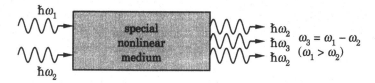

Fig. 12.7. Illustration of difference-frequency generation.

frequency generation is optical rectification (Fig 12.8). In this case, the incident photons are of equal frequency, yielding a difference of zero: a direct field is generated! Difference-frequency generation [12.9] and

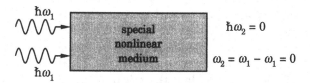

Fig. 12.8. Illustration of optical rectification.

optical rectification [12.10] were first realized experimentally in 1962. Difference-frequency generation may be used to generate infrared radiation [12.8, 12.9]. Optical rectification has been used, for example, to switch an optical shutter in ultrafast photography to capture light in flight [12.11]. Difference-frequency generation in effect is a parametric process in which the photon of energy $\hbar\omega_1$ is splitted into two photons of energies $\hbar\omega_2$ and $\hbar\omega_3$. Therefore, in Fig. 12.7, additionally to one photon of energy $\hbar\omega_3$ two photons of energy $\hbar\omega_2$ are present.

12.2.4 Optical Parametric Amplifier

In this case, a suitable medium is irradiated with a (strong) pump wave of frequency ω_p that decays into two light waves, one of frequency ω_s (signal wave) and the other of frequency ω_i (idler wave) (Fig. 12.9). An

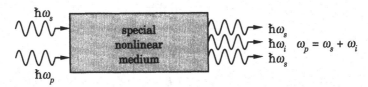

Fig. 12.9. Illustration of the optical parametric process in three-wave interaction.

optical parametric amplifier was first realized in 1965 [12.12]. It allows extremely low-noise light amplification when a signal wave ω_s is fed into the amplifier. The theory of parametric amplification [12.13] is discussed in Sect. 12.7.4 after the basics of three-wave interaction are introduced.

12.3 Four-Wave Interaction

The notion of four-wave interaction is mainly used in connection with parametric four-photon processes, in particular with phase conjugation in real time. Sum-frequency generation with three incident waves of frequencies ω_1, ω_2, and ω_3 generating a wave of frequency $\omega_4 = \omega_1 + \omega_2 + \omega_3$ is usually subsumed under the heading of multi-photon processes.

How phase conjugate waves (not in real time) can be generated with holograms has been discussed in Sect. 7.1.4. In the parametric four-photon process two photons of frequencies ω_1 and ω_2 are converted into two new photons of frequencies ω_3 and ω_4 (Fig. 12.10). As in the cases before, the nonlinear medium plays the role of a catalyst and emerges from the reaction unchanged. This process can be used to generate a wave which is phase conjugate to a given wave, that is, a wave with inverted k-vector. Because of its ability to correct aberrations, optical phase conjugation is intensely investigated [12.14]. The materials presently available (photorefractive crystals such as $BaTiO_3$, $LiNbO_3$, $Bi_{12}SiO_{20}$ (BSO)),

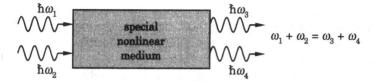

Fig. 12.10. Illustration of the parametric four-photon process.

however, have long response times, so that the potential of phase conjugation cannot yet be fully used.

12.4 Multi-photon Interaction

In multi-photon processes usually more than three photons are engaged in the interaction process. When many waves of different frequencies participate, and thus correspondingly many different types of photons, a large number of combinations are possible. The most important ones are discussed in the following.

12.4.1 Frequency Multiplication

In frequency multiplication, n photons of identical or different frequencies simultaneously interact with an atom that has levels of suitable energy difference. A new photon of frequency $\omega_m = n\omega_1$ or, most generally, $\omega_m = \omega_1 + \omega_2 + \ldots + \omega_n$ is created (Fig. 12.11). Special cases are third-harmonic

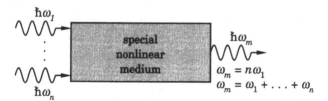

Fig. 12.11. Illustration of frequency multiplication.

generation (THG) with $\omega_m = 3\omega_1$ [12.15] and fourth-harmonic generation (FHG) with $\omega_m = 4\omega_1$. These processes are used to produce coherent light at short wavelengths. An effective conversion demands strong light fields. They may soon lead to damage of the nonlinear medium, a fact that hampers high multiplication factors. Moreover, the necessary transmittance may be lacking, thus stopping the multiplication process, or it may not be possible to meet the phase-matching condition (see Sect. 12.7.1). All the more surprising is the fact that high-harmonic generation (HHG) has been observed up to about the 300th harmonic [12.16].

This is highly improbable according to quantum mechanical perturbation theory. The description, covering most aspects of the experimental observations, consists of ionized electron motion, the electron returning to the parent atom it was stripped off after acquiring energy in the strong laser field. Upon recombination the total energy is released as a high-harmonic photon [12.17].

12.4.2 Multi-photon Absorption and Ionization

In the process of multi-photon absorption, as in the process of high-harmonic generation, n photons participate in the absorption process (Fig. 12.12). A first realization with three photons being absorbed succeeded in 1964 [12.18].

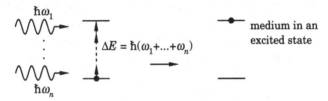

Fig. 12.12. Illustration of multi-photon absorption.

The process of multi-photon ionization is completely analogous to multi-photon absorption. The only difference consists in the ionization of the atom, as the sum of the energy of the interacting photons is larger than the binding energy of the atom for an electron (Fig. 12.13). There is an interesting connection to high-harmonic generation as multi-photon ionization seems to be the first step in this phenomenon [12.19].

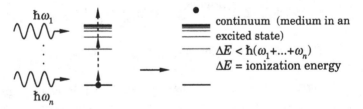

Fig. 12.13. Illustration of multi-photon ionization.

12.5 Further Nonlinear Optical Phenomena

Besides the fundamental nonlinear wave-interaction processes discussed, which were long known with acoustic waves and electronic oscillations (when applicable), a number of additional phenomena are encountered with intense light. We list some of them here without detailed description. More information may be found in the literature given at the end of this chapter.

- Intensity dependent index of refraction (Kerr effect)
  (see also the chapter on fiber optics).
- Self-phase modulation
  This effect leads to spectral broadening. Together with the optical Kerr effect it is a prerequisite for solitons in glass fibers.
- Intensity dependent absorption, self-induced transparency (bleachable absorbers)
  It is different from multi-photon absorption in that the energy level difference is $\Delta E = \hbar \omega_1$. It is used to switch the quality of a laser resonator to achieve high-intensity short pulses.
- Self-focusing, self-defocusing, and self-guiding [12.20]
  These spatial effects are caused by the intensity dependent index of refraction leading to plane-wave instability. They are detrimental to large-area laser amplifiers as a laser beam may break up into many filaments (in the case of self-focusing) of extremely high intensity, damaging the laser slabs.
- Stimulated Raman scattering
  It is caused by the contribution of vibrational modes of molecules taking part in absorption and emission processes. It may lead to laser frequency modulation with a total bandwidth extending over the infrared, visible, and ultra-violet spectral regions, a prerequisite for ultrashort pulse generation [12.21].
- Stimulated Brillouin scattering
  It is caused by the contribution of phonons to the interaction of light with matter. In the process, a sound wave is generated by the incident light at which the light is scattered.
- Coherent pulse propagation
  Examples are π pulses, 2π pulses, photon echoes.
- Optical shock-wave formation
- Optical breakdown (see Fig. 7.16)
- Light–plasma interaction
  In the interaction process, subharmonics and ultraharmonics have been observed, for instance, $\omega_2 = 1/2\, \omega_1$. High harmonics appear. Even nuclear fusion has been achieved with table-top lasers focused into a gas of deuterium clusters [12.22].

- Material processing with light
  Light can be used for cutting, drilling, welding, hardening. In eye surgery it is used to pin the retina down and for cutting and removing unwanted biological material.
- Light–electron interaction
  (Rayleigh scattering, Thomson scattering, Compton scattering.)
- Light–light interaction
  It is predicted by quantum field theory. The interaction cross section is very small, so that it has not yet been realized experimentally.

As the list makes evident, nonlinear optics is rich in effects. Its realm is by far larger than that of linear optics and is still largely unexplored. It is awaiting its prospectors.

12.6 Nonlinear Potentials

In the preceding chapters we have considered only the part of optics that is valid for waves in a vacuum or for waves of a relatively low electric field amplitude in transparent media. When the electric field strength in the medium is increased, new phenomena, as presented in the preceding sections, may occur as a result of the nonlinearity of the medium. What does it mean to say a medium is nonlinear? The simplest, classical description is as follows. When a light wave propagates in matter the electric field excites the electrons to oscillations (Lorentz model). For small amplitudes, the restoring forces are proportional to the elongation. The electrons then perform harmonic oscillations according to a linear differential equation for the harmonic oscillator:

$$\ddot{x} + \frac{1}{\tau}\dot{x} + \omega_0^2 x = \frac{e}{m}E(t) .$$ (12.1)

Here, x is the elongation of the electron, τ the decay constant of the oscillation (radiation damping, interaction with the lattice), ω_0 the resonance frequency (the medium may have many), e the charge of the electron, m the mass of the electron, and $E(t)$ the electric field of the incident wave. With harmonic excitation, $E(t) = E_0 \cos \omega t$, the solutions are harmonic oscillations, too. They may be visualized as the motion of a particle in a parabolic potential $V(x) = \frac{m}{2}\omega_0^2 x^2$ (Fig. 12.14). Now, realistic potentials, for instance, those describing the binding forces of electrons to an atom, by no means are parabolic potentials also for larger elongations. To obtain a first approximation to a nonlinear potential, the potential is expanded into a power series,

$$V(x) = \frac{m}{2}\omega_0^2 x^2 + k_1 x^3 + k_2 x^4 + \dots ,$$ (12.2)

which is truncated after the second nonvanishing term. A symmetric potential, that is, one with $V(x) = V(-x)$ and thus $k_1 = 0$, then leads to a nonlinear restoring force $F(x)$ with a cubic force term (see Fig. 12.14):

$$F(x) = -(m\omega_0^2 x + m\alpha x^3).\tag{12.3}$$

We then get the differential equation of a nonlinear oscillator, a Duffing equation:

$$\ddot{x} + d\dot{x} + \omega_0^2 x + \alpha x^3 = \frac{e}{m}E(t).\tag{12.4}$$

Here, we have introduced the damping constant d instead of $1/\tau$. The Duffing equation with periodic external forcing, $E(t) = E_0 \cos\omega t$, corresponds to a nonlinearly bound electron in the field of a harmonic light wave. The solution of the driven Duffing equation cannot be written down analytically. Also, for nonlinear differential equations the linear superposition principle, which states that for two solutions $x_1(t)$ and $x_2(t)$ new solutions can be gained by linear combinations $ax_1 + bx_2$, is no longer valid. Thus,

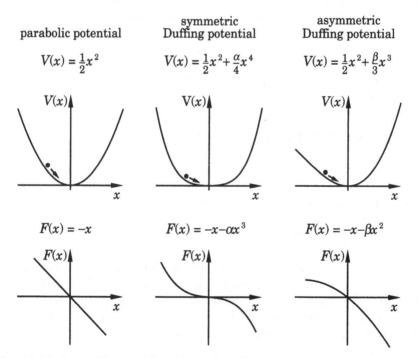

Fig. 12.14. An oscillator with a linear force law possesses a parabolic potential (*left*). A symmetric, to lowest-order nonlinear potential is given by $V(x) = x^2/2 + \alpha x^4/4$ (Duffing potential), having a restoring force $F(x) = -x - \alpha x^3$ (*middle*). An asymmetric, nonlinear potential $V(x) = x^2/2 + \beta x^3/3$ with its restoring force $F(x) = -x - \beta x^2$ is depicted to the right.

Fourier series and integrals are of little help for representing a general solution of (12.4).

To investigate nonlinear equations like the Duffing equation, numerical methods, methods of perturbation theory, or geometrical and topological methods must be employed. Despite formally being an almost trivial extension of the driven harmonic oscillator, the Duffing equation (12.4) may show a complicated dynamics, for instance, the chaotic behavior we have already come across when discussing the laser rate equations. If a parameter such as the forcing frequency ω is altered, period doubling to chaos, multiple resonances that can largely be classified, and horn-like bifurcation sets are observed in the parameter plane spanned by the electric field amplitude E_0 and the frequency ω.

Duffing equations and a number of nonlinear oscillators with different potentials (among them the Morse potential describing the interaction in diatomic molecules) have been investigated in the framework of the theory of dynamical systems. As no general theory for solving non-linear dynamical equations exists, at first each oscillator model must be studied separately in depth. Only then can a comparison be made to discover whether there exist similarities and conformities between different nonlinear models on a qualitative or topological level. Indeed, such similarities seem to exist between different nonlinear driven oscillators, for instance, with respect to their bifurcation set. To find and formulate the ordering principles is a task of present-day research [12.23].

12.7 Interaction of Light Waves

The classical treatment of nonlinear wave interactions in optics starts with Maxwell's equations. The classical description is a phenomenological theory that takes the existence of nonlinear phenomena from experiment and constructs a theoretical description with the help of reasonable assumptions. No statements are made about the 'microscopic' events, that is, events on an atomic level.

The Maxwell equations have already been presented in the chapter on the fundamentals of wave optics. We assume that the optical medium has the following properties: there exist no free charges ($\rho = 0$), the medium cannot be made magnetic ($M = 0$), and the medium is an isolator ($\sigma = 0$). Then, with $B_m = B$, the equations simplify to

$$\operatorname{div} E_m = 0, \tag{12.5}$$

$$\operatorname{div} B = 0, \tag{12.6}$$

$$\operatorname{curl} E = -\frac{\partial B}{\partial t}, \tag{12.7}$$

$$\operatorname{curl} B = \varepsilon_0 \mu_0 \frac{\partial E_m}{\partial t}, \tag{12.8}$$

along with the matter equations

$$E_{\mathrm{m}} = E + \frac{1}{\varepsilon_0} P,$$ (12.9)

$$B_{\mathrm{m}} = B.$$ (12.10)

We split the polarization P of the medium into a part depending linearly on E and a remaining part depending nonlinearly on E:

$$P = \varepsilon_0 \chi_{\mathrm{L}} E + P_{\mathrm{NL}}.$$ (12.11)

In nonisotropic media the susceptibility χ_{L} is given by a tensor. Later we shall assume an isotropic medium in which P (and also P_{NL}) is directed parallel to the internal field. Inserting the matter equations (12.9) and (12.10) into the field equations (12.5) to (12.8) and introducing the abbreviation

$$\varepsilon = \varepsilon_0 (1 + \chi_{\mathrm{L}}),$$ (12.12)

we obtain the new field equations:

$$\operatorname{div} E_{\mathrm{m}} = 0,$$ (12.13)

$$\operatorname{div} B = 0,$$ (12.14)

$$\operatorname{curl} E = -\frac{\partial B}{\partial t},$$ (12.15)

$$\operatorname{curl} B = \mu_0 \varepsilon \frac{\partial E}{\partial t} + \mu_0 \frac{\partial P_{\mathrm{NL}}}{\partial t}.$$ (12.16)

The elimination of B leads to the nonlinear wave equation which is also valid for a nonisotropic medium:

$$\operatorname{curl} \operatorname{curl} E + \mu_0 \varepsilon \frac{\partial^2 E}{\partial t^2} = -\mu_0 \frac{\partial^2 P_{\mathrm{NL}}}{\partial t^2}.$$ (12.17)

This equation cannot be simplified further in the general case of a nonisotropic medium, since then div $E_{\mathrm{m}} = 0$, but generally div $E \neq 0$. This means that $E \not\perp k$ (k = wave vector of the E_{m} field).

For simplicity, we now assume an isotropic medium ($E \perp k$) for the considerations to follow. The nonlinear wave equation then reads, along with div $E = 0$ and thus curl curl $E = \operatorname{grad} \operatorname{div} E - \Delta E = -\Delta E$,

$$\Delta E - \mu_0 \varepsilon \frac{\partial^2 E}{\partial t^2} = \mu_0 \frac{\partial^2 P_{\mathrm{NL}}}{\partial t^2}.$$ (12.18)

We would like to solve this nonlinear wave equation together with appropriate boundary conditions. Comparing this wave equation with the linear case we detect that the second time derivative of the nonlinear part of the polarization, $\partial^2 P_{\mathrm{NL}}/\partial t^2$, acts as a source for the light field. Analytic solutions of general type, as for the linear wave equation, do not exist. We therefore must look out for specific, yet typical, situations and find solutions for these special cases. A standard problem is given by three-wave interaction with quadratic nonlinearity.

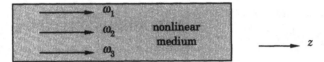

Fig. 12.15. Interaction of three plane waves propagating in the z-direction.

12.7.1 Three-Wave Interaction

In this section we derive the three-wave interaction equations for a medium with a quadratic nonlinearity. The problem is simplified by a series of assumptions:

1. The propagation proceeds in the z-direction with plane waves, that is, we have $\partial/\partial x = 0$ and $\partial/\partial y = 0$. Then

$$\boldsymbol{E}(z,t) = (E_x(z,t),\, E_y(z,t),\, 0)\,, \tag{12.19}$$

and the wave equation (12.18) takes the form

$$\frac{\partial^2 \boldsymbol{E}(z,t)}{\partial z^2} - \mu_0 \varepsilon \frac{\partial^2 \boldsymbol{E}(z,t)}{\partial t^2} = \mu_0 \frac{\partial^2 \boldsymbol{P}_{\mathrm{NL}}}{\partial t^2}\,. \tag{12.20}$$

This equation contains the additional assumption that $\boldsymbol{P}_{\mathrm{NL}}$ solely depends on z and t and has components only in the x- and y-directions. To sum up, we only consider transverse waves, and their interactions shall generate transverse waves only.

2. The nonlinear part of the polarization $\boldsymbol{P}_{\mathrm{NL}}$ is given by a quadratic nonlinearity, that is, by a tensor of third order (d_{ijk}):

$$(\boldsymbol{P}_{\mathrm{NL}})_i = d_{ijk} E_j E_k\,, \qquad i,j,k \in \{x,y\}\,, \tag{12.21}$$

whereby a summation has to be performed over repeated indices.

3. The interaction is a pure three-wave interaction, that is, only three waves of frequencies ω_1, ω_2, and ω_3 are present (see Fig. 12.15), which are related by

$$\omega_1 + \omega_2 = \omega_3\,. \tag{12.22}$$

In real notation, as needed for nonlinear problems, the three waves are given by

$$E_i^{(\omega_1)}(z,t) = \frac{1}{2}\left[E_i^{(1)}(z)\exp\left(ik_1 z - i\omega_1 t\right) + \text{c.c.}\right],$$

$$E_k^{(\omega_2)}(z,t) = \frac{1}{2}\left[E_k^{(2)}(z)\exp\left(ik_2 z - i\omega_2 t\right) + \text{c.c.}\right], \tag{12.23}$$

$$E_j^{(\omega_3)}(z,t) = \frac{1}{2}\left[E_j^{(3)}(z)\exp\left(ik_3 z - i\omega_3 t\right) + \text{c.c.}\right].$$

Here, as before, the indices i, j, or k may be x or y.

The solution ansatz for the nonlinear wave equation (12.20) in the case of the interaction of three waves is then as follows:

$$E_{x,y}(z,t) = \sum_{\nu=\pm 1}^{\pm 3} \frac{1}{2} E_{x,y}^{(\nu)}(z) \exp\left(ik_\nu z - i\omega_\nu t\right) . \tag{12.24}$$

Since the light-wave amplitudes are real quantities, we have

$$E_{x,y}^{(n)}(z) = E_{x,y}^{(-n)*}(z), \quad \omega_{-n} = -\omega_n, \quad k_{-n} = -k_n, \quad n = 1,2,3. \tag{12.25}$$

Referring back to the linear dispersion law, we may write for each of the waves:

$$k_n^2 = \mu_0 \varepsilon_n \omega_n^2, \quad n = 1,2,3. \tag{12.26}$$

Herein, $\varepsilon_n = \varepsilon(\omega_n)$, which has to be taken from (12.12) at the frequency ω_n. This ansatz, of course, can only be an approximation, as the (quadratic) nonlinearity also gives rise to all combination frequencies

$$\omega_{l_1,l_2,l_3} = l_1\omega_1 + l_2\omega_2 + l_3\omega_3, \quad l_1,l_2,l_3 = 0,\pm 1,\pm 2,\dots \tag{12.27}$$

Therefore, the validity of the ansatz has to be checked carefully in a given case. The ansatz, to be sure, will be a good approximation if, as assumed, $\omega_1 + \omega_2 = \omega_3$ holds and the wave with ω_3 is of very small amplitude. Then the additional combination frequencies with the frequency ω_3 as the source term will be of even smaller amplitude and can therefore be neglected.

If we take one of the three partial waves of frequency ω_ν, only that part of the nonlinear polarization that has the same frequency will be a source for this partial wave. This part will be called $P_{NL}^{(\omega_\nu)}(z,t)$. The remaining frequencies of the nonlinear polarization do not contribute, on average. With extremely short pulses, this approximation may lose its validity. Each partial wave, according to this approximation, has to obey the nonlinear wave equation,

$$\frac{\partial^2 E^{(\omega_\nu)}(z,t)}{\partial z^2} - \mu_0 \varepsilon(\omega_\nu) \frac{\partial^2 E^{(\omega_\nu)}(z,t)}{\partial t^2} = \mu_0 \frac{\partial^2 P_{NL}^{(\omega_\nu)}(z,t)}{\partial t^2} , \tag{12.28}$$

for $\nu = 1,2,3$.

The source term on the right-hand side should be a small perturbation for the wave $E^{(\omega_\nu)}$ that, undisturbed by $P_{NL}^{(\omega_\nu)}$, would obey the linear wave equation. The nonlinear part of the polarization, $P_{NL}^{(\omega_\nu)}$, again is a plane wave of the form

$$P_{NL}^{(\omega_\nu)} = \frac{1}{2}\left(P_{NL}^{(\nu)}(z)\exp\left[i(k_\nu z - \omega_\nu t)\right] + \text{c.c.}\right), \tag{12.29}$$

giving

$$\frac{\partial^2 P_{NL}^{(\omega_\nu)}}{\partial t^2} = -\omega_\nu^2 P_{NL}^{(\omega_\nu)} . \tag{12.30}$$

Similarly, we have

$$\frac{\partial^2 \boldsymbol{E}^{(\omega_\nu)}(z,t)}{\partial t^2} = -\omega_\nu^2 \boldsymbol{E}^{(\omega_\nu)}(z,t).$$ (12.31)

The system of partial differential equations (12.28) then is transformed into a system of ordinary differential equations, yielding

$$\frac{\mathrm{d}^2 \boldsymbol{E}^{(\omega_\nu)}(z,t)}{\mathrm{d}z^2} + \mu_0 \varepsilon(\omega_\nu)\omega_\nu^2 \boldsymbol{E}^{(\omega_\nu)}(z,t) = -\mu_0 \omega_\nu^2 \boldsymbol{P}_{\mathrm{NL}}^{(\omega_\nu)}(z,t)$$ (12.32)

for $\nu = 1, 2, 3$.

We now calculate the source term $\boldsymbol{P}_{\mathrm{NL}}^{(\omega_\nu)}(z,t)$ in the case of a quadratic nonlinearity. We consider the frequency $\omega_1 = \omega_3 - \omega_2$:

$$\left[\boldsymbol{P}_{\mathrm{NL}}^{(\omega_1)}(z,t)\right]_i = \frac{1}{2}\left(d_{ijk}E_j^{(3)}(z)E_k^{(2)*}(z)\exp\left[i(k_3 - k_2)z - i(\omega_3 - \omega_2)t\right] + \mathrm{c.c.}\right).$$ (12.33)

The nonlinear equation for the partial wave at the frequency ω_1 then reads:

$$\frac{\mathrm{d}^2 E_i^{(\omega_1)}(z,t)}{\mathrm{d}z^2} + \mu_0 \varepsilon(\omega_1)\omega_1^2 E_i^{(\omega_1)}(z,t)$$

$$= \frac{-\mu_0 \omega_1^2 d_{ijk}}{2}\left(E_j^{(3)}(z)E_k^{(2)*}(z)\exp\left[i(k_3 - k_2)z - i(\omega_3 - \omega_2)t\right] + \mathrm{c.c.}\right).$$ (12.34)

We now introduce further simplifications which often prove true experimentally; namely that the complex amplitudes $E_i^{(\nu)}$ of all three partial waves are slowly varying in the z-direction. Then the second derivative can be neglected compared to the first. This approximation is called the slowly varying amplitude approximation (SVA approximation). Written for the frequency ω_1, we have:

$$\frac{\mathrm{d}^2 E_i^{(\omega_1)}(z,t)}{\mathrm{d}z^2} = \frac{1}{2}\frac{\mathrm{d}^2}{\mathrm{d}z^2}\left(E_i^{(1)}(z)\exp\left[i(k_1 z - \omega_1 t)\right] + \mathrm{c.c.}\right)$$

$$= \frac{1}{2}\left[\frac{\mathrm{d}^2 E_i^{(1)}(z)}{\mathrm{d}z^2} + 2ik_1\frac{\mathrm{d}E_i^{(1)}(z)}{\mathrm{d}z} - k_1^2 E_i^{(1)}(z)\right]\exp\left[i(k_1 z - \omega_1 t)\right] + \mathrm{c.c.}$$

$$\approx -\frac{1}{2}\left[k_1^2 E_i^{(1)}(z) - 2ik_1\frac{\mathrm{d}E_i^{(1)}(z)}{\mathrm{d}z}\right]\exp\left[i(k_1 z - \omega_1 t)\right] + \mathrm{c.c.},$$ (12.35)

$$\text{with} \quad \left|\frac{\mathrm{d}^2 E_i^{(1)}(z)}{\mathrm{d}z^2}\right| \ll k_1\left|\frac{\mathrm{d}E_i^{(1)}(z)}{\mathrm{d}z}\right|.$$ (12.36)

Similar equations are valid for ω_2 and ω_3. For ease of derivation, it is recommended to cyclically permute the indices $1, 2, 3$ along with i, j, k $(1 \rightarrow 2 \rightarrow 3 \Longleftrightarrow i \rightarrow k \rightarrow j)$; that is, to set $E_k^{(\omega_2)}$ and $E_j^{(\omega_3)}$. In the SVA approximation the differential equation for $E_i^{(\omega_1)}(z,t)$ reads:

$$\left(\frac{k_1^2}{2}E_i^{(1)}(z) - ik_1\frac{dE_i^{(1)}(z)}{dz}\right)\exp\left[i(k_1z - \omega_1 t)\right] + \text{c.c.}$$

$$-\frac{1}{2}\mu_0\varepsilon(\omega_1)\omega_1^2 E_i^{(1)}(z)\exp\left[i(k_1z - \omega_1 t)\right] + \text{c.c.}$$

$$= +\frac{\mu_0\omega_1^2}{2}d_{ijk}E_j^{(3)}(z)E_k^{(2)*}(z)\exp\left[i(k_3 - k_2)z - i(\omega_3 - \omega_2)t\right] + \text{c.c.} \quad (12.37)$$

The two terms $k_1^2 E_i^{(1)}/2$ and $-\mu_0\varepsilon(\omega_1)\omega_1^2 E_i^{(1)}/2$ cancel, since the dispersion relation $k_1^2 = \mu_0\varepsilon(\omega_1)\omega_1^2$ is assumed to be valid (see (12.26)).

Equation (12.37) is an equation for real quantities, since the 'complex conjugate' is included. To further simplify the equations, we now proceed with complex quantities. This procedure is backed by an approximation called the rotating wave approximation (RWA). Only corotating waves (seen in the complex plane) interact strongly. We then have two equations each:

$$\frac{dE_i^{(1)}(z)}{dz} = +\frac{i\mu_0\omega_1^2}{2k_1}d_{ijk}E_j^{(3)}(z)E_k^{(2)*}(z)\exp\left[i(k_3 - k_2 - k_1)z\right] \quad (12.38)$$

and a second one for the complex conjugate. The second one, however, contains no further information. Now the time dependence is no longer present because of $\omega_3 - \omega_2 - \omega_1 = 0$.

Before giving the equations for the remaining two waves, we introduce Δk, the phase mismatch,

$$\Delta k = (k_1 + k_2) - k_3 . \quad (12.39)$$

Further, with the help of the dispersion relation

$$k_1^2 = \mu_0\varepsilon_1\omega_1^2 , \quad (12.40)$$

we write ($\varepsilon_1 = \varepsilon(\omega_1)$)

$$\frac{\mu_0\omega_1^2}{k_1} = \omega_1\sqrt{\frac{\mu_0}{\varepsilon_1}} . \quad (12.41)$$

Then, after having derived the corresponding equations for ω_2 and ω_3 according to (12.38), we obtain the following system of equations:

$$\frac{dE_i^{(1)}}{dz} = +\frac{i\omega_1}{2}\sqrt{\frac{\mu_0}{\varepsilon_1}}d_{ijk}E_j^{(3)}E_k^{(2)*}\exp(-i\Delta kz) , \quad (12.42)$$

$$\frac{dE_k^{(2)*}}{dz} = -\frac{i\omega_2}{2}\sqrt{\frac{\mu_0}{\varepsilon_2}}d_{kij}E_i^{(1)}E_j^{(3)*}\exp(i\Delta kz) , \quad (12.43)$$

$$\frac{dE_j^{(3)}}{dz} = +\frac{i\omega_3}{2}\sqrt{\frac{\mu_0}{\varepsilon_3}}d_{jki}E_k^{(2)}E_i^{(1)}\exp(i\Delta kz) . \quad (12.44)$$

This system describes the collinear propagation of three waves in the z-direction, with $\omega_1 + \omega_2 = \omega_3$ and quadratic nonlinearity of the medium.

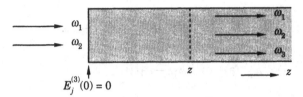

Fig. 12.16. Geometry and initial conditions for sum-frequency generation.

As i, j, k may be either x or y, we get six equations for the complex field amplitudes of the three waves (each having two polarization directions). The system describes second-harmonic generation, sum and difference-frequency generation including optical rectification, up- and down-conversion and the parametric amplifier. It may appear peculiar to have many different names associated with the three-wave interaction process. The reason is that they denote a number of particular cases which differ in which wave is the incident one and which one is generated, and also in which wave has a large amplitude and which one should stay small.

The application of the three-wave interaction equations (12.42) to (12.44) will now be demonstrated by way of example. We take sum-frequency generation with the input wave staying constant during interaction (Fig. 12.16).

Two waves of frequency ω_1 and ω_2 are radiated into a suitable nonlinear crystal. Let the amplitude of the wave at the sum frequency $\omega_3 = \omega_1 + \omega_2$ be zero at the surface of the crystal:

$$E_j^{(3)}(z = 0) = 0. \tag{12.45}$$

The problem of the coupling of the two waves, at frequencies ω_1 and ω_2, into the crystal is ignored. We further assume that the two incident waves retain their intensities. This means a small conversion rate, as often is the case:

$$E_i^{(1)} = \text{const} \quad \text{or} \quad \frac{dE_i^{(1)}}{dz} = 0, \tag{12.46}$$

$$E_k^{(2)*} = \text{const} \quad \text{or} \quad \frac{dE_k^{(2)*}}{dz} = 0. \tag{12.47}$$

Then, from the three equations (12.42) to (12.44) only the third one remains:

$$\frac{dE_j^{(3)}(z)}{dz} = +\frac{i\omega_3}{2}\sqrt{\frac{\mu_0}{\varepsilon_3}}d_{jki}E_k^{(2)}E_i^{(1)}\exp(i\Delta kz), \tag{12.48}$$

where $\Delta k = k_1 + k_2 - k_3$. This has to be solved with the boundary condition (12.45). The differential equation (12.48) is of the type

$$\frac{dy}{dx} = a\exp(ibx),$$

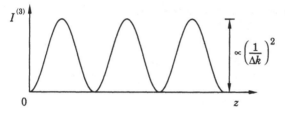

Fig. 12.17. Intensity of the sum frequency in dependence on the length of the crystal at $\Delta k = \text{const} \neq 0$.

whose solution can immediately be given. Therefore, we have

$$E_j^{(3)}(z) = \frac{i\omega_3}{2}\sqrt{\frac{\mu_0}{\varepsilon_3}}d_{jki}E_i^{(1)}E_k^{(2)}\frac{\exp(i\Delta kz) - 1}{i\Delta k}. \tag{12.49}$$

The intensity of the newly obtained wave of frequency ω_3 is given by

$$I^{(3)}(z) \propto E_j^{(3)}(z)E_j^{(3)*}(z) = \omega_3^2\frac{\mu_0}{\varepsilon_3}d_{jki}^2\left|E_i^{(1)}\right|^2\left|E_k^{(2)}\right|^2\frac{\sin^2[(\Delta k/2)z]}{(\Delta k)^2}, \quad \Delta k \neq 0. \tag{12.50}$$

To arrive at this equation, the relation

$$\left[\exp(i\Delta kz) - 1\right]\left[\exp(-i\Delta kz) - 1\right] = 2 - 2\cos(\Delta kz) = 4\sin^2\left(\frac{\Delta k}{2}z\right)$$

has been used. When $\Delta k \neq 0$, the intensity of the newly generated wave increases and decreases periodically with the path length in the crystal. In the approximation of constant pump waves at frequency ω_1 and ω_2 the intensity variation has the form of a squared sine function (Fig. 12.17). The maximum intensity is smaller, the larger the phase mismatch Δk:

$$I_{\max}^{(3)} \propto \left(\frac{1}{\Delta k}\right)^2. \tag{12.51}$$

The marked periodic dependence of the intensity on the length of the crystal was observed in 1962 [12.24] in the case of second-harmonic generation ($\omega_1 = \omega_2$).

For $\Delta k = 0$ (phase matching) the intensity of the sum-frequency wave increases quadratically. This can be valid for sufficiently small z only. For larger z it must be taken into account that the incident waves $E^{(1)}$ and $E^{(2)}$ no longer stay constant. Then, the full system of differential equations must be solved.

The example discussed also contains the case of difference-frequency generation, because the formalism for difference-frequency generation is similar to that for sum-frequency generation. It has only to be assumed that ω_1 and ω_3 are the frequencies of the pump waves and that $\omega_2 = \omega_3 - \omega_1$ is the frequency of the wave generated.

12.7.2 Scalar Three-Wave Interaction

The three-wave interaction equations (12.42) to (12.44) may be simplified when further assumptions are made. To begin, we set

$$A^{(\nu)} = \sqrt{\frac{n_\nu}{\omega_\nu}}\, E^{(\nu)}, \qquad \nu = 1,2,3, \tag{12.52}$$

where n_ν is the index of refraction at frequency ω_ν. Then Eq. (12.42) reads:

$$\sqrt{\frac{\omega_1}{n_1}}\frac{dA_i^{(1)}}{dz} = +\frac{i\omega_1}{2}\sqrt{\frac{\mu_0}{\varepsilon_1}}\,d_{ijk}\sqrt{\frac{\omega_2\omega_3}{n_2n_3}}A_j^{(3)}A_k^{(2)*}\exp(-i\Delta kz)\,. \tag{12.53}$$

Replacing ε_1 by

$$\varepsilon_1 = n_1^2\varepsilon_0\,, \tag{12.54}$$

we obtain

$$\frac{dA_i^{(1)}}{dz} = +\frac{i}{2}\sqrt{\frac{\mu_0}{\varepsilon_0}}\,d_{ijk}\sqrt{\frac{\omega_1\omega_2\omega_3}{n_1n_2n_3}}A_j^{(3)}A_k^{(2)*}\exp(-i\Delta kz)\,. \tag{12.55}$$

Introducing the abbreviation

$$C = \sqrt{\frac{\mu_0}{\varepsilon_0}}\sqrt{\frac{\omega_1\omega_2\omega_3}{n_1n_2n_3}}\,, \tag{12.56}$$

and taking the complete system (12.42) to (12.44), we finally get

$$\frac{dA_i^{(1)}}{dz} = +\frac{i}{2}Cd_{ijk}A_j^{(3)}A_k^{(2)*}\exp(-i\Delta kz)\,, \tag{12.57}$$

$$\frac{dA_k^{(2)*}}{dz} = -\frac{i}{2}Cd_{kij}A_i^{(1)}A_j^{(3)*}\exp(i\Delta kz)\,, \tag{12.58}$$

$$\frac{dA_j^{(3)}}{dz} = +\frac{i}{2}Cd_{jki}A_k^{(2)}A_i^{(1)}\exp(i\Delta kz)\,. \tag{12.59}$$

We now further assume that the polarization direction of each individual wave is not altered in the interaction process. For example, $A_x^{(1)}$ and $A_y^{(1)}$ simultaneously alter in such a way that no change in the polarization direction occurs. Then, instead of two equations for $A_i^{(1)}$, $i = x, y$, we only need one to describe the event, and similarly for the other waves: for each wave, we consider only one polarization direction. We may then simplify the notation:

$$A_i^{(1)} = A_1\,, \tag{12.60}$$

where now the index 1 denotes the wave of frequency ω_1 and its corresponding polarization direction. Similarly, we write:

$$A_k^{(2)} = A_2 \quad \text{and} \quad A_j^{(3)} = A_3\,. \tag{12.61}$$

The coefficients of the nonlinear susceptibilities (Eq. (12.21)) then take the form

$$d_{ijk} \leftrightarrow d_{132}, \quad d_{kij} \leftrightarrow d_{213}, \quad d_{jki} \leftrightarrow d_{321}. \tag{12.62}$$

Our assumption has simultaneously eliminated the summation over the indices, as now we consider the polarization directions as being independent from each other.

For lossless media, Kleinman's rule applies (see [12.25]) which states that all coefficients of the susceptibility tensor (d_{ijk}) with permutated indices are equal:

$$d_{132} = d_{321} = d_{213} = d. \tag{12.63}$$

We additionally assume d to be independent of frequency and finally obtain:

$$\frac{dA_1}{dz} = +iKA_2^* A_3 \exp(-i\Delta kz), \tag{12.64}$$

$$\frac{dA_2^*}{dz} = -iKA_1 A_3^* \exp(+i\Delta kz), \tag{12.65}$$

$$\frac{dA_3}{dz} = +iKA_2 A_1 \exp(+i\Delta kz), \tag{12.66}$$

with

$$\omega_1 + \omega_2 = \omega_3, \tag{12.67}$$

$$\Delta k = k_1 + k_2 - k_3, \tag{12.68}$$

$$K = \frac{1}{2}d\sqrt{\frac{\mu_0}{\varepsilon_0}}\sqrt{\frac{\omega_1\omega_2\omega_3}{n_1 n_2 n_3}}, \tag{12.69}$$

$$A = \sqrt{\frac{n}{\omega}}E \quad \text{and} \quad E \propto \exp\left[i(kz - \omega t)\right]. \tag{12.70}$$

12.7.3 Second-Harmonic Generation

We now take the set of equations (12.64) to (12.66) to describe second-harmonic generation. In this case we have

$$A_1 = A_2, \quad \omega_1 = \omega_2, \quad k_1 = k_2, \quad \omega_3 = 2\omega_1, \quad \Delta k = 2k_1 - k_3. \tag{12.71}$$

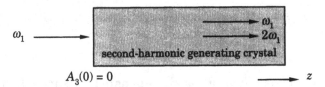

$A_3(0) = 0$ $\longrightarrow$ z

Fig. 12.18. Geometry and initial conditions for second-harmonic generation.

From the three equations only two remain for obvious reasons:

$$\frac{dA_1}{dz} = +iKA_1^* A_3 \exp(-i\Delta kz) \,, \tag{12.72}$$

$$\frac{dA_3}{dz} = +iKA_1^2 \exp(+i\Delta kz) \,. \tag{12.73}$$

This is a system of nonlinear differential equations describing second-harmonic generation, for instance, in KH_2PO_4 (named KDP).

We set out to solve the equation in the case of phase matching ($\Delta k = 0$). To do so, we also need initial conditions. We take $A_1(0)$ to be a real nonvanishing number (Fig. 12.18) and $A_3(0) = 0$.

The equations now read:

$$\frac{dA_1}{dz} = +iKA_1^* A_3 \,, \tag{12.74}$$

$$\frac{dA_3}{dz} = +iKA_1^2 \,. \tag{12.75}$$

We detect that $A_1(z)$ is real and $A_3(z)$ is purely imaginary. Therefore, we set

$$A_3 = i\tilde{A}_3 \,, \tag{12.76}$$

with $\tilde{A}_3$ real, and get, because of $A_1 = A_1^*$,

$$\frac{dA_1}{dz} = -K\tilde{A}_3 A_1 \,, \tag{12.77}$$

$$\frac{d\tilde{A}_3}{dz} = KA_1^2 \,. \tag{12.78}$$

Thus, we arrive at a real, nonlinear system of differential equations, which can be solved in closed form. To begin, we multiply the first equation by A_1 and the second by $\tilde{A}_3$ and add both equations:

$$\frac{d}{dz}\left(A_1^2 + \tilde{A}_3^2\right) = 0 \,. \tag{12.79}$$

This is an equation expressing conservation of energy. Since at the beginning ($z = 0$) we have only the wave $A_1(0)$, the integration constant must be

$$\left(A_1^2(z) + \tilde{A}_3^2(z)\right) = A_1^2(0) \,, \tag{12.80}$$

and therefore

$$\frac{d\tilde{A}_3(z)}{dz} = KA_1^2(z) = K\left(A_1^2(0) - \tilde{A}_3^2(z)\right).$$ (12.81)

The solution is given by

$$\tilde{A}_3(z) = A_1(0)\tanh\left(KA_1(0)z\right),$$ (12.82)

as can easily be verified, remembering $d\tanh z/dz = 1 - \tanh^2 z$.
Again we only observe the intensity:

$$I_3(z) \propto \tilde{A}_3^2(z) = A_1^2(0)\tanh^2\left(KA_1(0)z\right).$$ (12.83)

The decrease in intensity of the incident wave is immediately obtained
from (12.80):

$$I_1(z) \propto A_1^2(z) = A_1^2(0) - \tilde{A}_3^2(z) = A_1^2(0)\left[1 - \tanh^2\left(KA_1(0)z\right)\right]$$

$$= A_1^2(0)\operatorname{sech}^2\left(KA_1(0)z\right).$$ (12.84)

The solutions are plotted in Fig. 12.19.

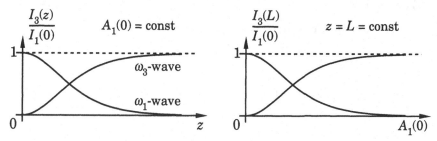

Fig. 12.19. Graph of the solutions for second-harmonic generation in the case of
phase matching ($\Delta k = 0$).

A discussion of the solution yields the following interesting statements:

- In the case of phase matching it is obviously possible to convert almost
  the total incident intensity into the newly generated, second-harmonic
  wave. For that, we need a large interaction length ($z \to \infty$) or a high
  intensity ($A_1^2(0) \to \infty$).
- Furthermore, it is remarkable that the process of converting light to
  its second harmonic needs no threshold and no noise to start with.
- The validity of conservation of energy as expressed in (12.79) is astonishing since approximations have been made when deriving the equations.

12.7.4 Optical Parametric Amplifier

Three-wave interaction in a nonlinear medium may be used to amplify a light signal at a frequency ω_1. To this end, an intense pump wave of frequency $\omega_3 > \omega_1$ must be applied together with the signal. Then, necessarily, a further light wave is generated at $\omega_2 = \omega_3 - \omega_1$: the idler wave (Fig. 12.20). This case, too, can be solved with the three-wave interaction equations (12.64–12.66).

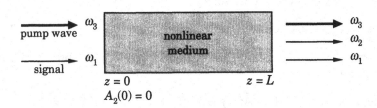

Fig. 12.20. Geometry and initial conditions for the optical parametric amplifier.

We consider the case of a constant pump wave of real amplitude:

$$A_3(z) = A_3(0). \tag{12.85}$$

Then only two equations are left:

$$\frac{dA_1}{dz} = +iK_p A_2^* \exp(-i\Delta kz), \tag{12.86}$$

$$\frac{dA_2^*}{dz} = -iK_p A_1 \exp(+i\Delta kz), \tag{12.87}$$

where

$$K_p = KA_3(0) \tag{12.88}$$

and $\Delta k = k_1 + k_2 - k_3$ as before. This system may be solved in closed form. Here, we just consider the case of phase matching, $\Delta k = 0$. Then we arrive at the linear system:

$$\frac{dA_1}{dz} = +iK_p A_2^*, \tag{12.89}$$

$$\frac{dA_2^*}{dz} = -iK_p A_1. \tag{12.90}$$

As initial conditions we take a weak signal $A_1(0) \neq 0$ and a zero idler wave $A_2(0) = 0$. Then the solution reads:

$$A_1(z) = A_1(0)\cosh(K_p z), \tag{12.91}$$

$$A_2^*(z) = -iA_1(0)\sinh(K_p z), \tag{12.92}$$

as can easily by verified by substitution into (12.89) and (12.90).

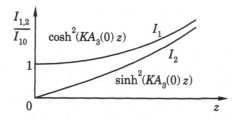

Fig. 12.21. Solution curves of the optical parametric amplifier, $\Delta k = 0$.

The intensity of the waves is given by

$$I_1(z) = A_1(z)A_1^*(z) = A_1(0)A_1^*(0)\cosh^2\left(K_p z\right) = I_{10}\cosh^2\left(KA_3(0)z\right),$$
(12.93)

$$I_2(z) = A_2(z)A_2^*(z) = A_1(0)A_1^*(0)\sinh^2\left(K_p z\right) = I_{10}\sinh^2\left(KA_3(0)z\right).$$
(12.94)

It can be detected that at first the signal wave at frequency ω_1 is amplified only moderately, whereas the idler wave at frequency ω_2 is newly generated (Fig. 12.21). The optical parametric amplifier, besides being a noiseless signal amplifier, may be used for generating coherent light at new frequencies that are not bound to a corresponding atomic transition.

12.7.5 Optical Parametric Oscillator

An amplifier for a certain frequency can always be used also for the construction of an oscillator. All that has to be done is to provide a phase-matched feedback and to overcome the unavoidable losses. In optics, phase-matched feedback is obtained by placing the amplifier into a resonator that is tuned to the wavelength to be generated. Exactly as in the case of a usual laser that makes use of atomic transitions between energy levels, an oscillator is obtained, called the optical parametric oscillator (Fig. 12.22). The frequencies ω_1 and ω_2 are initiated from noise. They do not have to correspond to an atomic transition in the nonlinear medium, but only have to obey the condition

$$\omega_1 + \omega_2 = \omega_3,$$
(12.95)

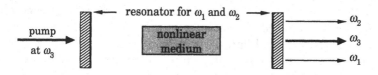

Fig. 12.22. The optical parametric oscillator.

the energy conservation law, and, for effective operation,

$$k_1 + k_2 = k_3 , \tag{12.96}$$

the phase matching condition (conservation of momentum). Both conditions can be met with suitable nonlinear media. Oscillators of this kind for coherent light production can be tuned in frequency or wavelength by changing the feedback conditions. Any frequency out of a continuous range of frequencies may be generated, depending on the geometrical setting. They are commercially available, mostly in connection with Ti:sapphire lasers that themselves are tunable over a broad frequency region, giving continuous coverage from the far infrared to deep into the ultraviolet.

12.7.6 Three-Wave Interaction in the Photon Picture

The equations (12.64)–(12.66), despite their underlying approximations, contain a general relation valid between the intensities of the three waves. This relation is known as the Manley–Rowe relation from communication science [12.26]. In the case of three-wave interaction and in our notation with $\omega_1 + \omega_2 = \omega_3$ it reads:

$$\frac{d}{dz}(A_1 A_1^*) = \frac{d}{dz}(A_2 A_2^*) = -\frac{d}{dz}(A_3 A_3^*) . \tag{12.97}$$

The proof is straightforward. We just have to introduce the corresponding expressions from (12.64)–(12.66) into the individual terms above:

$$\frac{d}{dz}(A_1 A_1^*) = A_1 \frac{dA_1^*}{dz} + A_1^* \frac{dA_1}{dz}$$
$$= A_1[-iKA_2 A_3^* \exp(+i\Delta kz)] + A_1^*[iKA_2^* A_3 \exp(-i\Delta kz)] , \tag{12.98}$$

$$\frac{d}{dz}(A_2 A_2^*) = A_2 \frac{dA_2^*}{dz} + A_2^* \frac{dA_2}{dz}$$
$$= A_2[-iKA_1 A_3^* \exp(+i\Delta kz)] + A_2^*[iKA_1^* A_3 \exp(-i\Delta kz)] , \tag{12.99}$$

$$-\frac{d}{dz}(A_3 A_3^*) = -A_3 \frac{dA_3^*}{dz} - A_3^* \frac{dA_3}{dz}$$
$$= -A_3[-iKA_2^* A_1^* \exp(-i\Delta kz)] - A_3^*[iKA_2 A_1 \exp(+i\Delta kz)] . \tag{12.100}$$

It turns out that all three terms look the same and thus are equal.

Since the intensities $A_n A_n^*$, $n = 1, 2, 3$, are proportional to the photon flux at the corresponding frequency ω_n, the Manley–Rowe relation can be interpreted in the photon picture in the following way. With each photon of frequency ω_3 created, a photon of frequency ω_1 and a photon of frequency ω_2 must be annihilated, whereby $\omega_1 + \omega_2 = \omega_3$. Conversely, with each photon of frequency ω_3 that disappears, a photon of frequency ω_1 and a photon of frequency ω_2 must appear, whereby the condition $\omega_3 = \omega_1 + \omega_2$ holds (Fig. 12.23).

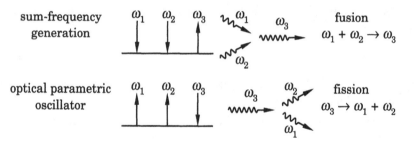

Fig. 12.23. Illustration of the two basic types of three-wave interaction in the photon picture.

In the classical relation of (12.97) the quantum nature of light is reflected, and indeed must be reflected therein. When light can interact only in discrete energy humps, then in a nonlinear frequency conversion process this condition must be met photon by photon. Thus, besides a wave being generated from a given one there must always be a further wave of different frequency, except in the degenerate case where the two waves coincide in frequency. An example for an important degenerate case is second-harmonic generation (Fig. 12.24).

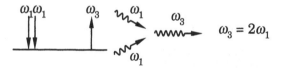

Fig. 12.24. Degenerate case of three-wave interaction: second-harmonic generation.

Nonlinear optics, in the form of three-wave interaction, supplies us with a means to merge and split photons. Basically, there are only two three-wave interaction processes: fusion ($\omega_1 + \omega_2 \to \omega_3$) and fission ($\omega_3 \to \omega_1 + \omega_2$) of photons. Viewed in this way, the nonlinear processes discussed find a simple interpretation and classification.

Problems

12.1 Consider the sinusoidally driven Duffing oscillator,

$$\ddot{x} + d\dot{x} + \omega_0^2 x + \alpha x^3 = E_0 \cos(\omega t). \tag{12.101}$$

Show that by making an ansatz according to the SVA approximation,

$$x(t) = A(t) \exp(-i\omega t) + A^*(t) \exp(i\omega t), \tag{12.102}$$

and neglecting the term d^2A/dt^2, the envelope function $A(t)$ obeys the following differential equation:

$$(d - 2i\omega)\frac{dA}{dt} + (\omega_0^2 - \omega^2 - id\omega + 3\alpha|A|^2)A = \frac{E_0}{2}. \tag{12.103}$$

Assuming that $A(t)$ does not vary with time, derive an equation for determining the quantity $|A|^2$.

12.2 A Duffing oscillator is driven by two harmonic oscillations having different frequencies ω_1 and ω_2:

$$\ddot{x} + d\dot{x} + \omega_0^2 x + \alpha x^3 = E_1 \cos(\omega_1 t) + E_2 \cos(\omega_2 t). \tag{12.104}$$

Let the oscillation amplitude of the oscillator be sufficiently small. By substituting the ansatz, $x^{(1)}(t) = A_1 \exp(i\omega_1 t) + A_2 \exp(i\omega_2 t) + \text{c.c.}$, into (12.104) determine the amplitudes of oscillation, A_1 and A_2, to a first approximation (the term αx^3 is to be neglected). Which new frequency components are generated (to first order) by the nonlinear term? How many coefficients have to be included in a second-order approximation of $x(t)$?

12.3 The (linearly polarized) beam of a ruby laser ($\lambda = 694.3$ nm), having a power density of $S = 100$ MW cm^{-2}, enters a KDP crystal. Take the scalar three-mode interaction theory for calculating second-harmonic generation. Of what length must the crystal be for the amplitude of the second harmonic to reach a fraction $1/\sqrt{2}$ of the input amplitude of the fundamental wave? The refractive index of the crystal is $n = 1.507$, the constant $K = 1.2 \cdot 10^{-6}$ V^{-1}. Assume index matching, $\Delta k = 0$.
Hints: the relation between the power density and field strength reads: $S = \frac{1}{2}nc\varepsilon_0 E_0^2$. Assume the losses upon entry of the beam into the crystal to be negligible.

12.4 Verify that the functions

$$A_1(z) = A_1(0) \exp(-i\Delta kz/2) \left(\cosh(bz) - i\frac{\Delta k}{2b} \sinh(bz) \right)$$

$$+ i\frac{K_p}{b} A_2^*(0) \sinh(bz) \exp(-i\Delta kz/2),$$

$$A_2^*(z) = A_2^*(0) \exp(i\Delta kz/2) \left(\cosh(bz) - i\frac{\Delta k}{2b} \sinh(bz) \right)$$

$$- i\frac{K_p}{b} A_1(0) \sinh(bz) \exp(i\Delta kz/2),$$

with $b = \sqrt{K_p^2 - (\Delta k)^2/4}$, solve the differential equations (12.86) and (12.87) for the optical parametric amplifier with a constant pump wave.

13. Fiber Optics

Information has been transmitted by optical means since ancient times. Just remember the smoke signals of Indians or signal flags and flashlight signals of different kinds. Today, lighthouses serve as sign posts with their light signals, and differently colored light (red, yellow, green) is used to control the traffic. Optics, however, has not played a role in the beginnings of communication engineering. It is true, wireless communication owes its being to the discovery of electromagnetic waves by *Heinrich Hertz*, and light, too, is an electromagnetic wave, but the coherent generation of such waves at first succeeded in a relatively low-frequency range only (radio, TV). With the development of the laser, in this context mainly the semiconductor laser, coherent waves are now readily available in the optical range, too. This opens up totally new possibilities for communication engineering due to the high communication bandwidth. However, in the Earth's atmosphere, optical propagation is often disturbed by the atmospheric conditions. It was not until the development of glass fibers with extremely low damping that optical communication came to a breakthrough. Present-day systems now are no longer limited in their performance by the fiber but by the electronics for modulating the laser and by the detector. A coarse scheme of a fiber-optic communication system is given in Fig. 13.1.

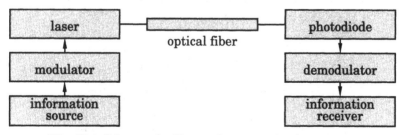

Fig. 13.1. Scheme of a fiber-optic communication system.

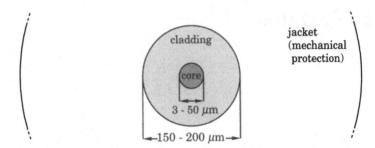

Fig. 13.2. Cross section through a glass fiber (core, cladding, jacket).

13.1 Glass Fibers

In this section, we will be occupied with part of this system, the fiber link. The propagation of light in fibers, quite naturally, is different from the propagation in free space, as the waves are guided and may travel along quite sinuous paths according to the bends of the fiber [13.1].

13.1.1 Profile

To begin, we consider the profile of a glass fiber (Fig. 13.2). It mainly consists of a core and a cladding of glass (SiO_2) which have different refractive indices, n_1 and n_2, respectively. To increase the index of refraction, GeO_2 is used as a dopant, for instance. Depending on the transition of the refractive index between the core and cladding, step-index and graded-index glass fibers are distinguished (Fig. 13.3). In the step-index fiber the transition of the refractive index proceeds in a step, in the graded-index fiber it proceeds gradually. The advantage of the graded-index fiber will become clear soon.

A special role is played by single-mode (monomode) fibers. They have an extremely small core diameter of 3–5 µm, so that only one transverse mode can propagate at a suitable wavelength of the light.

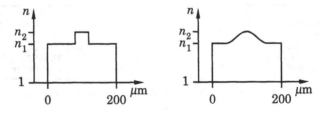

Fig. 13.3. Change of the refractive index across the fiber for a step-index fiber (*left*) and a graded-index fiber (*right*).

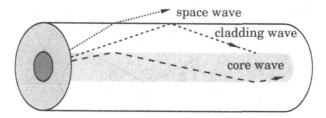

Fig. 13.4. Types of waves in a fiber.

13.1.2 Guided Waves

In connection with fibers, three types of waves are distinguished. We call them core waves, cladding waves, and space waves (Fig. 13.4). Light propagation over large distances occurs only with waves in the core. Excitation of the other waves leads to losses. Therefore, we take every effort to couple light only into the core. To this end, the angle of incidence should not be made too large, otherwise part of the light induces cladding and space waves (Fig. 13.4). This is particularly difficult for single-mode fibers, with their 3–5 μm core. To focus light onto the face of a core, one needs a microscope objective. These, however, have a large numerical aperture.

In a step-index fiber the core waves are retained in the interior by total internal reflection, in the graded-index fiber this is done by gradual inclination of the propagation direction of the wave. In Fig. 13.5, the light propagation is shown for fibers with a step-index change, a graded-index change, and for a monomode fiber.

We now will have a closer look at a step-index fiber (Fig. 13.6). To simplify the discussion, we take a plane model of a waveguide with a thickness of $2a$. At a given wavelength λ, the interference of the waves being reflected back and forth means that propagation is not allowed for all angles $\Theta < \Theta_c$ (Θ_c being the critical angle of total reflection) but only for discrete ones. Besides on λ these angles depend on the thickness $2a$ and the refractive indices n_1 and n_2. As illustrated in Fig. 13.6, after two reflections at the core–cladding interface the planes of equal phase must coincide for constructive interference. Elimination of $\overline{AB}$ from the

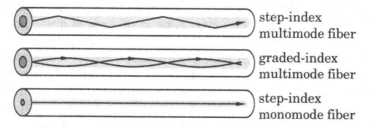

Fig. 13.5. Light propagation in different types of fibers.

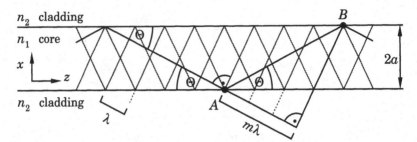

Fig. 13.6. Plane model of a light waveguide. The shaded, inclined lines indicate wave fronts of the mode considered.

two equations $\sin \Theta = 2a/\overline{AB}$ and $\cos 2\Theta = m\lambda/\overline{AB}$ (m integer) yields the quadratic equation:

$$\sin^2 \Theta + \frac{m\lambda}{4a} \sin \Theta - \frac{1}{2} = 0. \tag{13.1}$$

Its solutions are

$$\sin \Theta_m = -\frac{m\lambda}{8a} \pm \sqrt{\left(\frac{m\lambda}{8a}\right)^2 + \frac{1}{2}}, \tag{13.2}$$

with the additional condition $\Theta_m < \Theta_c$. The maximum number M of modes allowed depends on the parameters λ, $2a$, n_1, n_2, and the length of the fiber. The waves at the discrete angles Θ_m are the propagating modes of the fiber. For $M = 1$, a single-mode fiber is present, since only a single mode can be propagated.

The different modes possess different propagation velocities along the fiber, as the corresponding waves travel along different paths. This effect is called mode dispersion. When a signal is coupled into a multimode fiber, usually all modes are excited simultaneously. Therefore, a short pulse at the input broadens along the fiber, depending on the path length (Fig. 13.7). This limits the highest allowed bit rate or clock rate of communication. The problem of mode dispersion can be circumvented with the single-mode fiber. Because of the small core diameter, technical difficulties arise due to the necessary precision elements. Therefore often graded-index fibers are used in less demanding applications. Because the refractive index decreases with the distance from the axis, modes with larger Θ_m propagate a longer time in an area of lower refractive index, that is, they propagate at a higher velocity c_0/n, than do near-axis modes. That way, the larger path lengths can be almost compensated for by a higher velocity. With a suitable radial dependence of the refractive index, mode dispersion can be considerably reduced: from typically about 50 ns km^{-1} with step-index fibers to typically 0.5 ns km^{-1} with graded-index fibers. Besides mode dispersion, the usual chromatic dispersion is present, which is given by the dependence of the propagation velocity on

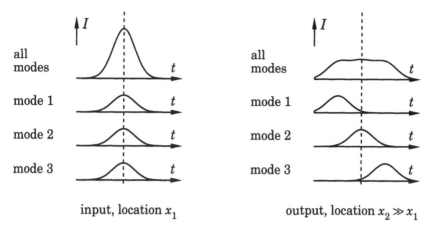

input, location x_1 output, location $x_2 \gg x_1$

Fig. 13.7. Illustration of pulse broadening by mode dispersion. Due to the different propagation velocities of the modes along the fiber, the different modes of a pulse reach the output at different times and superimpose to form a broadened pulse.

the wavelength. It is usually much smaller than the mode dispersion of a multimode fiber. Taking advantage of nonlinear effects, we may almost completely compensate for the chromatic dispersion. How that works is discussed in Sect. 13.3 on solitons.

13.1.3 Attenuation

An important quantity is the attenuation of the light wave as it propagates along the fiber. Fibers only become interesting for communication purposes at extremely low optical losses. Values as low as 0.2 dB km^{-1} at $\lambda \approx 1.55$ μm have been reached. The peculiar unit has practical reasons. The decibel (dB) is a logarithmic unit most valuable with exponential laws and large ranges. It expresses the ratio of two quantities on a logarithmic scale. For intensities or powers it is defined by

$$\frac{I_1}{I_2} \, \mathrm{dB} = 10 \, \log_{10} \frac{I_1}{I_2}. \tag{13.3}$$

The factor 10 in the definition occurs by historical convention. We easily see that a factor of two in intensity ($I_1 = 2I_2$) gives 3 dB, a factor of 1/2 gives −3 dB. No loss in intensity upon propagation is expressed by 0 dB. Equipped with this knowledge, we immediately see that 0.2 dB km^{-1} is equivalent to saying that the intensity of the light drops to half its value after 15 km of travel along the fiber (attenuation constants are expressed as positive numbers).

We now take a short look at the attenuation mechanisms. Figure 13.8 shows the optical losses of a Ge-doped multimode fiber versus the wavelength λ. The strong attenuation in the short wavelength region is due to

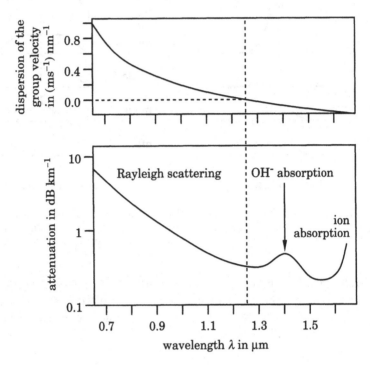

Fig. 13.8. Attenuation and group velocity dispersion(GVD, see Sect. 13.3.1 below) of a Ge-doped multi-mode fiber in dependence on wavelength.

Rayleigh scattering, which increases as $1/\lambda^4$ with decreasing wavelength. In the visible, the attenuation is already large, too large for optical communication over large distances. At $\lambda \approx 1.4\,\mu$m a local maximum of attenuation is found. It is caused by vibrational modes of OH$^-$ ions that tend to be present as impurities in the fiber. The increase in absorption in the infrared region $\lambda > 1.6\,\mu$m is also caused by ion vibrations. Thus, between these regions of large damping, two local minima are found, one at $\lambda = 1.55\,\mu$m with attenuation down to 0.2 dB km^{-1} (depending on the degree of OH$^-$ contamination) and a second, less pronounced one at $\lambda = 1.3\,\mu$m with attenuation down to about 0.6 dB km^{-1}.

Thus, optical communication in fibers has at its disposal just two narrow wavelength regions around $\lambda = 1.3\,\mu$m and $\lambda = 1.55\,\mu$m. Special semiconductor lasers (GaAs, GaAlAs) have been developed which optimally emit at these wavelengths. Source–fiber systems optimized that way reach extremely high bit rates of more than 100 Gbit s^{-1} over many kilometers of fiber.

Today, fibers already transport a large part of telephone signals. Many countries have connected their big cities with fiber links. Computer networks progressively use fibers due to the demands of high transmission capacity. The FDDI network (fiber distributed data interface) for connecting computers operates at a data rate of 1 Gbit s^{-1}, a factor of 10 up on the conventional Ethernet but short of the potential available optically. Networks with a data rate of 10 Gbit s^{-1} are under construction. By combining 160 signals of 10 Gbit s^{-1} each at closely spaced carrier wavelengths between 1.53 μm and 1.60 μm (a technology called dense wavelength-division multiplexing, DWDM) data transmission rates as high as 1.6 TBit s^{-1} have been achieved on a single fiber. In Europe field tests with 160 GBit s^{-1} were successfully completed as communication links. Networks of more than 1 Tbit s^{-1} for a single channel are presently being tested. They are based on optical solitons in fibers. These are stable-wave pulses with durations down to a few femtoseconds (see Sect. 13.3 on optical solitons). The theoretical limit in the optimal wavelength range (1.2 μm – 1.6 μm) is about 50 TBit s^{-1} [13.2].

13.2 Fiber Sensors

Fibers are suited not only for communication with light, but also as sensors for measuring quantities that influence the light propagation in the fiber (intrinsic fiber sensors) or between two fibers (extrinsic fiber sensors) (Fig. 13.9). Usually, the change in amplitude (intensity) or phase of the light is used for measurement, less frequently the wavelength, polarization, or spatial modes are used. Quantities as different as pressure, temperature, rotation, voltage, and electric and magnetic fields may be measured without problems. The art of measurement has thus gained a new thrust by fiber optics.

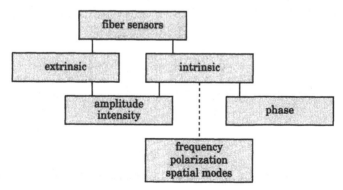

Fig. 13.9. Classification of fiber-optic sensors according to the characteristic quantity of light influenced.

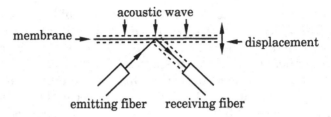

Fig. 13.10. Principle of an optical microphone by modulation of the intensity coupled in into a receiving fiber.

An example of an extrinsic fiber-optic sensor is given in Fig. 13.10: an optical microphone. A thin membrane is set into oscillation by a sound wave. The back of the membrane is illuminated with laser light via a fiber. The reflected laser beam is coupled into a receiving fiber. When the membrane is displaced, the light beam is deflected and causes a corresponding intensity variation in the receiving fiber as now less light is coupled in. This arrangement may be miniaturized, allowing the production of tiny microphones. Moreover, the optical microphone may be incorporated into integrated optical circuits as a sensor to the outside world for robot applications.

The membrane of the optical microphone can be dispensed with when measuring sound or shock waves in liquids. The plane-cut end of a bare fiber forms the sensitive area of a fiber-optic hydrophone [13.3], shown in Fig. 13.11. Its principle of operation is as follows. Laser light is coupled into fiber 1 and guided via a fiber coupler (port 2) to the end of the fiber probe. There it is partly reflected, the reflection coefficient R depending

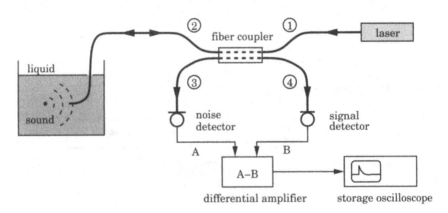

Fig. 13.11. Basic layout of a fiber-optic hydrophone.

on the refractive index of the liquid, n_L, and of the fiber core, n_c:

$$R = \left(\frac{n_c - n_L}{n_c + n_L} \right)^2 . \tag{13.4}$$

As the refractive index of the liquid n_L, in turn, depends on the density and thus on the pressure in the liquid, the intensity of the wave reflected back into the fiber is a measure of the pressure at the fiber tip. This wave is directed to a photodiode (signal detector) at port 4 of the fiber coupler. To reduce errors from fluctuations and noise of the laser, a part of the incoming laser light is coupled to port 3 where it is detected by a second photodiode (noise detector). The difference of the two detector signals is amplified and displayed on an oscilloscope, for example. After suitable calibration of the hydrophone and deconvolution of the recorded signal, the pressure $p(t)$ at the fiber tip can be determined.

While a fiber-optic hydrophone is not particularly sensitive, compared to conventional hydrophones, it features a fast response (mainly limited by the bandwidth of the detectors and electronics) and small detector area. Thus it is well suited to measure high-frequency ultrasound and acoustic shock waves. An example is shown in Fig. 13.12. A pulsed laser beam is focused into a cuvette containing water. By optical breakdown a bubble is generated from the plasma focus. The bubble expands and, after about 450 μs, collapses sharply. The hydrophone's fiber tip is placed at a short distance to the focus. Both the explosive generation of the bubble and the violent collapse are accompanied by the emission of an acoustic shock wave (compare Fig. 7.16) that is detected by the device (peaks B and C).

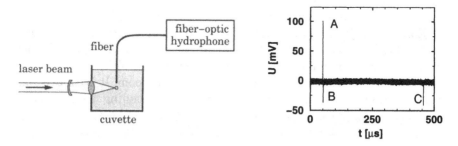

Fig. 13.12. *Left:* Experimental arrangement for investigation of cavitation bubbles in a liquid produced by optical breakdown; *right:* unprocessed detector signal from the fiber-optic hydrophone showing the generating laser pulse (peak A) and the acoustic shock waves emitted at plasma spot formation (peak B) and bubble collapse (peak C).

Further methods of acoustic-field sensing with optical fibers, for instance, those utilising the phase changes in a Mach–Zehnder interferometer, may be found in [13.4]. An overview on fiber sensors is given in [13.5, 13.6].

13.3 Optical Solitons

In analogue communications we attempt to realize transmission lines with properties as linear as possible. Nonlinearities lead to the interaction of waves of different frequencies and thus to crosstalk. The signals can no longer be transmitted independently of each other in different frequency channels.

However, nonlinear properties of a medium in conjunction with the dispersion usually present may lead to waves of quite unusual properties, called solitons. This name has come into use for a pulse-like wave that propagates without significant change of shape in a nonlinear, dispersive medium. The notion soliton suggests particle-like properties, similarly to the notions photon and electron. Solitons show an astonishing stability. In a suitable soliton-bearing medium a single soliton propagates without broadening and also keeps its form after interaction with a second or more solitons. This property of solitons was first found in numerical calculations on the behavior of surface waves in shallow water described by the Korteweg–de Vries equation [13.7]. Solitons, then called the great wave of translation, were observed in 1844 by *John Scott Russell* (1808–1882) in the form of surface waves in a canal. Today, solitons are an important part of nonlinear physics. In optics, they are found in fibers as envelope solitons and are proposed as the unit of information in digital optical communication and information processing [13.8, 13.9, 13.10].

Solitons owe their existence to the combined effects of dispersion and nonlinearity. Dispersion alone leads to broadening of a wave packet, nonlinearity alone provides for a steepening of a wave front. An acoustic example for the action of nonlinearity is the sonic boom. It is conceivable that a suitable combination of broadening and steepening, that is, of dispersion and nonlinearity, may lead to a wave with a stable form. How this phenomenon is cast into theoretic form will now be discussed.

13.3.1 Dispersion

It is known that optical media show chromatic dispersion. By this we mean that the speed of light, c, depends on its frequency ω: $c = c(\omega)$. This formulation implies the use of harmonic waves that possess, according to their frequency, their own propagation velocity $c(\omega)$, the phase velocity.

We came across this in the chapter on the fundamentals of wave optics. For the phase velocity of a harmonic wave we always have

$$c = \nu\lambda = \frac{\omega}{k}. \tag{13.5}$$

This relation, quite naturally, also holds when a different phase velocity c has to be assigned to each frequency ν or circular frequency ω. The dependence

$$\omega = c(\omega)k \quad \text{or} \quad k = \frac{\omega}{c(\omega)} = k(\omega) \tag{13.6}$$

is called the dispersion relation of the material in question. There exists a unique connection between the dispersion relation and the propagation equation belonging to it. Taking a harmonic wave,

$$E(z,t) = E_0 \exp\left[i(kz - \omega t)\right], \tag{13.7}$$

we have

$$\frac{\partial E}{\partial z} = ikE \quad \text{and} \quad \frac{\partial E}{\partial t} = -i\omega E, \tag{13.8}$$

and thus may relate

$$\frac{\partial}{\partial z} \leftrightarrow ik \quad \text{and} \quad \frac{\partial}{\partial t} \leftrightarrow -i\omega. \tag{13.9}$$

The wave equation,

$$\frac{\partial^2 E}{\partial z^2} - \frac{1}{c^2}\frac{\partial^2 E}{\partial t^2} = 0, \tag{13.10}$$

then corresponds to the dispersion relation

$$\omega^2 = c^2(\omega)k^2 \quad \text{or} \quad k^2 = \frac{\omega^2}{c^2(\omega)} \tag{13.11}$$

and vice versa. This relation is equivalent to (13.6) when only positive quantities are considered, that is, when we have propagation of the wave in the positive z-direction as assumed in (13.7).

The refractive index, n, of a medium is given by the ratio of the speed of light in a vacuum, c_0, to the speed of light (phase velocity) in the medium, c:

$$n = \frac{c_0}{c}. \tag{13.12}$$

When the speed of light in the medium depends on the frequency, this implies a dependence of the refractive index on the frequency of the light wave:

$$n(\omega) = \frac{c_0}{c(\omega)}. \tag{13.13}$$

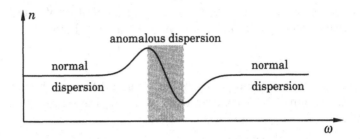

Fig. 13.13. Typical dependence of a dispersion curve $n(\omega)$.

We detect that with an increasing refractive index the phase velocity decreases. A typical dependence $n(\omega)$ is shown in Fig. 13.13. The form of the curve will not be derived here, but is considered as a given experimental fact. In those regions where the refractive index n increases with frequency, that is, where the phase velocity c decreases, we speak of normal dispersion, whereas in the region with a decreasing refractive index and correspondingly increasing phase velocity, we speak of anomalous dispersion.

We now consider the action of dispersion upon the propagation of a wave mixture of different frequencies. We are interested in pulses – no information can be transmitted by a harmonic wave – and take a wave pulse with a Gaussian envelope (Fig. 13.14), as we did in Chapter 11.

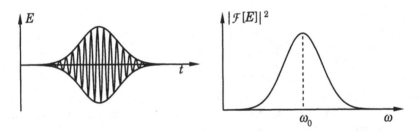

Fig. 13.14. A Gaussian wave pulse and its power spectrum.

Its power spectrum also has a Gaussian profile with its maximum at the circular frequency ω_0. This frequency is called the carrier frequency of the pulse. A wave pulse can be considered as a modulated harmonic wave and may be formulated as

$$E(z,t) = E_e(z,t) \exp\left[i(k_0 z - \omega_0 t)\right] . \tag{13.14}$$

Here, $E_e(z,t)$ is usually, compared to the carrier wave, a slowly varying amplitude function, called an envelope function.

We write $E(z,t)$ as a Fourier transform, ignoring a prefactor of $(1/2\pi)^2$, and set $\Delta k = k - k_0$, $\Delta\omega = \omega - \omega_0$:

$$E(z,t) = \int\limits_{-\infty}^{+\infty} \int\limits_{-\infty}^{+\infty} E_0(\omega,k) \exp\left[i(kz - \omega t)\right] \, \mathrm{d}\omega \, \mathrm{d}k,$$

$$= \int\limits_{-\infty}^{+\infty} \int\limits_{-\infty}^{+\infty} E_{0\Delta}(\Delta\omega, \Delta k) \exp\left[i(\Delta kz - \Delta\omega t)\right] \exp\left[i(k_0 z - \omega_0 t)\right] \, \mathrm{d}(\Delta\omega) \, \mathrm{d}(\Delta k),$$

$$= \left(\int\limits_{-\infty}^{+\infty} \int\limits_{-\infty}^{+\infty} E_{0\Delta}(\Delta\omega, \Delta k) \exp\left[i(\Delta kz - \Delta\omega t)\right] \, \mathrm{d}(\Delta\omega) \, \mathrm{d}(\Delta k) \right) \exp\left[i(k_0 z - \omega_0 t)\right],$$

$$= E_e(z,t) \exp\left[i(k_0 z - \omega_0 t)\right] . \tag{13.15}$$

We detect that $E_e(z,t)$, too, can be given by a Fourier transform with the spectral amplitudes $E_{0\Delta}(\Delta\omega, \Delta k)$. Suppose that the envelope consists of one spectral component $E_{0\Delta}(\Delta\omega, \Delta k)$ only. Then $E_e(z,t)$ is just the harmonic wave

$$E_e(z,t) = E_{0\Delta}(\Delta\omega, \Delta k) \exp\left[i(\Delta kz - \Delta\omega t)\right] . \tag{13.16}$$

Its propagation velocity is given by

$$c_g = \frac{\Delta\omega}{\Delta k} = \frac{\mathrm{d}\omega}{\mathrm{d}k}, \tag{13.17}$$

when $\Delta\omega \ll \omega_0$ and $\Delta k \ll k_0$. This velocity is called the group velocity c_g. When c_g does not depend on the frequency, every wave group given by (13.15) propagates without a change in shape. This holds true because with

$$\frac{\partial E_e}{\partial z} = i\,\Delta k E_e \quad \text{and} \quad \frac{\partial E_e}{\partial t} = -i\,\Delta\omega E_e \tag{13.18}$$

we see that $E_e(z,t)$ obeys the linear wave equation with (13.17) as the dispersion relation. Thus, the group velocity is the phase velocity of the envelope of a wave packet. When chromatic dispersion is present, ω is not proportional to k and the phase velocity $c = \omega/k$ does not equal the group velocity $c_g = \mathrm{d}\omega/\mathrm{d}k$. However, when $c = \text{const}$, then the phase and group velocities coincide, $c = c_g$.

The refractive index is defined by $n = c_0/c$. When working with short pulses and trying to measure n, we obtain a different value in the case $c_g \neq c$. Therefore a group velocity refractive index is defined:

$$n_g = \frac{c_0}{c_g}. \tag{13.19}$$

The group velocity, c_g, again may depend on the frequency, as can be seen from the graph of the refractive index in Fig. 13.13. Neither n, nor $\mathrm{d}n/\mathrm{d}\omega$ is constant. To obtain solitons, this contribution to dispersion has also to be taken into account. It is called the group velocity dispersion, abbreviated as GVD. It is of utmost importance for ultrashort pulse generation and propagation. In Fig. 13.8 above, we have already plotted the group velocity dispersion for a fiber.

To arrive at a theoretical expression for the group velocity dispersion, we expand the dispersion relation in the form $k = k(\omega)$ at ω_0 and stop after the second term:

$$
\begin{aligned}
k - k_0 &= \left.\frac{dk}{d\omega}\right|_{\omega_0} (\omega - \omega_0) + \frac{1}{2}\left.\frac{d^2k}{d\omega^2}\right|_{\omega_0} (\omega - \omega_0)^2 \\
&= k'(\omega - \omega_0) + \frac{1}{2}k''(\omega - \omega_0)^2 .
\end{aligned}
\tag{13.20}
$$

Since the envelope $E_e(z,t)$ may be represented as a superposition of harmonic waves, we proceed as before and set

$$
\Delta k \leftrightarrow -i\frac{\partial}{\partial z} \quad \text{and} \quad \Delta\omega \leftrightarrow i\frac{\partial}{\partial t}
\tag{13.21}
$$

to obtain the wave equation corresponding to the dispersion relation (13.20). Because of $k' = 1/c_g$ we have

$$
i\left(\frac{\partial}{\partial z} + \frac{1}{c_g}\frac{\partial}{\partial t}\right) E_e(z,t) - \frac{1}{2}k''\frac{\partial^2}{\partial t^2}E_e(z,t) = 0 .
\tag{13.22}
$$

This equation attains a simpler form in a coordinate system moving with the group velocity c_g of the wave, that is, upon the coordinate transformation

$$
\zeta = z, \quad \text{and} \quad \tau = t - \frac{1}{c_g}z .
\tag{13.23}
$$

Then the propagation equation for the envelope of a wave pulse reads

$$
i\frac{\partial E_e}{\partial \zeta} - \frac{k''}{2}\frac{\partial^2 E_e}{\partial \tau^2} = 0 .
\tag{13.24}
$$

The solutions of this equation broaden when $k'' \neq 0$. The quantity k'' can be considered as a measure of the group velocity dispersion, since

$$
k''(\omega_0) = \frac{d}{d\omega}(k'(\omega))|_{\omega_0} = \frac{d}{d\omega}\left(\frac{1}{c_g(\omega)}\right)\Bigg|_{\omega_0} = -\frac{1}{c_g^2(\omega_0)}\left.\frac{dc_g(\omega)}{d\omega}\right|_{\omega_0} .
\tag{13.25}
$$

From this relation it follows that with $k'' > 0$ the group velocity decreases with frequency and, conversely, increases when $k'' < 0$.

Analogous to the notions of normal and anomalous dispersion of phase velocity, the notions normal and anomalous dispersion of the group velocity are introduced. When $k''(\omega) > 0$ we speak of normal dispersion of the group velocity, when $k''(\omega) < 0$, of anomalous dispersion of the group velocity.

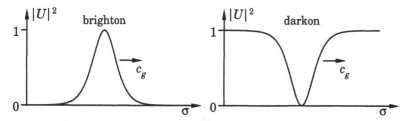

Fig. 13.15. Two special kinds of solitons, the brighton ($\text{sech}^2\sigma$) and the darkon ($\tanh^2\sigma$). A quantity $|U|^2$ which is proportional to the light intensity is plotted.

As we will see below, only in the region of anomalous dispersion of the group velocity do light pulses exist on an otherwise dark background, propagating as solitons. In the region of normal dispersion only solitons that propagate as pulses of reduced intensity exist on an otherwise bright background (Fig. 13.15). We call them brightons and darkons as a shortened form of bright solitons and dark solitons, respectively.

13.3.2 Nonlinearity

The refractive index, n, of a dielectric, transparent medium not only depends on frequency but also on the electric field. In connection with solitons in fibers the optical Kerr effect is of importance, that is, the intensity dependence of the refractive index:

$$n(\omega, |E_0|^2) = n_0(\omega) + n_2|E_0|^2 . \tag{13.26}$$

The Kerr constant, n_2, for fused silica is of the order of $n_2 = 10^{-22}$ m^2 V^{-2}. The Kerr effect is caused by the deformation of the electron cloud of atoms or molecules by the electric field of the light wave. It is extremely fast, of the order of a period of the light wave [13.11].

The nonlinearity (13.26) can be introduced into the dispersion relation (13.20). For that, we need the relation between k and ω in a formulation containing the refractive index n. Because of $n = c_0/c = c_0 k/\omega$, we have

$$k(\omega) = \frac{n\omega}{c_0} = \frac{1}{c_0}\left(n_0(\omega) + n_2|E_0|^2\right)\omega. \tag{13.27}$$

For the carrier wave we set

$$k_0 = \frac{n_0(\omega_0)}{c_0}\omega_0 . \tag{13.28}$$

For $\omega = \omega_0$ the nonlinear dispersion law (13.27), along with (13.28), yields:

$$k(\omega_0) = k_0 + k_0\frac{n_2}{n_0(\omega_0)}|E_0|^2 = k_0 + k_0^{(NL)} . \tag{13.29}$$

The nonlinear term

$$k_0^{(NL)} = k_0 \frac{n_2}{n_0(\omega_0)} |E_0|^2 \tag{13.30}$$

leads, in Δk, to an additional deviation from k_0 of just the amount $k_0^{(NL)}$. Because of the speed of the Kerr effect it is supposed to be instantaneous (adiabatic approximation). When this nonlinear approximation is included, with $|E_0|^2$ changed to $|E_e|^2$, the dispersion equation (13.20) reads:

$$k - k_0 = k_0 \frac{n_2}{n_0(\omega_0)} |E_e|^2 + k'(\omega - \omega_0) + \frac{1}{2} k''(\omega - \omega_0)^2. \tag{13.31}$$

Changing from the dispersion relation to the differential equation by applying rules (13.21) we obtain the propagation equation for the envelope $E_e(z,t)$ of a wave pulse:

$$i\left(\frac{\partial}{\partial z} + \frac{1}{c_g(\omega_0)} \frac{\partial}{\partial t}\right) E_e - \frac{1}{2} k'' \frac{\partial^2}{\partial t^2} E_e = -k_0 \frac{n_2}{n_0(\omega_0)} |E_e|^2 E_e. \tag{13.32}$$

Again we change to a coordinate system which moves with the group velocity $c_g(\omega_0)$. In analogy to (13.24) we obtain for $E_e = E_e(\zeta, \tau)$:

$$i\frac{\partial E_e}{\partial \zeta} - \frac{1}{2} k'' \frac{\partial^2 E_e}{\partial \tau^2} = -k_0 \frac{n_2}{n_0(\omega_0)} |E_e|^2 E_e. \tag{13.33}$$

We simplify the notation by setting

$$U = \sqrt{\frac{k_0 n_2}{n_0(\omega_0)}} E_e, \quad \sigma = \frac{\tau}{\pm\sqrt{|k''(\omega_0)|}}, \tag{13.34}$$

to arrive at the normalized form of the propagation equation for $U = U(\zeta, \sigma)$:

$$i\frac{\partial U}{\partial \zeta} \pm \frac{1}{2} \frac{\partial^2 U}{\partial \sigma^2} + |U|^2 U = 0, \tag{13.35}$$

or, even shorter:

$$iU_\zeta \pm \frac{1}{2} U_{\sigma\sigma} + |U|^2 U = 0. \tag{13.36}$$

The positive sign goes with $k''(\omega) < 0$, the negative one with $k''(\omega) > 0$. This equation has the form of a Schrödinger equation, as known from quantum mechanics, but with the quantity $|U|^2$ instead of a potential. Therefore, it is called the nonlinear Schrödinger equation, often for short, NLS. The equation with positive sign,

$$i\frac{\partial U}{\partial \zeta} + \frac{1}{2} \frac{\partial^2 U}{\partial \sigma^2} + |U|^2 U = 0, \tag{13.37}$$

possesses a solution propagating without change of shape, the fundamental soliton solution:

$$U(\zeta,\sigma) = a\,\text{sech}(a\sigma)\exp(ia^2\zeta/2), \qquad a > 0. \tag{13.38}$$

This may easily be verified by the substitution of (13.38) into (13.37). For that, helpful relations are $(d/d\sigma)\text{sech}\sigma = -\text{sech}\sigma\,\tanh\sigma$, $(d/d\sigma)\tanh\sigma = \text{sech}^2\sigma$ and $\tanh^2\sigma = 1 - \text{sech}^2\sigma$. A particularly simple form is obtained for $a = 1$:

$$U(\zeta,\sigma) = \text{sech}(\sigma)\exp(i\zeta/2). \tag{13.39}$$

The solution (13.38) is a 'bright' soliton or brighton (Fig. 13.15). Its width decreases with increasing amplitude a. Figure 13.16 shows some

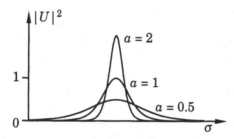

Fig. 13.16. Brightons of different amplitude and width.

brightons of different amplitudes. Therein, the normalized intensity of the soliton, $|U|^2$, is plotted versus σ, the normalized time.

Besides the fundamental soliton (13.38), there exist interacting multi-soliton solutions. Unlike the fundamental solitons they alter in time, but always recur in the same forms. Thus they also are stable. These solitons are called 'breathing' or, equivalently, bound soliton states.

The nonlinear Schrödinger equation (13.35) with negative sign,

$$i\frac{\partial U}{\partial \zeta} - \frac{1}{2}\frac{\partial^2 U}{\partial \sigma^2} + |U|^2\,U = 0, \tag{13.40}$$

possesses more complex solutions compared with the equation with positive sign (13.37). An elementary one-soliton solution is given by

$$U(\zeta,\sigma) = a\tanh(a\sigma)\exp(ia^2\zeta). \tag{13.41}$$

In its simplest form with the amplitude $a = 1$, it reads:

$$U(\zeta,\sigma) = \tanh(\sigma)\exp(i\zeta). \tag{13.42}$$

This solution is plotted in Fig. 13.15. It represents a pulse of reduced intensity and therefore is called a 'dark' soliton or darkon. The darkon, too, gets narrower with increasing amplitude. To increase the amplitude a, the background intensity has to be increased. At the pulse minimum the light intensity is zero.

The Schrödinger solitons of the form (13.38) and (13.41), brightons and darkons, propagate with the group velocity c_g, independently of the amplitude a. However, there exist solutions at any velocity independently of the amplitude. As the shapes $U(\zeta, \sigma)$ of (13.38) and (13.41) are just envelopes of the wave pulses, Schrödinger solitons are also called envelope solitons. In analogy to wave pulses they also might be called wave solitons, as they have an inner wave structure. It is customary to simply speak of 'optical solitons'.

Optical solitons in fibers have a great potential for applications, in particular for fast communication. Electronic systems may stop at a bandwidth of about 40 GHz, whereas a fiber, just in the region of 1.3 µm to 1.6 µm, reaches a bandwidth of 40 THz, that is, one thousandfold of that of an electronic system. To make use of this bandwidth effectively, fast, purely optical switches are needed to keep up with data rates in the Tbit/s region. Such switches should be possible with virtual optical transitions due to their almost instantaneous response. There are no obstacles from physical reasons. At present, the main technical difficulties arise from switching energies that are too high and from problems of cascadibility.

Already at a switching energy of 1 pJ/bit a data rate of 1 Tbit/s requires 1 Watt of power. Lasers with these powers for high bit rates have yet to be developed for the fiber wavelength region. To compensate for losses, amplifiers are needed. This may be achieved with erbium-doped fibers that are pumped to amplify the light pulses traveling along them [13.5, 13.12].

Cascadibility is an important aspect for the building of complex systems out of modular units. The output pulses of photonic devices should have the same properties as the input pulses for data to be processed in a long series of steps. Solitons are the method of choice. They may be switched as a unit and do not suffer from the problem of pulse splitting [13.10]. The fundamental soliton, the brighton, may serve as the unit of information, as 1 bit.

13.4 Fiber-Optic Signal Processing

Light is well suited for (almost) interaction-free measurement of many physical quantities. An example for the measurement of velocity is laser Doppler anemometry (see Sect. 6.4). Two examples for measuring sound pressures have been given in Sect. 13.2. As these quantities are present in the form of modulated light signals, the question may be posed whether the optical signals have immediately to be transformed to electronic signals for further processing or may be processed optically. In some cases this indeed can be done. In particular, fiber-optic sensors are well suited for this purpose, as in the course of the development of telecommunication technology components have been manufactured that may also be used for optical processing. Among them are directional couplers for dividing and combining beams (in free-space optics they are called beam splitters), delay elements, and optical-fiber amplifiers.

With these elements, linear filters and even processors for matrix multiplication can be built. As analogue devices they suffer from known restrictions concerning accuracy and stability. Light that is as incoherent as possible is used to avoid interference effects. That restricts the filter process, since only positive real values can be used. Moreover, fiber techniques are one-dimensional and do not make use of the power inherent in optics, its multidimensionality.

The restrictions concerning the accuracy may be overcome by digitizing the operations. Fiber-optic digital techniques developed in this respect have yet to come a long way. The nonlinear elements necessary for digital operations suffer from technical difficulties [13.12]. In feasibility studies, the mathematical operations of binary addition, subtraction, and multiplication in monomode fibers have been demonstrated.

Problems

13.1 Consider the two-dimensional model of a step-index fiber treated in the text. Derive an expression for the smallest mode number, $N > 0$, for which the corresponding mode can still propagate through the fiber. Calculate N for a fiber with a cladding refractive index of $n_1 = 1.52$, with a core refractive index of $n_2 = 1.62$, and a core diameter of $D = 2a = 50$ µm at the wavelength $\lambda = 1.3$ µm. For such a fiber, what is the maximum propagation time difference per km between different modes?

13.2 A fiber 2 m long is to be used for interferometric sensing of temperature changes. The coefficient of linear thermal expansion of the fiber material is $\alpha = 0.6 \cdot 10^{-6}$ K^{-1}, its refractive index is $n = 1.46$, and the change of refractive index with temperature is given by the coefficient $\mathrm{d}n/\mathrm{d}T = 13 \cdot 10^{-6}$ K^{-1}. Determine the smallest temperature change that can be detected if a phase variation of 0.1 rad of the light wave through the fiber can be resolved.

13.3 Determine the spatial extent of an envelope soliton (brighton) having a duration of 25 ps (FWHM) in a fiber with a refractive index of $n = 1.5$ in units of the vacuum wavelength λ_0, $\lambda_0 = 1.55$ µm. Conversely, what is the pulse duration Δt_h of a soliton having an extension of 100λ? Upon optical transmission of digital data, the detection window for one soliton (1 bit) ought to be about seven times the soliton duration. Give the maximum data transmission rate that can be achieved with solitons of width Δt_h and the associated bandwidth. What distance can such solitons travel in a fiber with a loss of 0.12 dB km^{-1} until their width has doubled?

13.4 A Gaussian wave packet propagates in a medium having the dispersion relation

$$k(\omega) = k_0 + \left.\frac{\mathrm{d}k}{\mathrm{d}\omega}\right|_{\omega_0} (\omega - \omega_0) + \frac{1}{2}\left.\frac{\mathrm{d}^2 k}{\mathrm{d}\omega^2}\right|_{\omega_0} (\omega - \omega_0)^2 \dots .$$

Upon propagation, the width of the packet increases according to

$$t(z) = t_0 \sqrt{1 + 4z^2 (d^2k/d\omega^2)^2 / t_0^4},$$

where t_0 denotes the initial pulse width and $t(z)$ the pulse width at location z. Calculate the distance a wave packet of initial width $t_0 = 10$ ps has traveled in a fiber with $d^2k/d\omega^2 \approx 27$ ps^2 km^{-1} until its width has spread tenfold, $t(z) = 10 t_0$.

13.5 Verify that the brighton (13.38) is a solution of the nonlinear Schrödinger equation (13.37), and also that the darkon (13.41) is a solution of (13.40).

13.6 Write down the brighton solution (13.38) in a non-scaled form for the electric field amplitude as a function of time. A brighton of duration $t_0 = 10$ ps at wavelength $\lambda = 1.5$ μm propagates in a monomode fiber having $|d^2k/d\omega^2| = 20$ ps^2 km^{-1}, a core diameter of $D = 10$ μm, an index of refraction $n_0 = 1.5$, and a Kerr coefficient $n_2 = 1.2 \cdot 10^{-28}$ km^2 V^{-2}. Calculate the maximum amplitude of the electric field of the soliton pulse and its maximum power.

Hint: the maximum intensity is given by $I_{max} = (1/2)\varepsilon_0 n_0 c |E_{max}|^2$.

13.7 Show that the quantity

$$I_0 = \int_{-\infty}^{+\infty} |U|^2 \, d\sigma \tag{13.43}$$

is invariant in time according to the nonlinear Schrödinger equations (13.37) and (13.40). Thereby, assume that the solution U satisfies the boundary conditions $\lim_{\sigma \to \pm\infty} U(\zeta, \sigma) = 0$ und $\lim_{\sigma \to \pm\infty} \partial U(\zeta, \sigma)/\partial\sigma = 0$.

Hint: multiply the NLS equation by U^* and subtract the complex conjugate of the resulting expression.

A. The Fourier Transform

In this appendix, the most important facts pertaining to the Fourier transform are surveyed without proof. For a more detailed exposition of Fourier theory and its applications the reader is referred to one of the numerous textbooks ([A.1]–[A.6]).

A.1 One-Dimensional Fourier Transform

The one-dimensional Fourier transform plays an important part in the analysis of time signals, for example, of the electric field amplitude $E(t)$ at some point in space. The idea is to represent a given signal $f(t)$ as a superposition of harmonic oscillations $\exp(i2\pi\nu t)$. The weighting factors that appear in the sum constitute the complex amplitude spectrum or complex spectral function $H(\nu)$ of the signal. This spectrum is given by the Fourier transform of f:

$$H(\nu) = \mathcal{F}[f(t)](\nu) = \int_{-\infty}^{+\infty} f(t)\exp(-2\pi i\nu t)\,\mathrm{d}t. \qquad (A.1)$$

Conversely, given the spectral function $H(\nu)$ the Fourier representation of the signal reads:

$$f(t) = \int_{-\infty}^{+\infty} H(\nu)\exp(2\pi i\nu t)\,\mathrm{d}\nu. \qquad (A.2)$$

This integral inverts the mapping $f \to H$ of (A.1); that is, it defines the inverse Fourier transform. It differs from (A.1) only by the sign of the argument of the exponential function. In some texts, prefactors of the form $1/\sqrt{2\pi}$ or $1/2\pi$ may be found in the definition of the Fourier transform or its inverse. In the formulae given above such factors do not occur because we integrate over the frequency ν.

Here, the function f has been expanded into a sum of complex exponentials that form an orthonormal function system. To do the same in real notation, the functions 'sin' and 'cos' are used, resulting in the sine and cosine transform, respectively. However, the complex notation appears to be more elegant and also more advantageous, because it leads straight to the notion of the analytic signal.

The Fourier transform can be understood as an invertible, linear map (a linear operator) between function spaces. The existence of an inverse operator is assured by the Fourier integral theorem which states that the relation

$$\mathcal{F}^{-1}(\mathcal{F}[f]) = \mathcal{F}(\mathcal{F}^{-1}[f]) = f \tag{A.3}$$

holds at every point where the function f is continuous.

Mathematical questions not being of our concern here, we suppose in the following that all functions involved have properties (for instance, absolute integrability) sufficient to guarantee the existence of the integrals (A.1) and (A.2). With real signals of finite energy, represented by ordinary, smooth functions, no difficulties arise. Care has to be exercised with other, idealized signals described by not absolutely integrable functions or distributions (the harmonic wave being an elementary example). Then, the relations given above have to be extended by properly including limiting processes, leading to the definition of the generalized Fourier transform. For simplicity, we shall ignore this delicacy in the following and, wherever possible, manipulate all functions as though they were ordinary, finite functions.

Often, in an experiment the power spectrum $W(v)$ of a time signal $f(t)$ is measured. It is defined by

$$W(v) = |\mathcal{F}[f](v)|^2 . \tag{A.4}$$

The power spectrum contains information only on the amplitudes of the harmonic components, the phase information is lost.

A.2 Two-Dimensional Fourier Transform

The Fourier transform of functions depending on two variables x, y is defined analogously to the one-dimensional case (A.1):

$$\tilde{f}(v_x, v_y) = \mathcal{F}[f(x,y)](v_x, v_y) = \int\limits_{-\infty}^{+\infty} \int\limits_{-\infty}^{+\infty} f(x,y) \exp\left[-2\pi i(v_x x + v_y y)\right] \, \mathrm{d}x \, \mathrm{d}y. \tag{A.5}$$

Up to the sign of the argument of the exponential the inverse transformation $\mathcal{F}^{-1}$ has the same form:

$$f(x,y) = \mathcal{F}^{-1}[\tilde{f}(v_x, v_y)](x,y)$$
$$= \int\limits_{-\infty}^{+\infty} \int\limits_{-\infty}^{+\infty} \tilde{f}(v_x, v_y) \exp\left[+2\pi i(x v_x + y v_y)\right] \, \mathrm{d}v_x \, \mathrm{d}v_y. \tag{A.6}$$

If the function f can be factorized, that is, if it can be written as a product $f(x,y) = f_1(x)f_2(y)$, then also the Fourier transform (A.5) decomposes into a product of two one-dimensional transforms:

$$\mathcal{F}[f_1(x)f_2(y)](\nu_x, \nu_y) = \mathcal{F}[f_1(x)](\nu_x)\mathcal{F}[f_2(y)](\nu_y). \tag{A.7}$$

One of the two functions may be a distribution, but the product of two distributions in general is not defined. Therefore, the decomposition of a distribution into a product is not correct in a strict sense but often is done formally to facilitate calculations. For example, the two-dimensional δ function sometimes is written as a product of two one-dimensional δ functions.

As before, the power spectrum of a function f is given by

$$W(\nu_x, \nu_y) = |\mathcal{F}[f(x,y)]|^2 (\nu_x, \nu_y). \tag{A.8}$$

For example, the intensity distribution in the back focal plane of a positive lens is given by the power spectrum of the field distribution f present in the front focal plane.

A.3 Convolution and Autocorrelation

The convolution operation assigns two given functions f and g another function h. In two dimensions, the definition of the convolution product reads:

$$h(\xi, \eta) = [f(x,y) * g(x,y)](\xi, \eta)$$
$$= \int\limits_{-\infty}^{+\infty} \int\limits_{-\infty}^{+\infty} f(x,y)g(\xi - x, \eta - y)\,\mathrm{d}x\,\mathrm{d}y. \tag{A.9}$$

As before, we assume that the integral exists.

The convolution operation is commutative, $(f_1 + f_2) * g = f_1 * g + f_2 * g$, and distributive, $(f_1 + f_2) * g = f_1 * g + f_2 * g$. The convolution product is of great significance for linear systems theory. The output signal $a(t)$ of a linear system having an impulse response function h and being fed with an input signal $e(t)$ (see Fig. A.1) is given by

$$a(t) = (e * h)(t). \tag{A.10}$$

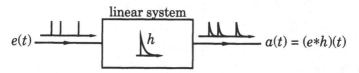

Fig. A.1. The output signal $a(t)$ of a linear system corresponds to the convolution product of the input signal $e(t)$ and the impulse response function h.

Many optical systems, even a region of empty space traversed by a light wave, can be considered as a linear system and thus can be assigned an impulse response function (see the chapter on Fourier optics).

The autocorrelation of a function f, written symbolically as $f \otimes f$, is defined by

$$(f \otimes f)(\xi, \eta) = \int\limits_{-\infty}^{+\infty} \int\limits_{-\infty}^{+\infty} f(x,y) f^*(x - \xi, y - \eta) \, dx \, dy. \tag{A.11}$$

This expression is related to the convolution product (A.9), because for a real-valued function we have $f \otimes f = (f * f)^{(-)}$. Here, $f^{(-)}$ denotes the function obtained from f by reflecting the coordinates at the origin, $f^{(-)}(x,y) = f(-x,-y)$. A connection between the autocorrelation of a function and its Fourier transform is established by the Wiener–Khinchin theorem. It states that the autocorrelation and the power spectrum of a function are Fourier transforms of one another.

A.4 Properties of the Fourier Transform

The Fourier transform is closely related to the convolution product. Also, certain rules apply concerning its interchangeability with other operations like translation, scaling or differentiation of a function. In the following, written for the two-dimensional Fourier transform, these properties are briefly reviewed.

1. Linearity
   The Fourier transform is linear (a linear operator). For arbitrary real or complex constants a, b it holds that

   $$\mathcal{F}[a \cdot f + b \cdot g](\nu_x, \nu_y) = a\mathcal{F}[f](\nu_x, \nu_y) + b\mathcal{F}[g](\nu_x, \nu_y). \tag{A.12}$$

2. Parseval's Identity
   The total energy of a function, measured as the integral over $|f|^2$, is equal to the total energy of its spectrum, given as the integral over the power spectrum $|\mathcal{F}[f]|^2$:

   $$\int\limits_{-\infty}^{+\infty} \int\limits_{-\infty}^{+\infty} |f(x,y)|^2 \, dx \, dy = \int\limits_{-\infty}^{+\infty} \int\limits_{-\infty}^{+\infty} |\mathcal{F}[f](\nu_x, \nu_y)|^2 \, d\nu_x \, d\nu_y. \tag{A.13}$$

3. Complex Conjugation
   The prescription of how to interchange Fourier transform and complex conjugation reads:

   $$\mathcal{F}[f^*(x,y)](\nu_x, \nu_y) = (\mathcal{F}[f(x,y)])^*(-\nu_x, -\nu_y). \tag{A.14}$$

4. Shift Theorem
   A shift of the function f along the x- and y-axes by an amount Δx and Δy, respectively, results in a phase factor in the amplitude spectrum

of the shifted function:

$$\mathcal{F}[f(x + \Delta x, y + \Delta y)](\nu_x, \nu_y)$$

$$= \exp\left[2\pi i(\nu_x \Delta x + \nu_y \Delta y)\right] \mathcal{F}[f(x,y)](\nu_x, \nu_y). \tag{A.15}$$

The shift does not affect the power spectrum.

5. Scaling Theorem

A scaling of the independent variables x, y of a function f effects a reciprocal scaling of the Fourier spectrum. Because of Parseval's identity, the amplitude of the spectrum also changes,

$$\mathcal{F}[f(ax, by)](\nu_x, \nu_y) = \frac{1}{|a| \cdot |b|} \mathcal{F}[f(x,y)]\left(\frac{\nu_x}{a}, \frac{\nu_y}{b}\right). \tag{A.16}$$

6. Convolution Theorems

The Fourier transform of a product of two functions f and g equals the convolution product of the individual spectra:

$$\mathcal{F}[f(x,y) \cdot g(x,y)](\nu_x, \nu_y) = \mathcal{F}[f(x,y)] * \mathcal{F}[g(x,y)]. \tag{A.17}$$

Conversely, the amplitude spectrum of a convolution product of two functions is identical to the product of their spectra,

$$\mathcal{F}[f(x,y) * g(x,y)](\nu_x, \nu_y) = \mathcal{F}[f(x,y)](\nu_x, \nu_y) \cdot \mathcal{F}[g(x,y)](\nu_x, \nu_y). \tag{A.18}$$

7. Correlation Theorems

The correlation theorems connect the Fourier transform of a product of two functions f and g with the crosscorrelation function of their spectra:

$$\mathcal{F}[f^*(x,y) \cdot g(x,y)](\nu_x, \nu_y) = \mathcal{F}[f(x,y)] \otimes \mathcal{F}[g(x,y)]. \tag{A.19}$$

Conversely, we have:

$$\mathcal{F}[f(x,y) \otimes g(x,y)](\nu_x, \nu_y) = \mathcal{F}[f(x,y)](\nu_x, \nu_y) \cdot \mathcal{F}[g(x,y)]^*(\nu_x, \nu_y). \tag{A.20}$$

8. Differentiation

In the Fourier spectrum, the operator $\partial/\partial x$ carries over to a multiplication by $i2\pi\nu_x$,

$$\mathcal{F}[\frac{\partial}{\partial x} f(x,y)](\nu_x, \nu_y) = i2\pi\nu_x \mathcal{F}[f(x,y)](\nu_x, \nu_y). \tag{A.21}$$

Similarly, the operator $\partial/\partial y$ corresponds to a multiplication of the spectrum by $i2\pi\nu_y$.

These rules allow us to effectively calculate the Fourier transforms of functions that are composed of or can be derived from other, simpler functions.

In such cases, results previously obtained can be used and the Fourier integral can be computed more efficiently. We take advantage of this fact in the chapter on Fourier optics.

A.5 Selected Functions and Their Fourier Transforms

The temporal or spatial dependence of physical quantities often is described mathematically by idealized wave forms. For example, a purely harmonic wave does not exist in nature. Nevertheless, in many cases the functions $\exp(-i\omega t)$ or $\exp[i(kx - \omega t)]$ are very good approximations to reality. The Dirac delta distribution δ is used to describe, in an idealized form, events or objects localized in time or space, such as a short pulse (one-dimensional) or a point source of light (two-dimensional). It may be helpful to visualize the δ function as a function that is infinite at one single point and is zero everywhere else. However, it should be kept in mind that, strictly speaking, the symbol δ or the formal integration $\int \delta(x)f(x)\,dx$ is just a shorthand notation for the functional mapping

$$\delta(x){:}f \mapsto f(0) = \int \delta(x)f(x)\,dx. \tag{A.22}$$

Because of the (distribution) identity,

$$\delta(\nu) = \int_{-\infty}^{+\infty} \exp(i2\pi\nu t)\,dt, \tag{A.23}$$

the generalized Fourier transform of the δ functional yields a constant spectrum, the constant being one. From δ functions, other useful functions can be composed, for instance, the comb function. It is well suited for representing periodic structures by a convolution product.

In Fig. A.5 a selection of one-dimensional functions occurring in this book is presented together with their Fourier transforms. Some examples of two-dimensional Fourier transforms are worked out in the chapter on Fourier optics.

Problems

A.1 Verify the expression given in Fig. A.5 for the Fourier transform of the Gaussian function $f(t) = \exp(-\pi t^2)$. From this, determine the Fourier transform of a rescaled Gaussian pulse, $f_a(t) = \exp(-a\pi t^2)$, $a > 0$.

A.2 Calculate the amplitude spectrum of the triangular function

$$\Lambda(t) = \begin{cases} 1 - |t| & \text{for} \quad |t| \le 1, \\ 0 & \text{otherwise.} \end{cases} \tag{A.24}$$

A.3 Determine the autocorrelation function of a square pulse, rect(t). For this example, verify the Wiener–Chintchin theorem using the result of the previous problem and the Fourier transform of a square pulse as given in Fig. A.5.

A.4 Prove the convolution theorems for the one-dimensional Fourier transform and convolution.

A.5 Determine the amplitude spectrum of a cosine oscillation modulated by a sine function: $f(t) = A\sin(2\pi\nu_1 t)\cos(2\pi\nu_2 t)$. Use the table of Fig. A.5 and the convolution theorem.

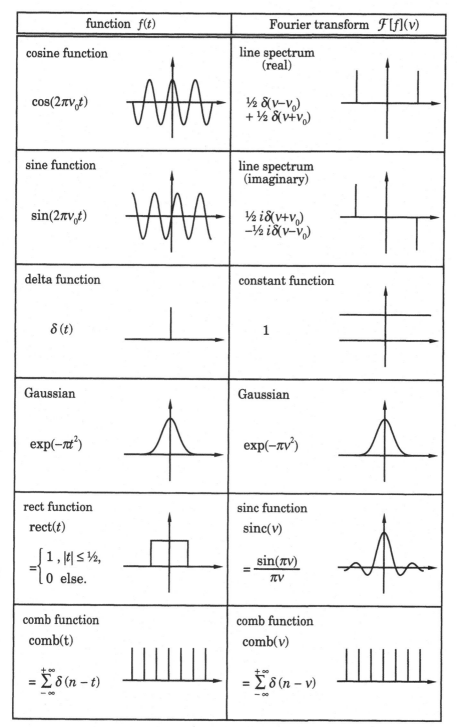

function $f(t)$	Fourier transform $\mathcal{F}[f](\nu)$
cosine function $\cos(2\pi\nu_0 t)$	line spectrum (real) $\tfrac{1}{2}\,\delta(\nu-\nu_0)$ $+\tfrac{1}{2}\,\delta(\nu+\nu_0)$
sine function $\sin(2\pi\nu_0 t)$	line spectrum (imaginary) $\tfrac{1}{2}\,i\delta(\nu+\nu_0)$ $-\tfrac{1}{2}\,i\delta(\nu-\nu_0)$
delta function $\delta(t)$	constant function 1
Gaussian $\exp(-\pi t^2)$	Gaussian $\exp(-\pi \nu^2)$
rect function $\mathrm{rect}(t)$ $=\begin{cases}1\,,\ \lvert t\rvert \le \tfrac{1}{2},\\ 0\ \text{else.}\end{cases}$	sinc function $\mathrm{sinc}(\nu)$ $=\dfrac{\sin(\pi\nu)}{\pi\nu}$
comb function $\mathrm{comb}(t)$ $=\displaystyle\sum_{-\infty}^{+\infty}\delta(n-t)$	comb function $\mathrm{comb}(\nu)$ $=\displaystyle\sum_{-\infty}^{+\infty}\delta(n-\nu)$

Fig. A.2. Some frequently occurring functions and their Fourier transforms.

B. Solutions of Problems

Chapter 1

1.1. Using $\nu = c/\lambda$, $E = hc/\lambda$, $p = h/\lambda$, and $m = h/(\lambda c)$ we obtain for a photon of wavelength $\lambda = 550$ nm: $\nu = 5.45 \cdot 10^{14}$ Hz, $E = 3.61 \cdot 10^{-19}$ J, $p = 1.20 \cdot 10^{-27}$ Ns, and $m = 4.02 \cdot 10^{-36}$ kg.

1.2. (a) Substituting the relations $\lambda = c/\nu$ and $|d\lambda| = (c/\lambda^2)|d\nu|$ into (1.2) yields:

$$\rho(\lambda)\,d\lambda = \frac{8\pi hc}{\lambda^5} \frac{1}{\exp\left(hc/\lambda kT\right) - 1}\,d\lambda.$$

(b) At an extremum (here: maximum) of the function $\rho(\nu)$ its first derivative must vanish,

$$\frac{d\rho(\nu)}{d\nu} = \frac{8\pi h}{c^3} \frac{\nu^2}{\left[\exp\left(h\nu/kT\right) - 1\right]^2} \left[\exp\left(\frac{h\nu}{kT}\right)\left(3 - \frac{h\nu}{kT}\right) - 3\right] = 0.$$

Since the solution $\nu = 0$ is not of interest here, the right-hand term in brackets should vanish. With the abbreviation $z = h\nu/kT$ we obtain $e^z(3 - z) = 3$, and thus, using the root $z = 2.821$ given in the text, we find:

$$\nu_{max} = 2.821\frac{kT}{h} = (5.88 \cdot 10^{10} \text{ K}^{-1} \text{ s}^{-1})\cdot T.$$

A similar calculation for $\rho(\lambda)$, this time with $z = hc/\lambda kT$, leads to the condition $e^z(5 - z) = 5$. Substituting the root $z = 4.965$ into it leads to the usual form of *Wien's* displacement law:

$$\lambda_{max} = \frac{hc}{4.965 \cdot kT} = (2.898 \cdot 10^{-3} \text{ Km})\cdot \frac{1}{T}.$$

The fact that ν_{max} does not correspond to λ_{max}, that is, $\nu_{max} \neq c/\lambda_{max}$, is due to the nonlinearity of the coordinate transformation $\nu \leftrightarrow \lambda$.

(c) Integrating $\rho(\nu)$ yields:

$$\rho_t = \int_0^\infty \rho(\nu)\,d\nu = \frac{8\pi h}{c^3}\left(\frac{kT}{h}\right)^4 \int_0^\infty \frac{z^3}{\exp(z) - 1}\,dz = \frac{8\pi^5 k^4}{15c^3 h^3}T^4.$$

Thus, the total energy density ρ_t of the radiation field grows as the fourth power of the temperature. The Stefan–Boltzmann radiation law is obtained by multiplying ρ_t by $c/4$:

$$M = \frac{c}{4}\rho_t = \frac{2\pi^5 k^4}{15c^2 h^3}T^4 = (5.67 \cdot 10^{-8} \text{ J K}^{-4} \text{ m}^{-2} \text{ s}^{-1})\cdot T^4.$$

It gives the radiant emittance of a black body into half space.

Chapter 3

3.1. The superposition of both waves again yields a harmonic wave of frequency ω, written in real form as

$$E(z,t) = (E_1 \cos\varphi_1 + E_2 \cos\varphi_2) \sin(kz - \omega t)$$
$$+ (E_1 \sin\varphi_1 + E_2 \sin\varphi_2) \cos(kz - \omega t).$$

Setting $E(z,t) = E_0 \sin(kz - \omega t + \varphi_0)$, expanding this expression, and comparing coefficients leads to:

$$E_0 = \sqrt{E_1^2 + E_2^2 + 2E_1 E_2 \cos(\varphi_2 - \varphi_1)} \,,$$

$$\tan\varphi_0 = \frac{E_1 \sin\varphi_1 + E_2 \sin\varphi_2}{E_1 \cos\varphi_1 + E_2 \cos\varphi_2}.$$

In complex notation, the superposition reads:

$$E(z,t) = (\tilde{E}_1 + \tilde{E}_2) \exp\left[i(kz - \omega t)\right] = \tilde{E}_0 \exp\left[i(kz - \omega t)\right].$$

The complex amplitudes are related to the real quantities by $\tilde{E}_1 = E_1 \exp(i\varphi_1')$, $\tilde{E}_2 = E_2 \exp(i\varphi_2')$ with $\varphi_{1,2}' = \varphi_{1,2} + \pi/2$. For the complex total amplitude $\tilde{E}_0$ and its phase angle φ_0 we obtain, as before,

$$|\tilde{E}_0| = \sqrt{|\tilde{E}_1|^2 + |\tilde{E}_2|^2 + 2\mathrm{Re}\,\{\tilde{E}_1 \tilde{E}_2^*\}},$$

$$\tan\varphi_0 = \frac{\mathrm{Im}\{\tilde{E}_0\}}{\mathrm{Re}\{\tilde{E}_0\}}.$$

3.2. Frequency $\nu = 5$, period of oscillation $T = 1/\nu = 0.2$, wave number $k = 6\pi$, wavelength $\lambda = 1/3$, amplitude $E_0 = \sqrt{4^2 + 3^2} = 5$, phase velocity $v = \omega/k = \nu\lambda = 5/3$. The propagation is in the positive z-direction. The complex amplitude is given by $\tilde{E}_0 = 3 - 4i$.

3.3. Let φ be the angle at which the radius $r = 0.5$ cm of the circular area appears, as seen from the light source S at the distance R. This angle being small, we have $\varphi = r/R$ and $\cos\varphi = 1 - \varphi^2/2$. Thus, the path length difference of rays from S to the center of the circular area and of rays from S to its border is $\Delta = R(1 - \cos\varphi) = R\varphi^2/2 = r^2/2R$. Setting $\Delta = \lambda/10 = 55$ nm, $r = 0.5$ cm we obtain $R = r^2/2\Delta = 227.3$ m.

3.4. Insert the ansatz $\exp\left[i(kz - \omega t)\right]$ into the differential equation and the dispersion relation follows at once: $\omega(k) = c\,|k|\,\sqrt{1 - \eta k^2}$.

3.5. Written in cylindrical coordinates (ρ, Θ, z), with $\rho = \sqrt{x^2 + y^2}$ and $\Theta = \arctan(y/x)$, a monofrequency wave with cylindrical symmetry about the z-axis does not depend on z and $\Theta : E(r,t) = E(\rho,t) = E_R(\rho)\exp(-i\omega t)$. The function $E_R(\rho)$ describes the radial dependence of the wave amplitude. For $\rho \gg \lambda$ it is to be expected that the field E can be approximated by plane harmonic waves. Thus, the power radiated by the wave per unit length of the z-axis will be $\propto E^2(\rho)$ and $\propto 1/(2\pi\rho)$ ($=$ 1/surface area of a cylinder of height one). From energy conservation we then have $E_R(\rho) \propto 1/\sqrt{\rho}$. Using the ansatz given above in the scalar wave equation yields, in cylindrical coordinates, the differential equation

$$\frac{dE_R}{d\rho}(\rho) + \rho\frac{d^2 E_R}{d\rho^2}(\rho) + \rho k^2 E_R(\rho) = 0 \,.$$

As may be verified easily, the Bessel function $E_R(\rho) = J_0(k\rho)$ is a solution. From the asymptotic expansion of J_0, $J_0(k\rho) \approx \sqrt{2/(\pi k\rho)}\cos(k\rho - \pi/4)$ for $k\rho \gg 1$, the $1/\sqrt{\rho}$ dependence of the amplitude is confirmed.

3.6. Applying the approximate relations given in Fig. 3.3 the diameter of the axicon is obtained: $L = 2R = 2z_B n(1-\gamma) = 26.2\,\text{cm}$.

3.7. By introducing the function (3.65) into the three-dimensional scalar wave equation and eliminating the common factor $E_0 \exp\left[-i(\omega t - \beta z)\right]$ we obtain

$$\left[\frac{J_1(\alpha\rho)}{\alpha\rho} - J_0(\alpha\rho)\right]\alpha^2 + \frac{dJ_0(z)}{dz}\Big|_{z=\alpha\rho}\frac{\alpha}{\rho} + J_0(\alpha\rho)\left(\frac{\omega^2}{c^2} - \beta^2\right) = 0.$$

This relation is true since $\omega^2/c^2 = k^2 = \alpha^2 + \beta^2$ and $dJ_0(z)/dz = -J_1(z)$. For $\alpha = \pm k$ the Bessel wave turns into a cylindrical wave $E_0 J_0(\pm k\rho)\exp(-i\omega t)$ (compare problem 3.5). By inserting the integral representation of J_0 into (3.65),

$$E(\boldsymbol{r},t) = \frac{E_0}{2\pi}\exp(-i\omega t)\int_0^{2\pi}\exp\left[i(\alpha\rho\cos\varphi + \beta z)\right]d\varphi\,,$$

it is seen that the Bessel wave may be considered as a superposition of plane harmonic waves of equal amplitude. The associated wave vectors $\boldsymbol{k}$ also have equal magnitude and a common z-component, $k_z = \beta$. Thus, the radial components of their $\boldsymbol{k}$-vectors have constant magnitude, too. This means that the $\boldsymbol{k}$-vectors of all these partial waves are lying on a cone with the angle $2\Theta = 2\arcsin(\alpha/k)$. This explains why Bessel waves may be generated by refraction at a circular glass cone. For small angles, the angle Θ obeys $\sin(\Theta + \gamma)/\sin\gamma \approx 1 + \Theta/\gamma = n$ (law of refraction, γ: axicon angle, see Fig. 3.3).

3.8. The sought-after intensity is $I = 8\,\text{W/m}^2$. If we average over the instantaneous intensity of a harmonic wave, in (3.85) a prefactor of $1/2$ is introduced. Thus, the electric field amplitude is given by $E = \sqrt{2I/(\varepsilon_0 c)} = 77.6\,\text{V m}^{-1}$.

3.9. Along with $\overline{\omega} = (\omega_1 + \omega_2)/2$ and $\Delta\omega = (\omega_1 - \omega_2)/2$ we have

$$E(z,t) = E_0\exp(-i\overline{\omega}t)\left[\exp(-i\Delta\omega t) + \exp(+i\Delta\omega t)\right] = 2E_0\exp(-i\overline{\omega}t)\cos(\Delta\omega t).$$

Here, the complex function $E(z,t)$, instead of $\text{Re}\{E(z,t)\}$, may be decomposed into a product of functions because one of the factors is real. For determining the instantaneous intensity $I(t)$, the real field amplitude has to be squared,

$$I(t) \propto \left(\text{Re}\{E(z,t)\}\right)^2 = 4E_0^2\cos^2(\overline{\omega}\,t)\cos^2(\Delta\omega t).$$

Thus, the instantaneous intensity oscillates, on a fast time scale, at twice the light frequency. However, if the definition (3.94) of intensity is applied directly to the wave $E(z,t)$ we obtain

$$I'(t) = E(z,t)E^*(z,t) = 2E_0^2\left[1 + \cos(\omega_1 - \omega_2)t\right] = 4E_0^2\cos^2(\Delta\omega t).$$

The intensity now varies according to the slow modulation of $I(t)$ due to the beats. The averaging over the faster time scale, corresponding to the frequency $\overline{\omega}$, is already incorporated in the definition (3.94). Thus, the quantity I' is a short-term intensity with an averaging time $T_m \ll T_s = 2\pi/|\omega_1 - \omega_2| = \pi/|\Delta\omega|$, but $T_m \gg T_0 = \pi/\overline{\omega}$. For $T_0 \ll T_m$ we get the following form of the short-term intensity (3.96),

$$\overline{I}(t) = \frac{1}{T_m}\int_{t-T_m/2}^{t+T_m/2} I'(t')\,dt' = 2E_0^2\left[1 + \text{sinc}\left(\frac{T_m}{T_s}\right)\cos(\omega_1 - \omega_2)t\right].$$

In the integrand of (3.96), as for $I'(t)$ given above, the averaging over the fast oscillation $\propto \cos(\overline{\omega}\, t)$ is already contained. For $T_0 \ll T_m \ll T_s$, the sinc term is practically unity, and the short-term intensity follows the modulation of the oscillation, $\overline{I}(t) = I'(t)$. On the other hand, for $T_m \gg T_s$ the short-term intensity hardly varies and the intensity of the wave is recovered, $I_\infty = 2E_0^2$. For $T_m \approx T_s$ the amplitude of variation of the short-term intensity strongly depends on the choice of the averaging interval.

Chapter 4

4.1. The superposition of the two oscillations is written as a real quantity,

$$E(t) = \tfrac{1}{2}[E_{01}\exp(-i\omega_1 t) + E_{02}\exp(-i\omega_2 t) + \text{c.c.}].$$

For simplicity, we assume both amplitudes E_{01} and E_{02} to be real. Intending to write the superposition as a product of harmonic oscillations, we set $\overline{E} = \tfrac{1}{2}(E_{01} + E_{02})$, $\Delta E = \tfrac{1}{2}(E_{01} - E_{02})$, $\overline{\omega} = \tfrac{1}{2}(\omega_1 + \omega_2)$, and $\Delta\omega = \tfrac{1}{2}(\omega_1 - \omega_2)$, and obtain

$$E(t) = \tfrac{1}{2}\overline{E}\exp(-i\overline{\omega}t)[\exp(-i\Delta\omega t) + \exp(+i\Delta\omega t)]$$
$$\qquad + \tfrac{1}{2}\Delta E\exp(-i\overline{\omega}t)[\exp(-i\Delta\omega t) - \exp(+i\Delta\omega t)] + \text{c.c.}$$
$$= 2\overline{E}\cos(\overline{\omega}t)\cos(\Delta\omega t) - 2\Delta E\sin(\overline{\omega}t)\sin(\Delta\omega t).$$

This expression describes a beat signal that in general is not fully modulated (see the Figure). The self coherence function $\Gamma(\tau)$ of the complex oscillation $\tilde{E}(t)$

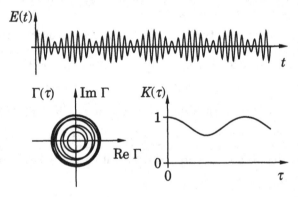

Fig. B.1. Sketch of the time function, the self coherence function and the contrast function of the beat signal.

corresponds to a superposition of harmonic oscillations (in τ)

$$\Gamma(\tau) = \langle \tilde{E}^*(t)\tilde{E}(t+\tau)\rangle = I_1\exp(-i\omega_1\tau) + I_2\exp(-i\omega_2\tau),$$

with $I_1 = |E_{01}|^2$ and $I_2 = |E_{02}|^2$. The complex degree of coherence thus is given by $\gamma(\tau) = \Gamma(\tau)/(I_1 + I_2)$. In the complex plane, the graph of $\gamma(\tau)$ is located between the circles $|\gamma| = 1$ and $\gamma = |I_1 - I_2|/(I_1 + I_2)$. The contrast function reads:

$$K(\tau) = |\gamma(\tau)| = \frac{\sqrt{I_1^2 + I_2^2 + 2I_1 I_2 \cos((\omega_1 - \omega_2)\tau)}}{I_1 + I_2}.$$

4.2. For signals of finite energy, the division by T is omitted in the definition (4.5) of the self coherence function,

$$\Gamma(\tau) = \int_{-\infty}^{+\infty} E^*(t)E(t+\tau)\,dt.$$

Hence, for the Gaussian wave packet the self coherence function is given by

$$\begin{aligned}
\Gamma(\tau) &= \frac{|E_0|^2}{2\pi\sigma^2} \int_{-\infty}^{+\infty} \exp\left(i\omega t - \frac{t^2}{2\sigma^2}\right) \exp\left[-i\omega(t+\tau) - \frac{(t+\tau)^2}{2\sigma^2}\right]dt \\
&= \frac{|E_0|^2}{2\pi\sigma^2} \exp\left(-\frac{\tau^2}{4\sigma^2}\right) \exp(-i\omega\tau) \int_{-\infty}^{+\infty} \exp\left[-\frac{(t+\tau/2)^2}{\sigma^2}\right]dt \\
&= \frac{|E_0|^2}{2\sqrt{\pi}\sigma} \exp\left(-\frac{\tau^2}{4\sigma^2}\right) \exp(-i\omega\tau).
\end{aligned}$$

Therefore, the contrast function also has the form of a Gaussian function,

$$K(\tau) = |\gamma(\tau)| = \left|\exp\left(-\frac{\tau^2}{4\sigma^2}\right)\exp(-i\omega\tau)\right| = \exp\left(-\frac{\tau^2}{4\sigma^2}\right).$$

The coherence time is $\tau_c = 2\sigma$, because $K(\tau_c) = K(2\sigma) = 1/e$.

4.3. Substituting

$$E(t) = \int_0^\infty E_0(\nu)\exp(-i2\pi\nu t)\,d\nu$$

into the definition of the coherence function and using

$$\lim_{T\to\infty} \frac{1}{T}\int_{-T/2}^{+T/2} \exp[i2\pi(\nu-\nu')t]\,dt = \delta(\nu-\nu')$$

we can verify relation (4.30) directly:

$$\begin{aligned}
\Gamma(\tau) &= \lim_{T\to\infty} \frac{1}{T}\int_{-T/2}^{+T/2} dt \int_0^\infty E_0^*(\nu)\exp(i2\pi\nu t)\,d\nu \int_0^\infty E_0(\nu')\exp[-i2\pi\nu'(t+\tau)]\,d\nu' \\
&= \int_0^\infty d\nu \int_0^\infty d\nu' E_0^*(\nu)E_0(\nu')\delta(\nu-\nu')\exp(-i2\pi\nu'\tau) \\
&= \int_0^\infty |E_0(\nu)|^2 \exp(-i2\pi\nu\tau)\,d\nu.
\end{aligned}$$

4.4. Applying the spatial coherence condition (4.53) we obtain $d \le d_{max} = \lambda R/2L$ $= \lambda/2\Theta$. The angular diameter of the sun being $\Theta = L/R = 30$ arc minutes, the hole separation should be smaller than $d_{max} = 31.5\ \mu m$.

4.5. The total intensity at a point (ξ, η) of the screen is obtained by incoherent superposition of the diffraction images, that is, by addition of intensities. Integrating (4.2) in polar coordinates (r, Θ) over the circular source area and using $\varphi_2 - \varphi_1 = kdx/L_1 = kdr\cos\Theta/L_1$ leads to:

$$I(\xi, \eta) \propto \int_0^R dr\, r \int_0^{2\pi} d\Theta\left[1 + \cos\left(\frac{kd}{L_1}r\cos\Theta + \frac{kd}{L_2}\xi\right)\right].$$

The cosine function in the integrand can be expanded by using a trigonometric identity. This yields two integrals, one of them being zero because of symmetry. It remains that:

$$I(\xi,\eta) \propto \pi R^2 + \cos\left(\frac{kd}{L_2}\xi\right) \int_0^R dr\, r \int_0^{2\pi} d\Theta \cos\left(\frac{kd}{L_1} r \cos\Theta\right)$$

$$= \pi R^2 + 2\pi \cos\left(\frac{kd}{L_2}\xi\right) \int_0^R dr\, r J_0\left(\frac{kd}{L_1} r\right)$$

$$= \pi R^2 \left[1 + 2\cos\left(\frac{kd}{L_2}\xi\right) \frac{J_1(kdR/L_1)}{kdR/L_1}\right].$$

4.6. The intensity of the superposition of two waves has the form $I = I_1 + I_2 + 2\,\mathrm{Re}\{\Gamma_{12}\}$. As with the discussion in this chapter we may write, for quasi-monochromatic light, $I_{\max} = I_1 + I_2 + 2|\Gamma_{12}|$ und $I_{\min} = I_1 + I_2 - 2|\Gamma_{12}|$. Along with the definition (4.58) we obtain relation (4.60) for the contrast function:

$$K_{12}(\tau) = \frac{4|\Gamma_{12}(\tau)|}{2(I_1+I_2)} = 2\frac{\sqrt{I_1 I_2}}{I_1+I_2}|\gamma_{12}(\tau)| = 2\frac{\sqrt{\Gamma_{11}(0)\Gamma_{22}(0)}}{\Gamma_{11}(0)+\Gamma_{22}(0)}|\gamma_{12}(\tau)|.$$

The mutual degree of coherence of a spherical wave $E(r,t) = (E_0/r)\exp[i(kr-\omega t)]$ thus is given by $\gamma_{12}(\tau) = \exp[ik(r_1-r_2)]\exp(-i\omega\tau)$, and the contrast, by

$$K_{12}(\tau) = \frac{2r_1 r_2}{r_1^2 + r_2^2}.$$

4.7. The Fourier spectrum of a square pulse, $f(t) = \mathrm{rect}\,(t)$, reads:

$$H(\nu) = \mathcal{F}[\mathrm{rect}](\nu) = \int_{-1/2}^{+1/2} \exp(-i2\pi\nu t)\, dt$$

$$= \frac{1}{i2\pi\nu}\left[\exp(i\pi\nu) - \exp(-i\pi\nu)\right] = \frac{\sin(\pi\nu)}{\pi\nu} = \mathrm{sinc}\,(\nu).$$

The analytic signal $\tilde{f}(t)$ may be calculated either by integrating over the positive part of the amplitude spectrum, or by application of the Hilbert transforms (4.71) and (4.72):

$$\mathrm{Re}\,\tilde{f}(t) = \mathrm{rect}\,(t), \quad \mathrm{Im}\,\tilde{f}(t) = \frac{1}{\pi}\int_{-1/2}^{+1/2}\frac{1}{t'-t}\,dt' = \frac{1}{\pi}\ln\left|\frac{t-\frac{1}{2}}{t+\frac{1}{2}}\right|.$$

4.8. For the Michelson stellar interferometer relation (4.89) reads:

$$I(\xi,\eta) \propto 1 + 2\cos(kd'\xi/L_2)\frac{J_1(kdR/L_1)}{kdR/L_1}.$$

Thereby, the star is supposed to be a circular source of constant surface brightness. This simple model does not include, for instance, a possible limb darkening of the stellar disk. The first minimum of the contrast occurs at the first root of J_1. Thus the angular diameter of Betelgeuse is $\varphi = 2R/L_1 = 2\cdot 3.83/kd = 1.22\lambda/d = 2.27\cdot 10^{-7}$ or $\varphi = 0.047$ arc seconds, and its linear diameter $2R = \varphi L_1 = 860\cdot 10^6$ km.

4.9. Since by measuring $R_{12}(\tau) \propto |\gamma_{12}(\tau)|^2$ essentially the contrast of the intensity distribution (4.89) is recorded the formula given in problem 4.8 applies. For an object in the zenith, the maximum base distance, $d_{max} = 188$ m, corresponds to a minimum angular diameter of a star still measurable of $\varphi_{min} = 3.83\lambda/\pi d_{max} = 1.22\lambda/d = 6.6 \cdot 10^{-4}$ arc seconds.

4.10. Let the interferogram ΔI_b (essentially being the autocorrelation function of the field amplitude) be recorded in the displacement interval $[0, \tau_{max}]$. This section of the complete interferogram, ΔI, may be represented by

$$\Delta I_b(\tau) = \Delta I \, \text{rect}\left(\frac{t}{\tau_{max}} - \frac{1}{2}\right).$$

Its power spectrum $W(\nu)$ is given by

$$W(\nu) = \mathcal{F}[\Delta I](\nu) * \mathcal{F}\left[\text{rect}\left(\frac{t}{\tau_{max}} - \frac{1}{2}\right)\right](\nu)$$

$$= \mathcal{F}[\Delta I](\nu) * \left[\tau_{max} \exp(-i\pi\nu)\text{sinc}(\tau_{max}\nu)\right].$$

Hence, due to the convolution with the sinc function, a spectral line at least has a width of $\Delta\nu \approx 1/\tau_{max}$; the frequency resolution therefore is $\Delta\nu \approx 1/\tau_{max}$. For the example given the time shift τ to be realized should at least be

$$\tau_{max} = \frac{1}{\Delta\nu_{min}} = \frac{\lambda_{max}A}{c} = 10^{-11} \text{ s}.$$

In the Michelson interferometer this corresponds to a mirror displacement of $l = c/(2\tau_{max}) = 1.5$ mm. The number of sampling points required to faithfully reconstruct the power spectrum is given by the sampling theorem. At maximum frequency, ν_{max}, the measuring interval is covered by the maximum number of interference fringes, $N_{max} = 2\tau_{max}\nu_{max} = 1925$. The number of samples should at least be twice as large, $N_s \geq 2N_{max} = 3850$. In this case, $N_s = 4096$ would be chosen since, for digital processing, commonly the fast Fourier transform (FFT) is used which operates on $N_s = 2^n$ data points.

4.11. The interferogram of the pulse, given by the autocorrelation function of the real signal, reads:

$$\Delta I(\tau) = \exp(-\sigma|\tau|)\frac{\omega_0^2}{4(\omega_0^2 + \sigma^2)}\left[\frac{1}{\sigma}\cos(\omega_0\tau) + \frac{1}{\omega_0}\sin(\omega_0|\tau|)\right].$$

The coherence time therefore is $\tau_c = 1/\sigma$, and the power spectrum,

$$W(\nu) = \frac{\omega_0^2}{(\omega_0^2 - \omega^2 + \sigma^2)^2 + 4\sigma^2\omega^2}.$$

For $\tau_c \gg T$ (the period of oscillation), that is, for $\omega_0 \gg \sigma$, the spectrum has a width (FWHM) of $\nu_h = 2\sigma$, being proportional to the reciprocal of the coherence time.

Chapter 5

5.1. The following values are obtained: mirror reflectivity $r = \sqrt{1 + \pi^2/4F^2} - \pi/2F \approx 1 - \pi/2F = 0.99984$, free spectral range $\Delta\nu = c/2L = 3 \cdot 10^9$ Hz, width (FWHM) of a spectral line $\nu_h = \Delta\nu/F = c/2LF = 3 \cdot 10^5$ Hz, spectral resolution $A = \nu/\nu_h = 2LF/\lambda = 2 \cdot 10^9$. The two spectral lines have the frequency separation

$\Delta \nu = c\Delta\lambda/\lambda^2 = 2.24 \cdot 10^{11}$ Hz. Since the free spectral range is about 100 times smaller than $\Delta \nu$ the spectral components cannot be resolved unambiguously.

5.2. The superposition of the partial waves reflected at the first mirror (first term) and those reflected at the second and transmitted by the first mirror (remaining terms) yields the total amplitude of the reflected wave:

$$E_r = E_e r \left[1 - t^2 \exp(ik2L) - t^2 r^2 \exp(ik4L) - t^2 r^4 \exp(ik6L) + \ldots\right].$$

Summing the geometric series gives the total amplitude reflectance

$$R = r\frac{1 - \exp(ik2L)}{1 - r^2 \exp(ik2L)}$$

and thus the intensity reflectance

$$|R|^2 = 2r^2 \frac{1 - \cos 2kL}{1 + r^4 - 2r^2 \cos 2kL} = \frac{K \sin^2 kL}{1 + K \sin^2 kL},$$

where $K = 4r^2/(1-r^2)^2$ is the finesse coefficient. From this, along with (5.17), the validity of $|R|^2 + |T|^2 = 1$ immediately follows.

5.3. Recall that $I(kd) = 2I_1(1 + \cos(2kd))$ for the output intensity of a Michelson interferometer; its free spectral range is found to be $\Delta \nu = c/2d$, and the full width at half maximum of a spectral line is $\nu_h = c/4d$. Thus, the finesse of a Michelson interferometer has the value $F = \Delta \nu/\nu_h = 2$.

5.4. The mirror displacement we are looking for in the case of the Fabry–Perot interferometer is given by

$$\Delta L = \frac{1}{k\sqrt{K}} = \frac{\lambda}{4F} = 1.58 \cdot 10^{-10} \text{ m}.$$

For the Michelson interferometer the displacement is $\Delta L = \lambda/8 = 79.1$ nm (see also the previous problem, $F = 2$). Thus the Fabry–Perot instrument, compared to the Michelson interferometer, is more sensitive to small optical path length variations by a factor of $F/2 = 500$, the ratio of the finesse parameters.

5.5. Let us denote the amplitude of the incident wave immediately in front of the first mirror, M_1, by E_e, the amplitude of the reflected wave there, by E_r, and the amplitude of the reflected wave immediately behind M_1, by E_1. Also, let E_2 denote the amplitude of the transmitted wave in front of the second mirror, M_2, and E_o, that of the outgoing wave immediately behind M_2 (compare Fig. 5.3). Assuming r, r', t, t' to be real we have

$$E_0 = E_2 t', \quad E_1 = E_2 r' \exp(ikL), \quad E_2 = (E_e t + E_1 r') \exp(ikL), \quad E_r = r E_e + t' E_1.$$

For maximum transmission, $\exp(ikL) = \pm 1$ and $E_r = 0$, thus $E_1 = (r'/t')E_e$, $E_2 = (1/t')\exp(ikL)E_e = \pm(1/t')E_e$. In the cavity of the Fabry–Perot interferometer the two counterpropagating waves of amplitude E_1 and $E_2 \exp(-ikL) = E_e/t'$, respectively, superimpose to yield a standing wave field with amplitude E_s,

$$E_s(z,t) = 2E_e\frac{r'}{t'}\cos(kz)\exp(-i\omega t),$$

and a forward-traveling wave

$$E_f(z,t) = E_e\frac{1-r'}{t'}\exp\left[i(kz - \omega t)\right].$$

For $t' \ll 1$ and $r \lesssim 1$, the propagating wave can be neglected and the field in the cavity is just a standing wave field of amplitude $\approx 2E_e/t'$ and of mean energy density

$$u = \frac{1}{cL} \int_0^L |E(z,t)|^2 \, dz \approx \frac{2I_e}{c|t'|^2} \gg \frac{I_e}{c}.$$

For minimum transmission, $\exp(ikL) = \pm i$ and thus $E_2 = it/(1+r'^2)E_e$, $E_1 = -tr'/(1+r'^2)E_e$. It follows that

$$E_s(z,t) = 2iE_e \frac{r't}{1+r'^2} \sin(kz) \exp(-i\omega t)$$

and

$$E_l(z,t) = E_e \frac{(1-r')t}{1+r'^2} \exp\left[i(kz - \omega t)\right].$$

Once again, for $t' \ll 1$ and $r' \lesssim 1$, neglecting the wave E_l we now have a standing wave field of energy density

$$u \approx \frac{2|E_e|^2 t^2}{c(1+r'^2)^2} \approx \frac{(t^2)I_e}{2c} \ll \frac{I_e}{c}.$$

5.6. The frequency separation of the longitudinal modes of a laser with resonator length L is given by $\Delta\nu = c/2nL$. For the He–Ne laser ($n \approx 1$) we find $\Delta\nu = 250$ MHz, for the semiconductor laser $\Delta\nu = 86$ GHz. Upon difference frequency analysis of the three laser modes the photo current $i(t)$ varies according to

$$i(t) \propto |E(t)|^2 \propto \left| \exp\left[-i(\omega - \Delta\omega)t\right] + 2\exp(-i\omega t) + \exp\left[-i(\omega + \Delta\omega)t\right] \right|^2$$
$$= 6 + 8\cos(\Delta\omega t) + 2\cos(2\Delta\omega t),$$

that is, we observe frequency components $\Delta\nu$ and $2\Delta\nu$ with relative amplitudes 4:1.

Chapter 6

6.1. Along with $I = |E|^2$, $dI = 2|E|\,d|E|$, and $\bar{E} = \sqrt{\langle I \rangle}$ the following distribution is obtained:

$$p(|E|) = 2\frac{|E|}{\bar{E}^2} \exp\left(-\frac{|E|^2}{\bar{E}^2}\right).$$

The mean value of $|E|$ is $\langle |E| \rangle = \sqrt{\pi}\,\bar{E}/2$.

6.2. (a) The smallest speckle grain size on the screen is $d_{min} = 1.2\lambda L_1/d = 0.38$ mm. (b) On the retina of the observer the images of those grains have a diameter of $d' = Vd_{min} = bd_{min}/L_2 = 2.6$ μm ($b = 1.7$ cm being the distance from the eye lens to the retina). (c) The minimum grain size of subjective speckles on the observer's retina is given by $d'' = 1.2\lambda b/D_{pup} = 2.6$ μm, which is accidentally the same as the corresponding diameters of the smallest objective speckles seen on the screen ($D_{pup} =$ diameter of pupil). The minimum size of the subjective speckles as perceived by the observer does not depend on his or her distance from the screen as long as the diameter of the pupil does not change due to adaptation. This is not the case for the image of objective speckles since with increasing distance from the screen their image on the retina correspondingly gets smaller. Subjective speckles are always seen sharply, also by a short-sighted observer, because they result from statistical interference of coherent waves entering the eye, which is not affected significantly by the eye's lens.

6.3. The smallest resolvable object displacement for $F = 2$ and magnification $V = 1$ is $d_{min} = 2 \cdot (2.4\lambda F) = 4.7$ μm. Using the magnification V upon recording, we require the resulting image displacement to be at least twice as large as the minimum speckle grain size,

$$d_{min} = 2 \cdot \left(1.2\frac{V+1}{V}\lambda F\right) = \left(1+\frac{1}{V}\right) \cdot 2.34 \text{ μm}.$$

Thus, with $V > 1$ no significant improvement of resolution can be achieved since the grain size, for large V, is proportional to V. For $V < 1$, resolution is degraded (but the image brightness is increased).

6.4. The integral occurring in (6.31) can be written, for an arbitrary oscillation $f(t)$, as

$$\int_0^{T_s} \exp\left[2\pi i\nu_x f(t)\right] dt = \int_{-\infty}^{+\infty} p(s)\exp(2\pi i\nu_x s) \, ds.$$

Here, a stationary signal has been assumed and the time average has been replaced by an average over the range of the function, $p(s)$ being the corresponding probability density. For the triangular signal the density $p(s)$ is constant in the interval $[-A/2, A/2]$, $p(s) = 1/A$, and zero elsewhere. Thus the integral evaluates to sinc($\nu_x A$), and the diffraction pattern upon specklegram analysis, apart from $(\nu_x, \nu_y) = (0, 0)$, has the form of a speckle pattern modulated with the diffraction pattern of a slit,

$$I \propto |\mathcal{F}[I(x,y)]|^2 (\nu_x, \nu_y) \text{sinc}^2(\nu_x A).$$

6.5. Using (6.38) we obtain $u = \lambda\nu_{burst}/(2\sin\alpha) = 1.0 \text{ m s}^{-1}$.

Chapter 7

7.1. The fringe separation in the resulting interference pattern is (compare problem 6.5)

$$d = \frac{\lambda}{2\sin\alpha} = 0.49 \text{ μm}.$$

This means that the photographic emulsion should have a resolution of at least 2000 lines per mm.

7.2. With $T = a - b'I - c'I^2$ and $I = (RR^* + SS^* + RS^* + R^*S)$, the following terms are found for the reconstructed wave RT:

$R(a - b'\|R\|^2 - c'\|R\|^4)$	zeroth order,
$R(b'\|S\|^2 + 4c'\|R\|^2\|S\|^2 + c'\|S\|^4)$	broadened zeroth order,
$S(b'\|R\|^2 + 2c'\|R\|^4)$	direct image (first order),
$S(2c'\|R\|^2\|S\|^2)$	broadened first order (direct image),
$S^*R^2(b' + 2c'\|R\|^2)$	conjugate image,
$S^*R^2(2c'\|S\|^2)$	broadened first order (conjugate image),
$S^2R^*(c'\|R\|^2)$	direct image (second order),
$(S^*)^2R^3c'$	conjugate image (second order).

Thus the nonlinearity of the photographic characteristic leads, among other effects, to additional images in higher diffraction orders.

7.3. The conjugate image appears at the angle $\beta = \arcsin(2\sin\alpha - \sin\gamma) = 51.8°$ with the optical axis. Reconstructing with the reference wave R', entering at a different angle α', the conjugate image disappears if $\alpha' \geq \arcsin(1 + \sin\gamma - \sin\alpha) = 59°$.

7.4. Superimposing a plane and a spherical harmonic wave, both having unit amplitude in the plane of the hologram, yields the intensity distribution $I = 2(1 + \cos(kl))$. Here, $l = d - d_0$ denotes the length difference of the straight path from the point source to the point considered on the hologram (d) and that from the source to the intersection of hologram and optical axis (d_0). We assume that at the central point there is an intensity maximum of the interference pattern. The edges of the black and white rings of the zone plate occur for $\cos(kl) = 0$. This leads to the condition $k(d_n - d_0) = (2n - 1)\pi/2$ for the nth edge, counted from the center. The corresponding radii on the hologram are given by

$$r_n = \frac{\lambda}{4}\sqrt{(2n-1)^2 + (2n-1)\frac{8d_0}{\lambda}}.$$

7.5. The dense layers of the developed hologram have a separation of $d = \lambda/2 = 0.316\,\mu\text{m}$ (recording wavelength $\lambda = 633\,\text{nm}$, index of refraction $n = 1$). Applying the Bragg condition for green light (reconstruction wavelength $\lambda' = 500$ nm) yields:

$$\sin\alpha = \frac{\lambda'}{2d} = \frac{\lambda'}{\lambda} \rightarrow \alpha = 52.2°.$$

Here, α denotes the angle between the incident or outgoing beam and the plane of the hologram (Fig. 7.18).

7.6. The following notation is used: location of the source of the reference wave upon recording, (x_R, y_R, z_R); corresponding location upon reconstruction, (x'_R, y'_R, z'_R). Let the first object point be at (x_{S1}, y_{S1}, z_{S1}) and the second one at (x_{S2}, y_{S2}, z_{S2}), their image points be at (x_{Q1}, y_{Q1}, z_{Q1}) and at (x_{Q2}, y_{Q2}, z_{Q2}), respectively. The ratio of reconstruction and recording wavelength is denoted by m, $m = \lambda_2/\lambda_1$. In the case (a) we have $-\infty < z_R, z'_R < 0$.

Lateral magnification: the coordinates of the object points are chosen as $x_{S2} = x_{S1} + \Delta x$, $y_{S2} = y_{S1}$, $z_{S2} = z_{S1}$. From the imaging equations, $y_{Q1} = y_{Q2}$ and $z_{Q1} = z_{Q2}$ for both the direct and conjugate image. The lateral magnification of the direct image, $M_{\text{lat,d}}$, is given by the difference in x_Q,

$$x_{Q1} - x_{Q2} = z_{Q1}m\frac{\Delta x}{z_{S1}} \rightarrow M_{\text{lat,d}} = \frac{x_{Q1} - x_{Q2}}{\Delta x} = \left(m + \frac{z_{S1}}{z'_R} - m\frac{z_{S1}}{z_R}\right)^{-1}.$$

The lateral magnification of the conjugate image is obtained by just inverting signs of z_R and z_{S1}:

$$M_{\text{lat,c}} = \left(m - \frac{z_{S1}}{z'_R} - m\frac{z_{S1}}{z_R}\right)^{-1}.$$

To obtain the longitudinal magnification, two object points with $x_{S2} = x_{S1}$, $y_{S2} = y_{S1}$, and $z_{S2} = z_{S1} + \Delta z$ are chosen and their images are calculated. Applying the imaging equations reveals that not only the z-coordinate but also the x- and y-coordinates of the image point change with a shift of the object point in the z-direction (distortion). This effect, depending quadratically on Δz, is ignored. For simplicity, we suppose that both object points are on the z-axis: $x_{S1} = x_{S2} = 0$, $y_{S1} = y_{S2} = 0$. Substituting this into the imaging equations and calculating the magnification we get:

$$M_{\text{lo,d}} = \frac{z_{Q1} - z_{Q2}}{\Delta z} = \frac{mz_R'^2 z_R^2}{\left(z_R z_{S1} - mz_R' z_{S1} + mz_R' z_R\right)^2}.$$

As before, the longitudinal magnification of the conjugate image follows by change of signs:

$$M_{\text{lo,c}} = \frac{mz_R'^2 z_R^2}{\left(z_R z_{S1} + mz_R' z_{S1} - mz_R' z_R\right)^2}.$$

Cancelling the common factor $m^2 z_R'^2 z_R^2$ we immediately verify the relation $M_{\text{lo}} = mM_{\text{lat}}^2$ for the conjugate as well as the direct image. In the case (b) (plane reconstruction wave) we have $z_R' = -\infty$. It follows that $M_{\text{lat,c}} = M_{\text{lat,d}}$ and $M_{\text{lo,c}} = M_{\text{lo,d}} = mM_{\text{lat}}^2$. The lateral magnification is independent of m.

7.7. The expression for the maximum distance, (7.55), follows immediately from $z_{\max}^2 + a^2 = (z_{\max} + \lambda/2)^2$ (see right part of Fig. 7.28). The minimum distance, (7.54), is derived from $z_{\min}^2 + a^2 = D^2$ and $z_{\min}^2 + (a-b)^2 = (D-\lambda)^2$ by eliminating D and solving for $z_{\min}$ (see right part of Fig. 7.28).

7.8. For an estimation of the computing time it is sufficient to determine the time consumption of the main loop of the algorithm. For L object points and a hologram size of MN points the expression $r_{ijl} = \sqrt{z_l^2 + (x_i - x_l)^2 + (y_j - y_l)^2}$ has to be calculated MNL times, along with as many integer operations to determine whether r_{ijl} is even or odd (square wave approximation). Using intermediate results stored in main memory (see the code given in solution 7.9) three integer operations (addition and multiplication) are necessary to obtain the argument of the root. Thus a total of $(3+1)MNL$ integer operations and MNL evaluations of the root are required, leading to a computation time of at least $T = MNL(4 \cdot 10 + 200)$ ns $= 960000$ s $= 266.6$ hours.

7.9. The digital hologram of a point source with a normally incident reference wave is just a Fresnel zone plate as given in Fig. 7.29. The following C code represents a simple implementation of the algorithm. For it to work, the reader should supply routines init_output(), output_point() and end_output() that initialize the graphics output device, set a black dot, and terminate the output, respectively.

```
#include <stdio.h>
#include <math.h>

/* ..... maximum number of object points ..... */
#define MAXPTS 1000

int   npts;
int   xc[MAXPTS], yc[MAXPTS], zc[MAXPTS], zcq[MAXPTS],dyv[MAXPTS];

/* ..... definition of the size
         of the digital hologram ..... */
```

```
int  nxpixels  =   5000;
int  nypixels  =   3200;

main()
{
  register int    i,j,k,dx,dy;
  double          sum;

  /* ..... the coordinates of the object points are read from
            stdin. Each input line should contain three numbers,
            the x-, y- and z-coordinate of an object point ..... */
  npts = 0;

  while( scanf(" %d %d %d",xc+npts,yc+npts,zc+npts) == 3 && npts<MAXPTS )
    npts++;

  /* ..... initialize output ..... */
  init_output(nxpixels,nypixels);

  /* ..... store squares of the z-coordinates in array zcq ..... */
  for(i=0; i<npts;i++) zcq[i] = zc[i]*zc[i];

  /* ..... loop over the lines of the image ..... */
  for(j=0; j<nypixels; j++)
  {
    /* ..... save the vector   (y[i] - j)^2 in array dyv[] ..... */
    for(i=0;i<npts;i++)
      { dy = yc[i] - j; dyv[i] = dy*dy; }
    /* ..... loop over columns of the image ..... */
    for(k=0; k<nxpixels; k++)
    {
      sum = 0.;
      /* .... loop over object points ..... */
      for(i=0;i<npts;i++)
      {
        dx = xc[i] - k;
        sum += cos(fmod(sqrt((double)(dx*dx + dyv[i] + zcq[i])),
                    6.28318530718));
      }
      /* ..... set point at (j,k) if sum>= 0. ..... */
      if(sum >= 0.) output_point(j,k);
    }
  }
  /* ..... terminate output ..... */
  end_output();
}
```

Chapter 8

8.1. The reconstructed object wave is of the form $R'R(S_1 + S_2)$, where R' denotes the reconstruction wave, R the reference wave upon recording and S_1, S_2 the object waves of the first and second exposure, respectively. The image intensity is given by $I = |R'|^2 |R|^2 |S_1 + S_2|^2 \propto \cos^2(\Delta\varphi/2)$, where $\varphi = \arg(S_1 - S_2)$ depends only on the object waves. Thus, by choosing a different wavelength for reconstruction, the coverage of the object image by interference fringes is not affected, but the image is scaled and possibly distorted (see the holographic imaging equations).

8.2. In the picture to the left, the oscillation of the lid has a single radial node line. Nine intensity minima are visible, thus an oscillation amplitude of

$$d_1 = \frac{z_9\lambda}{2\pi(\cos\alpha + \cos\beta)} = 1.47\ \mu m$$

can be inferred. The picture in the middle reveals two orthogonal, radial node lines and one node circle. Two intensity minima are visible, leading to an amplitude $d_2 = 0.3\,\mu m$ on using z_2. In the picture to the right, the node circle is missing, and ten minima are counted. Thus the amplitude is $d_3 = 1.64\,\mu m$.

8.3. Written for an arbitrary oscillation $s(t)$, relation (8.30) reads:

$$E_d \propto \int_0^{t_B} S(t)\,dt = \int_0^{t_B} A\exp[iks(t)]\,dt\,.$$

Taking for $s(t)$ a square wave signal with duty cycle γ, that is, a signal having the amplitude $-d$ in a fraction γ of the oscillation period T_s and having an amplitude $+d$ in the remaining time $(1-\gamma)T_s$, it follows for $t_b \gg T_s$ that

$$E_d \propto \gamma\exp[ik(-d)] + (1-\gamma)\exp(ikd)\,.$$

For the intensity we obtain:

$$I = E_d E_d^* \propto \gamma^2 + (1-\gamma)^2 + 2\gamma(1-\gamma)\cos(2kd)\,.$$

Thus, the duty cycle γ determines the contrast of the resulting interference pattern,

$$K = \frac{2\gamma(1-\gamma)}{\gamma^2 + (1-\gamma)^2}\,,$$

being largest for $\gamma = 1/2$ (a symmetric oscillation) and vanishing for the duty cycle $\gamma \to 1$ or $\gamma \to 0$.

Chapter 9

9.1. The transmission function of the two-hole aperture may be written as $\tau(x,y) = circ_a(x,y) * (\delta(x-b/2) + \delta(x+b/2))$, with $circ_a$ being the aperture function of a circular hole of radius a. Far off the diffraction screen the field distribution is given by

$$\mathcal{F}[E_0\tau](\nu_x,\nu_y) = E_0\mathcal{F}[circ_a](\nu_x,\nu_y)\mathcal{F}[\delta(x-b/2)+\delta(x+b/2)]$$

$$= \frac{2aE_0}{\sqrt{\nu_x^2+\nu_y^2}}J_1(2\pi a\sqrt{\nu_x^2+\nu_y^2})\cos b\pi\nu_x\,.$$

9.2. If $\tau = circ_{R_2} - circ_{R_1}$, the wave field in the Fraunhofer approximation has the form

$$\mathcal{F}[E_0\tau](\nu_x,\nu_y) = E_0\frac{1}{\sqrt{\nu_x^2+\nu_y^2}}\left[R_2 J_1\left(2\pi R_2\sqrt{\nu_x^2+\nu_y^2}\right) - R_1 J_1\left(2\pi R_1\sqrt{\nu_x^2+\nu_y^2}\right)\right]$$

$$\approx E_0(2\pi R_1\,\Delta R)J_0\left(2\pi R_1\sqrt{\nu_x^2+\nu_y^2}\right)\,.$$

9.3. To achieve best filtering without too large a loss of intensity the diameter of the pinhole should match the diameter of the central diffraction spot. The lens aperture being given by the diameter of the incoming beam, we obtain for the pinhole diameter:

$$d = \frac{2.4\lambda f}{D} = 6.2 \,\mu\text{m}.$$

The focal length of the collimation lens should be $f_c = f(d_a/d_e) = 125$ mm.

9.4. The action of the lens as a phase filter is due to the fact that light waves entering the lens at different distances $\rho = \sqrt{x^2 + y^2}$ from the optical axis (the z-axis) traverse paths of different length in the glass, that is, they have different optical path lengths. Let z_0 be the thickness of the lens on the optical axis, k the wavenumber, r_1 and r_2 the radii of curvature and n the refractive index of the glass. Then, the phase shift of a light wave entering at a distance ρ from the axis is given by

$$\Delta\varphi(\rho) = kz_0 n - k\frac{\rho^2}{2}\left(\frac{1}{r_1} - \frac{1}{r_2}\right)(n-1) = kz_0 n - k\frac{\rho^2}{2f}.$$

Thus, the transmission function of the lens is

$$\tau(\rho) = \exp\left[i\Delta\varphi(\rho)\right] = \exp(ikz_0 n)\exp\left[-ik\rho^2/(2f)\right].$$

Let $E_1(x,y)$ denote the field distribution immediately in front of the lens. Using the impulse response of free space, the field distribution $E_2(x',y')$ in the back focal plane is obtained with Eq. (9.11):

$$E_2(x',y') = \left(E_1(x,y)\tau(x,y)\right) * h_f(x',y')$$

$$= \frac{\exp(ikf)}{i\lambda f}\exp\left(i\frac{k(x'^2 + y'^2)}{2f}\right)\mathcal{F}[E_1]\left(\frac{x'}{\lambda f}, \frac{y'}{\lambda f}\right).$$

9.5. For a function g to be invariant under $\mathcal{F}$ we require $\mathcal{F}[g(x,y)](v_x, v_y) = g(v_x, v_y)$. It follows that an invariant function g necessarily is point symmetric, that is, $g(x,y) = g(-x,-y)$, since $\mathcal{F}[\mathcal{F}[g]] = g^{(-)}$. Now, from every point symmetric function g an invariant (also point symmetric) function g' may be constructed:

$$g'(x,y) = g(x,y) + \mathcal{F}[g](x,y).$$

9.6. Upon illumination with a plane monofrequency wave incident along the normal the field distribution behind the Fresnel zone plate (taking $E_0 = 1$) has the form

$$E(r) = T(r) = \sum_{n=-\infty}^{+\infty} 2\,\text{sinc}\left(\frac{n}{2}\right)\exp(in\alpha r^2)\text{circ}_a(r).$$

The field at a distance d behind the plate can be expressed by the Fresnel integral

$$E(x',y',d) = \frac{\exp(ikd)}{i\lambda d}\exp\left[\frac{ik(x'^2 + y'^2)}{2d}\right]\sum_{n=-\infty}^{+\infty} 2\,\text{sinc}\left(\frac{n}{2}\right)\cdot$$

$$\cdot\iint \exp\left[i\left(n\alpha + \frac{k}{2d}\right)r^2 - i\frac{k}{d}(x'x + y'y)\right]\text{circ}_a\left(\sqrt{x^2 + y^2}\right)\,\text{d}x\,\text{d}y.$$

For $d = d_n = -2n\alpha/k$, n being odd, the quadratic phase factor of the integrand disappears (for even n we have $\mathrm{sinc}(n/2) = 0$). The integral for the summation index n gives a focus on the optical axis at distance d_n, which has the form of an Airy disk (diffraction pattern of the corresponding aperture):

$$\iint \exp\left[-i\frac{k}{d}(x'x + y'y)\right] \mathrm{circ}_a\left(\sqrt{x^2 + y^2}\right) dx\,dy$$
$$= \frac{a\lambda d}{\sqrt{x^2 + y^2}} J_1\left(2\pi a \frac{\sqrt{x^2 + y^2}}{\lambda d}\right).$$

9.7. Amplitude grating: its Fourier spectrum is given by

$$\mathcal{F}[E_0 \tau](\nu_x, \nu_y) = ab\,\mathrm{comb}\,(a\nu_x)\,\mathrm{sinc}\,(b\nu_x)\delta(\nu_y).$$

Eliminating the zeroth order $(\nu_x, \nu_y) = (0,0)$ (dark field method) results in the following field and intensity distribution in the filtered image $(\gamma = a/b)$:

$$E'(x,y) = E_0(\tau(x,y) - \gamma)$$
$$I'(x,y) = |E_0|^2\left[\left(\gamma(\gamma - 1) + \frac{1}{2}\right) + \left(\tau(x,y) - \frac{1}{2}\right)(1 - 2\gamma)\right].$$

A modulation of the intensity of size $|E_0|^2(1 - 2\gamma)$ is observed which vanishes for a symmetric grating $(a = b)$. With the phase contrast method, the same result is obtained up to a greater brightness of the image background:

$$E'(x,y) = E_0[\tau(x,y) - (1 - i)\gamma]$$
$$I'(x,y) = |E_0|^2\left[\left(\gamma(2\gamma - 1) + \frac{1}{2}\right) + \left(\tau(x,y) - \frac{1}{2}\right)(1 - 2\gamma)\right].$$

Phase grating: we abbreviate the square wave function by $f(x)$, that is, $f(x) = (\mathrm{comb}_a(\xi) * \mathrm{rect}_b(\xi))(x)$. The dark field method then leads to the following distributions:

$$E'(x,y) = i\alpha E_0[f(x) - \gamma]$$
$$I'(x,y) = \alpha^2 |E_0|^2\left[\left(\gamma(\gamma - 1) + \frac{1}{2}\right) + \left(f(x) - \frac{1}{2}\right)(1 - 2\gamma)\right].$$

Because of the factor α^2 the intensity is small, the relative intensity modulation being the same as in the case of the amplitude grating. For the phase contrast method we get

$$E'(x,y) = E_0\left[\alpha\gamma + i\left(1 + \alpha[f(x) - \gamma]\right)\right]$$
$$I'(x,y) = |E_0|^2\left[(1 + \alpha(1 - 2\gamma)) + 2\left(f(x) - \frac{1}{2}\right)\alpha\right].$$

Here, terms of the order of α^2 have been neglected. The modulation of intensity has size 2α, the phase modulation of the wave has been converted into an amplitude modulation.

9.8. Each group of equidistant parallels present in the raster generates a corresponding line spectrum in the diffraction pattern that is oriented perpendicularly to the parallels. The separation of the points in the spectrum is inversely proportional to the separation of the parallel lines. The sampled image is contained in the lower frequency part of the spatial Fourier spectrum and thus is concentrated in the spectral plane around the spot of zeroth order and, with less intensity, around those of higher orders. Therefore, a smooth image can be recovered by means of a circular, square, or hexagonal aperture with a diameter of about the minimum separation between two spectral spots, located at $(\nu_x, \nu_y) = (0,0)$ in the Fourier plane (Fig. B.2).

Fig. B.2. Diffraction patterns of the grids shown in Fig. 9.27. The circles outline possible apertures as low pass filters in the Fourier plane for smoothing the images.

9.9. If we realize the reference wave by a point source and assume equal amplitudes for illumination wave and reference wave, the field distribution in the object plane is given by:

$$E(x,y) = R\delta(x,y) + R \operatorname{rect}\left(\frac{x}{a}\right)\operatorname{rect}\left(\frac{y}{b}\right).$$

The field in the plane of the hologram essentially is given by the Fourier transform of $E(x,y)$. It leads, by intensity formation, to the transmission function (C_1 and C_2 being suitable constants):

$$\tau(x',y') = C_1 - C_2|R|^2\left[a^2b^2\operatorname{sinc}^2(a\nu_x)\operatorname{sinc}^2(b\nu_y) + 2ab\operatorname{sinc}(a\nu_x)\operatorname{sinc}(b\nu_y)\right].$$

Upon reconstruction with the point source $R\delta(x,y)$ this τ gives, since $\mathcal{F}[\operatorname{sinc}^2](x) = \Delta(x)$ (triangular function, compare Problem A.2), the following field distribution in the image plane:

$$E(x'',y'') = RC_1\delta(x'',y'') - C_2R|R|^2\left[ab\Delta\left(\frac{x''}{a}\right)\Delta\left(\frac{y''}{b}\right) + 2\operatorname{rect}\left(\frac{x}{a}\right)\operatorname{rect}\left(\frac{y}{b}\right)\right].$$

Upon filtering, the reference wave is replaced by the Fourier transform of the shifted rectangular aperture, $\mathcal{F}[R\operatorname{rect}((x-x_0)/a)\operatorname{rect}((y-y_0)/b)]$. This leads, in the process of holographic filtering, to the following field distribution in the image plane:

$$
\begin{aligned}
E(x'',y'') = {} & C_1R\exp(i\varphi)\operatorname{rect}\left(\frac{x''}{a}\right)\operatorname{rect}\left(\frac{y''}{b}\right) \\
& - C_2R|R|^2 a^3b^3\exp(i\varphi)\mathcal{F}[\operatorname{sinc}^3(a\nu_x)\operatorname{sinc}^3(b\nu_y)](x'',y'') \\
& - 2C_2R|R|^2 ab\exp(i\varphi)\Delta\left(\frac{x''}{a}\right)\Delta\left(\frac{y''}{b}\right).
\end{aligned}
$$

The displacement of the aperture is reflected by the phase factor $\exp(i\varphi) = \exp[-2i\pi(x_0\nu_x/a + y_0\nu_y/b)]$. The first term of the right-hand side corresponds to the object image. The term given in the second line, not further elaborated, represents the broadened zeroth order. The term in the third line is just the sum of the autocorrelation function and the convolution of the object distribution that coincide for the object considered here. The displacement (x_0, y_0) of the object, only entering the phase factor, does not show up in the image intensity.

Chapter 10

10.1. Denote the populations of levels $E_0, \ldots E_3$ by $N_0, \ldots N_3$, respectively. Because of the assumptions $N_3 = 0$ and $N_1 = 0$, the total number of atoms taking part in the laser process is $N = N_0 + N_2$. Formulated for the population N_2 and the photon number Q the intensity equation reads

$$\frac{\mathrm{d}Q}{\mathrm{d}t} = -\gamma Q + WQN_2 + W_{sm}N_2$$

and the material equation is given by

$$\frac{\mathrm{d}N_2}{\mathrm{d}t} = -W_{20}N_2 - WQN_2 + W_{02}N_0 .$$

Here, the parameter W_{02} denotes the normalized transition rate from the ground state to the upper laser level (via the pump level), and W_{20} is the normalized rate for transitions from the upper laser level to the ground level by relaxation. Compared to the three-level system, essentially the term $+WN_1Q$ has vanished in the material equation. Disregarding spontaneous emission, $W_{sm}N_2$, and introducing normalized variables $q = WQ/\gamma$, $n = WN_2/\gamma$, $p = WW_{02}N/\gamma^2$, $b = (W_{02} + W_{20})/\gamma$, and the rescaled-time derivatve $(\cdot) = (1/\gamma)\mathrm{d}()/\mathrm{d}t$ leads to (10.28).

10.2. Setting $\dot{x} = \dot{y} = \dot{z} = 0$, we find a stationary point at the origin $(0,0,0)$ for all parameter values. From an evaluation of the characteristic equation of the Jacobian of the vector field it turns out that this solution is stable for $\rho < 1$ and unstable for $\rho > 1$. Furthermore, for $\rho > 1$ an additional (symmetric) pair of stationary points $(x_0, x_0, \rho - 1)$ exists with $x_0 = \pm\sqrt{\beta(\rho - 1)}$. A quite tedious calculation along the same lines yields stability for $\rho < [\sigma(\sigma + \beta + 3)]/(\sigma - \beta - 1)$ and instability beyond that value.

10.3. At first, try to reproduce the trajectories of Fig. 10.16. For the chaotic attractor given there, calculate the distance $d(t)$ and the quantity $\log(d(t))$ of two initially very close states in the course of time. Repeat this calculation for different, randomly chosen initial conditions on the attractor.

Chapter 11

11.1. Since the electric field $E(t) \propto \exp(-at^2)$ we have for the on-axis intensity (irradiance) on the beam axis

$$I(t) = I_{\max}\exp(-2at^2)$$

and for the beam power, which is the integral of I over the cross-section of the beam,

$$P(t) = P_{\max}\exp(-2at^2) .$$

The pulse energy is

$$W_P = \int_{-\infty}^{\infty} P(t)\,\mathrm{d}t = P_{\max} \int_{-\infty}^{\infty} \exp(-2at^2)\,\mathrm{d}t = \sqrt{\frac{\pi}{2a}}P_{\max} .$$

As W_P is given, and a can be calculated from $\Delta t_G = 30\,\mathrm{fs}$ with (11.2), $a = 2\ln 2/\Delta t_G^2 = 1.54 \cdot 10^{27}\,\mathrm{s}^{-2}$, the peak power follows immediately: $P_{max} = \sqrt{2a/\pi}\,W_P = 375.7\,\mathrm{kW}$. Taking the Gaussian intensity distribution at $t = 0$ (maximum intensity) $I(r, \varphi) = I_{max}\exp(-r^2/r_s^2)$ and integrating over the cross-section of the beam to obtain P_{max},

$$P_{max} = \int_0^\infty r\,dr \int_0^{2\pi} d\varphi\, I(r, \varphi),$$

(r and φ being the cylindrical coordinates), we get

$$P_{max} = I_{max}\int_0^\infty r\,dr \int_0^{2\pi} d\varphi \exp\left(-\frac{r^2}{r_s^2}\right) = 2\pi I_{max}\int_0^\infty r\exp\left(-\frac{r^2}{r_s^2}\right) dr$$

$$= \pi I_{max} r_s^2.$$

For $r_s = 1\,\mathrm{mm}$, the unfocused beam, a peak intensity $I_{max} = P_{max}/\pi r_s^2$ of $I_{max} = 11.96\,\mathrm{MW/cm^2}$ is attained; for the focused beam with $r_s = 5\,\mathrm{\mu m}$, it is $I_{max} = 478\,\mathrm{GW/cm^2}$.

11.2. (i) The rectangular oscillation pulse can be written as (see Sect. A)

$$E_{rp}(t) = E_0\exp(-i\omega_0 t)\,\mathrm{rect}(t/\tau).$$

For $\tau \gg 2\pi/\omega_0$ the pulse width Δt_{rp} can be identified with τ, the width of the envelope. Upon Fourier transform of $E_{rp}(t)$, needed to calculate the spectral width $\Delta\nu$, we get

$$\tilde{E}_{rp}(\nu) = \mathcal{F}\{E_{rp}\}(\nu) = E_0\,\delta(\nu + \nu_0) * \tau\,\mathcal{F}\{\mathrm{rect}\}(\tau\nu) = E_0\,\tau\,\mathrm{sinc}(\pi\tau(\nu + \nu_0)).$$

The spectral width $\Delta\nu$ (FWHM of the intensity) is given by $\mathrm{sinc}(\pi\tau\Delta\nu/2) = 1/\sqrt{2}$. With $\xi = 1.3916$ as the approximate solution of the equation $\sin\xi = \xi/\sqrt{2}$, we get $\pi\tau\Delta\nu/2 \approx 1.3916$ (see Sect. 11.2.1) or

$$\Delta\nu\,\Delta t_{rp} \approx 0.886$$

as the time-bandwidth product of the rectangular oscillation pulse.

(ii) As before, when $\gamma^{-1} \gg 2\pi/\omega$, the width of the exponentially decaying oscillation pulse is given by the width of its intensity envelope, $\Delta t_{ep} = \ln 2/(2\gamma)$. The Fourier transform of $E_{ep}(t)$ reads

$$\tilde{E}_{ep}(\nu) = E_0\int_0^\infty \exp(-i2\pi\nu_0 t)\exp(-\gamma t)\exp(-i2\pi\nu t)\,dt = \frac{E_0}{i2\pi(\nu + \nu_0) + \gamma}.$$

For the power spectrum $|\tilde{E}_{ep}|^2$ to drop to half of its maximum value (around $-\nu_0$), we have the condition $4\pi^2(\Delta\nu/2)^2 = \gamma^2$, or, with $\gamma = \ln 2/(2\Delta t_{ep})$,

$$\Delta\nu\,\Delta t_{ep} = \frac{\ln 2}{2\pi} = 0.11.$$

11.3. This is too easy. Taking the inverse Fourier transform of (11.8) we get

$$E_{lp}(t) = \mathcal{F}^{-1}\{\tilde{E}_{lp}(\omega)\}(t)$$

$$= E_0\sqrt{\frac{\pi}{a}}\int_{-\infty}^\infty \exp[-i2\pi\tau(\nu + \nu_0)]\exp\left(-\frac{\pi^2(\nu + \nu_0)^2}{a}\right)\exp(i2\pi\nu t)\,d\nu$$

$$= \exp(-i2\pi\tau\nu_0)\left[E_0\sqrt{\frac{\pi}{a}}\int_{-\infty}^\infty \exp(i2\pi\nu(t - \tau))\exp\left(-\frac{\pi^2(\nu + \nu_0)^2}{a}\right)d\nu\right].$$

Reading (11.3) backwards, we can identify the above term in square brackets with $\mathcal{F}^{-1}\{\tilde{E}_{lp}(\omega)\}(t-\tau)$, that is, (11.8) is obtained immediately:

$$E_{lp}(t) = \exp(-i\tau\omega_0)E_0\exp(-i\omega_0(t-\tau))\exp(-a(t-\tau)^2)$$
$$= E_0\exp(-i\omega_0 t)\exp(-a(t-\tau)^2).$$

11.4. From (11.21) we have

$$\frac{\Delta t'_G}{\Delta t_G} = \frac{3\,\text{ps}}{30\,\text{fs}} = 100 = \sqrt{1+16\alpha^2\beta^2} \quad \rightarrow \quad |\alpha\beta| = \frac{\sqrt{9999}}{4}.$$

Comparing (11.26) with (11.10), taking into account that (11.26) is expanded around $\omega_0 > 0$ while the center frequency of the complex pulse spectrum (11.10) is $-\omega_0 < 0$, i.e. transform $\omega \rightarrow -\omega$, we find that $\beta = \frac{1}{2}k''_0 z$ while α is given by (see (11.4) and (11.2)) $\alpha = 2\ln 2/(\Delta t_G)^2$. Thus, with the assumed numerical values we get

$$|k''_0| = \frac{\sqrt{9999}}{4z\ln 2}\Delta t_G^2 \approx \frac{\Delta t'^2_G}{4z\ln 2} = 3.246\,\frac{\text{ps}^2}{\text{km}}.$$

11.5. The amplitude E_r of the reflected wave is the sum of the amplitudes of all partial waves leaving the device through the entrance mirrror. With $2kd = \omega t_0$ it is given by:

$$E_r = E_i\left(r_1 + t_1 r_2 t'_1\exp(i\omega t_0) + t_1 r_2^2 r'_1 t'_1\exp(i2\omega t_0) + t_1 r_2^3 r'^2_1 t_2\exp(i3\omega t_0)...\right),$$

where E_i is the amplitude of the incident wave. The meaning of the constants r_1, r'_1, t_1 and t'_1 is demonstrated by the Figure given with the problem. Collecting terms, using $r_1 = -r'_1$, $t_1 t'_1 = 1 - r_1^2$, $r_2 = -1$ and summing the geometric series we obtain for the amplitude reflectance $r_{GT} = E_r/E_i$ of the interferometer:

$$r_{GT} = r_1 - t_1 t'_1\exp(i\omega t_0)\sum_{n=0}^{\infty}(-r'_1\exp(i\omega t_0))^n = r_1 - \frac{t_1 t'_1\exp(i\omega t_0)}{1+r'_1\exp(i\omega t_0)}$$
$$= \frac{r_1 - \exp(i\omega t_0)}{1 - r_1\exp(i\omega t_0)}.$$

By multiplying r_{GT} with its complex conjugate we verify that $|r_{GT}| = 1$:

$$r_{GT}r^*_{GT} = |r_{GT}|^2 = \frac{(r_1 - \exp(i\omega t_0))(r_1 - \exp(-i\omega t_0))}{(1 - r_1\exp(i\omega t_0))(1 - r_1\exp(-i\omega t_0))} = 1.$$

To obtain $\partial\varphi/\partial\omega$ we note that since $|r_{GT}| = 1$, we can write $r_{GT} = \exp(i\varphi)$ and

$$\frac{\partial r_{GT}}{\partial\omega} = i\frac{\partial\varphi}{\partial\omega}\exp(i\varphi) = i\frac{\partial\varphi}{\partial\omega}r_{GT} \quad \rightarrow \quad \frac{\partial\varphi}{\partial\omega} = \frac{1}{ir_{GT}}\frac{\partial r_{GT}}{\partial\omega}.$$

The last expression can be evaluated easily to give

$$\frac{\partial\varphi}{\partial\omega} = \frac{(1-r_1^2)t_0}{1 - 2r_1\cos(\omega t_0) + r_1^2}.$$

Differentiating once again w.r.t. ω is straightforward and obtains

$$\frac{\partial^2 \varphi}{\partial \omega^2} = -\frac{2r_1(1-r_1^2)t_0^2 \sin(\omega t_0)}{(1+r_1^2-2r_1\cos(\omega t_0)^2}.$$

11.6. (a) After substituting the pulse $E(t) = E_0(t)\cos(\omega_0 t)$[1] in (11.54) and expanding all the terms the average intensity reads

$$I_s(\tau) = \left\{ \frac{1}{2T} \int_{-T}^{T} E_0^4(t)\cos^4(\omega_0 t)\,dt + \frac{1}{2T} \int_{-T}^{T} E_0^4(t-\tau)\cos^4(\omega_0(t-\tau))\,dt \right\}$$

$$+4\frac{1}{2T} \int_{-T}^{T} E_0^3(t)E_0(t-\tau)\cos^3(\omega_0 t)\cos(\omega_0(t-\tau))\,dt$$

$$+6\frac{1}{2T} \int_{-T}^{T} E_0^2(t)E_0^2(t-\tau)\cos^2(\omega_0 t)\cos^2(\omega_0(t-\tau))\,dt$$

$$+4\frac{1}{2T} \int_{-T}^{T} E_0(t)E_0^3(t-\tau)\cos(\omega_0 t)\cos^3(\omega_0(t-\tau))\,dt.$$

The powers and products of trigonometric functions are expanded by means of trigonometric identities. The idea of the approximation is that, because $E_0(t)$ is slowly varying, the integrals containing oscillating terms of the form $\cos n\omega_0 t$ and $\sin n\omega_0 t$ are very small and can be neglected. For the first integral of the above equation, we have $\cos^4 \omega_0 t = 3/8 + (1/2)\cos 2\omega_0 t + (1/8)\cos 4\omega_0 t$. As the second integral is identical to the first (the time shift can be cancelled as the integrand vanishes sufficiently fast) we have from these two terms the contribution

$$2\frac{3}{8}\frac{1}{2T} \int_{-T}^{T} E_0^4(t)\,dt.$$

Similarly, because $\cos^3(\omega_0 t)\cos(\omega_0(t-\tau)) = \cos^4(\omega_0 t)\cos(\omega_0 \tau) + \cos^3(\omega_0 t)\sin(\omega_0 t)$ $\cdot \sin(\omega_0 \tau)$, the only term remaining for the third integral (second row) is

$$4\frac{3}{8}\cos(\omega_0 \tau)\frac{1}{2T} \int_{-T}^{T} E_0^3(t)E_0(t-\tau)\,dt.$$

Analogously, the contribution of the fifth integral (fourth row) is given by

$$4\frac{3}{8}\cos(\omega_0 \tau)\frac{1}{2T} \int_{-T}^{T} E_0(t)E_0^3(t-\tau)\,dt.$$

(make the substitution $t \to t+\tau$ and note that $\cos(\omega_0 \tau)$ is symmetric in τ). Finally, for the fourth integral (third row) we have $\cos^2(\omega_0 t)\cos^2(\omega_0(t-\tau)) = (1/4)+(1/8)\cos(2\omega_0 \tau) + $ oscillating terms, and it thus reduces to

$$6\left(\frac{1}{4}+\frac{1}{8}\cos(2\omega_0 \tau)\right)\frac{1}{2T} \int_{-T}^{T} E_0^2(t)E_0^2(t-\tau)\,dt.$$

All the given intermediate results combined yield equation (11.56).

(b) We have to calculate (11.56) for a Gaussian envelope function $E_0(t) = \exp(-at^2)$. For simplicity, we assume a single pulse (single-shot autocorrelation), let $T \to \infty$ and omit the prefactor $1/2T$. With

---

[1] the real part has to be taken because of the nonlinearity of the equation

$$\int_{-\infty}^{\infty} E_0^4(t)\,dt = \int_{-\infty}^{\infty} \exp(-4at^2)\,dt = \sqrt{\frac{\pi}{4a}}$$

$$\int_{-\infty}^{\infty} E_0^2(t)E_0^2(t-\tau)\,dt = \exp(-a\tau^2)\int_{-\infty}^{\infty} \exp(-[2\sqrt{a}(t-\tau/2)]^2)\,dt$$

$$= \frac{1}{2}\sqrt{\frac{\pi}{a}}\exp(-a\tau^2)$$

$$\int_{-\infty}^{\infty} E_0^3(t)E_0(t-\tau)\,dt = \int_{-\infty}^{\infty} E_0(t)E_0^3(t-\tau)\,dt = \exp(-\frac{3}{4}a\tau^2)$$

it follows that

$$I_s(\tau) = \frac{3}{8}\sqrt{\frac{\pi}{a}}\left\{1 + 2\exp(-a\tau^2) + 4\cos(\omega\tau)\exp(-\frac{3}{4}a\tau^2) + \cos(2\omega\tau)\exp(-a\tau^2)\right\}.$$

Chapter 12

12.1. The time derivatives of the solution ansatz,

$$\dot{x} = [\dot{A}(t) - i\omega A(t)]\exp(-i\omega t) + \text{c.c.},$$
$$\ddot{x} = [\ddot{A}(t) - i2\omega\dot{A}(t) - \omega^2 A(t)]\exp(-i\omega t) + \text{c.c.}$$

as well as

$$x^3 = A^3(t)\exp(-i3\omega t) + 3A^2 A^* \exp(-i\omega t) + \text{c.c.}$$

are substituted into the Duffing equation, the term $\ddot{A}$ being ignored:

$$\left[\dot{A}(d - 2i\omega) + \left(\omega_0^2 - \omega^2 - id\omega + 3\alpha|A|^2\right)A\right]\exp(-i\omega t) + A^3\exp(i3\omega t) + \text{c.c.}$$
$$= E_0\cos(\omega t).$$

Since the right-hand side may be represented by $(E_0/2)\exp(-i\omega t) + \text{c.c.}$, the differential equation (12.103) immediately follows from a comparison of coefficients. Set $\dot{A} = 0$ and a cubic equation for the quantity $|A|^2$ is deduced:

$$9\alpha^2|A|^6 + 6\alpha(\omega_0^2 - \omega^2)|A|^4 + \left[(\omega_0^2 - \omega^2)^2 - \omega^2 d^2\right]|A|^2 - \frac{E_0^2}{4} = 0.$$

12.2. Ignoring the term αx^3, which only contributes to the amplitudes A_i in the third order, we just have, as the first approximation to the solution, a superposition of two harmonic waves of frequency ω_1 and ω_2 and of amplitude A_1 and A_2, respectively:

$$A_i = \frac{E_i}{\sqrt{(\omega_0^2 - \omega_i^2)^2 + d^2\omega_i^2}}, \qquad i = 1, 2.$$

Taking now into account the nonlinear terms, we obtain as a second approximation the following expression:

$$0 = A_1\left[-\omega_1^2 - id\omega_1 + \omega_0^2 + \alpha(3A_1 A_1^* + 6A_2 A_2^*)\right]\exp[-i\omega_1 t]$$
$$+ A_2\left[-\omega_2^2 - id\omega_2 + \omega_0^2 + \alpha(3A_2 A_2^* + 6A_1 A_1^*)\right]\exp[-i\omega_2 t]$$
$$+ \alpha\left(A_1^3\exp[-3i\omega_1 t] + 3A_1^2 A_2\exp\left[-i(2\omega_1 + \omega_2)t\right]\right)$$

$$+ 3A_1^2 A_2^* \exp\left[-i(2\omega_1 - \omega_2)t\right] + 3A_1 A_2^2 \exp\left[-i(\omega_1 + 2\omega_2)t\right]$$

$$+ 3A_1(A_2^*)^2 \exp\left[-i(\omega_1 - 2\omega_2)t\right] + A_2^3 \exp[-3i\omega_2 t]\Big) + \text{c.c.}$$

This ansatz for the solution contains eight complex or sixteen real parameters. It is seen that indeed all possible combination frequencies occur that can be generated from three frequencies.

12.3. The amplitude of the fundamental wave upon entry into the crystal is $A_1(0) = \sqrt{2S/nc\varepsilon_0} = 22.36 \cdot 10^6 \text{ V m}^{-1}$. From (12.83) the sought-after length of the crystal is calculated:

$$z_{1/2} = \frac{\operatorname{arctanh}(1/\sqrt{2})}{KA_1(0)} = \frac{0.88}{KA_1(0)} = 3.27 \text{ cm}.$$

12.4. Substituting the derivatives

$$\frac{dA_1}{dz}(z) = A_1(0)\exp\left(-i\Delta kz/2\right)\left[-i\Delta k\cosh(bz) + \left(b + \frac{(\Delta k)^2}{4b}\right)\sinh(bz)\right]$$

$$+ A_2^*(0)\exp\left(-i\Delta kz/2\right)\left[iK_p\cosh(bz) + \frac{\Delta k}{2b}\sinh(bz)\right],$$

$$\frac{dA_2^*}{dz}(z) = A_2^*(0)\exp\left(i\Delta kz/2\right)\left(b - i\frac{\Delta k}{2b}\right)\sinh(bz)$$

$$+ A_1(0)\exp\left(i\Delta kz/2\right)\left[-iK_p\cosh(bz) + \frac{\Delta k K_p}{2b}\sinh(bz)\right],$$

into the differential equations (12.86) and (12.87) leads to the desired result.

Chapter 13

13.1. With $\Theta_m \leq \Theta_T$ or $\sin\Theta_m \leq \sin\Theta_T$, where $\Theta_T = 90° - \Theta_c$ and Θ_c denotes the critcal angle for total internal reflection at the core–cladding interface, the relation (13.2) along with $\sin\Theta_T = \sqrt{1 - (n_2/n_1)^2}$ leads to

$$m \geq N = \frac{4a}{\lambda}\frac{\left|1/2 - (n_1/n_2)^2\right|}{\sqrt{1 - (n_1/n_2)^2}}.$$

In the model considered, for $l \to \infty$ infinitely many modes having a mode index of $m \geq N$ are able to propagate. Of course, the wave at $\Theta = 0$ (included in the ansatz (13.2) only as the limiting case $m \to \infty$) is a propagating mode. For a fiber with the properties stated we obtain $N = 85$. The maximum difference in propagation time exists for the wave at $\Theta_0 = 0$, running at the speed of light in the medium, $c_1 = c_0/n_1$, and the mode with index N, having an effective propagation velocity $c_{\min} \approx \cos\Theta_T(c_0/n_1) = \cos\Theta_T c_1$. In our example we get $c_{\min} = 0.938c_1$. Thus, due to mode dispersion the maximum difference in propagation time per kilometer is given by

$$\Delta T = \frac{1000 \text{ m}}{c_1}\left(\frac{1}{\sin\Theta_c} - 1\right) = \frac{1000 \text{ m}}{c_0}\frac{n_1}{n_2}(n_1 - n_2) = 0.355 \text{ μs}.$$

13.2. The smallest measurable change of temperature is:

$$\Delta T = \frac{\Delta \varphi_{\min}}{kL(\mathrm{d}n/\mathrm{d}t + n\alpha)} = 3.6 \cdot 10^{-4} \text{ K}.$$

13.3. The full width at half maximum covers ≈ 3224 wavelengths. A soliton having a width of 100 wavelengths lasts about $\Delta t_h = 775$ fs. Thus the bit rate achievable with such solitons is given by $b = 1/(7 \cdot 775 \text{ fs}) = 184$ Gbit/s. The spectrum of a periodic train of solitons at a maximum bit rate is a line spectrum, the lines having a separation of 184 GHz. The spectrum is modulated by a bell-shaped function, centered at the carrier frequency $\nu_0 = 1.93 \cdot 10^{14}$ Hz, that corresponds to the spectrum of a single soliton. The width (FWHM) of the spectral envelope is approximately $\Delta \nu_h \approx 1/(\pi \Delta t_h) = 411$ GHz $= 0.0021 \nu_0$. The pulse width being inversely proportional to the pulse amplitude, the soliton can travel a distance $l = -20 \log_{10}(1/2)/0.12$ km $= 50.2$ km until its width has doubled.

13.4. The width of the wave packet has increased tenfold after $l = 18.4$ km.

13.5. To verify the brighton solution, the derivatives

$$\frac{\partial U}{\partial \zeta} = i\frac{a^3}{2} \operatorname{sech}(a\sigma) \exp(ia^2 \zeta/2),$$

$$\frac{\partial^2 U}{\partial \sigma^2} = a^3 \left[\operatorname{sech}(a\sigma) \tanh^2(a\sigma) - \operatorname{sech}^3(a\sigma) \right] \exp(ia^2 \zeta/2),$$

$$|U|^2 U = a^3 \operatorname{sech}^3(a\sigma) \exp(ia^2 \zeta/2),$$

are substituted into the differential equation (13.37). To verify the darkon solution, the expressions

$$\frac{\partial U}{\partial \zeta} = ia^3 \tanh(a\sigma) \exp(ia^2 \zeta),$$

$$\frac{\partial^2 U}{\partial \sigma^2} = -2a^3 \operatorname{sech}^2(a\sigma) \tanh(a\sigma) \exp(ia^2 \zeta),$$

$$|U|^2 U = a^3 \tanh^3(a\sigma) \exp(ia^2 \zeta)$$

are substituted into (13.40).

13.6. The brighton solution, written in dimensional form, reads:

$$E(z,\tau) = \sqrt{\frac{|k_0''| n_0}{k_0 n_2}} \frac{1}{t_0} \operatorname{sech}\left(\frac{\tau}{t_0}\right) \exp\left[i\frac{|k_0''| z}{2t_0^2}\right],$$

t_0 being a measure of the pulse width. For a 10 ps pulse the maximum field strength amounts to $E_{\max} \approx 630$ kV m^{-1}. The area of the fiber core being $F = \pi D^2/4$, the maximum power of the pulse is given by $P_{\max} = (1/2)\varepsilon_0 n_0 c F |E_{\max}|^2 = 62$ mW.

13.7. We will show the invariance of I_0 for the solution of (13.37). In the case of (13.40) the calculations proceed analogously. By multiplying (13.37) with U^*, multiplying the complex conjugate equation with U, and subtracting the resulting expressions the nonlinear term is eliminated:

$$i\left(U^*\frac{\partial U}{\partial \xi} + U\frac{\partial U^*}{\partial \xi}\right) + \frac{1}{2}\left(U^*\frac{\partial^2 U}{\partial \sigma^2} + U\frac{\partial^2 U^*}{\partial \sigma^2}\right) = 0.$$

The first term equals $i(\partial/\partial \xi)[UU^*]$. Next, the equation is integrated over σ and the integral over the second term is eliminated by two-fold partial integration and by using the boundary conditions. This leads to the desired result,

$$\frac{\partial}{\partial \xi}\int_{-\infty}^{+\infty}|U|^2\,\mathrm{d}\sigma = 0.$$

Appendix A

A.1. The Fourier transform of $f(t) = \exp\left(-\pi t^2\right)$ is given by

$$\mathcal{F}[f](\nu) = \exp\left(-\pi \nu^2\right)\int_{-\infty}^{+\infty}\exp\left[\pi(t+i\nu)^2\right]\mathrm{d}t = \exp\left(-\pi \nu^2\right).$$

To arrive at this result, first a rectangle between $t = -C$ and $t = +C$ was chosen as integration path in the complex plane, its horizontal sides being located at $\nu = 0$ and ν. Exploiting the holomorphy of $\exp\left(z^2\right)$ and using $\int_{-\infty}^{+\infty}\exp\left(-\pi t^2\right)\mathrm{d}t = 1$, the relation is obtained in the limit $C \to \infty$. By means of the similarity theorem the Fourier transform of the rescaled Gaussian pulse follows at once,

$$\mathcal{F}[f_a](\nu) = a^{-1}\exp\left(-\frac{\pi \nu^2}{a}\right).$$

A.2. The amplitude spectrum of the triangular pulse,

$$\mathcal{F}[\Delta](\nu) = \int_{-\infty}^{+\infty}\Delta(t)\exp(-i2\pi \nu t)\,\mathrm{d}t,$$

is real since the function Δ is symmetric, that is,

$$\mathcal{F}[\Delta](\nu) = 2\int_0^1(1-t)\cos(2\pi \nu t)\,\mathrm{d}t$$

$$= \frac{1}{2\pi^2\nu^2}(1-\cos 2\pi \nu) = (\mathrm{sinc}\,\nu)^2.$$

A.3. Autocorrelation function: $\mathrm{rect}\otimes\mathrm{rect}\,(\tau) = \Delta(\tau)$ (triangular function). Because of $\mathcal{F}[\mathrm{rect}] = \mathrm{sinc}$, the Wiener–Khinchin theorem may be verified immediately:

$$\mathcal{F}[\mathrm{rect}\otimes\mathrm{rect}] = \mathcal{F}[\Delta] = |\mathcal{F}[\mathrm{rect}]|^2.$$

A.4. For the proof of the convolution theorems, the representation (A.23) of the Dirac δ distribution is used. In case of the first theorem we have:

$$\mathcal{F}[fg](v) = \int f(x)g(x)\exp(-i2\pi vx)\,dx$$

$$= \int \left[\int f(x')\delta(x-x')\,dx'\right]g(x)\exp(-i2\pi vx)\,dx$$

$$= \int \left(\int f(x')\left[\int \exp\left[i2\pi v'(x-x')\right]\,dv'\right]\,dx'\right)g(x)\exp(-i2\pi vx)\,dx$$

$$= \int \left[\int f(x')\exp(-i2\pi x'v')\,dx'\right]\left[\int g(x)\exp\left[-i2\pi x(v-v')\right]\,dx\right]\,dv'$$

$$= (\mathcal{F}[f] * \mathcal{F}[g])(v).$$

The proof of the second theorem proceeds analogously.

A.5. The Fourier transform of the beat signal is

$$\tilde{f}(v) = A\left[\frac{i}{2}\delta(v-v_1) - \frac{i}{2}\delta(v+v_1)\right] * \left[\frac{1}{2}\delta(v-v_2) + \frac{1}{2}\delta(v+v_2)\right]$$

$$= \frac{iA}{4}\left[\delta\left(v-(v_1+v_2)\right) - \delta\left(v+(v_1+v_2)\right)\right.$$

$$\left. +\delta\left(v-(v_1-v_2)\right) - \delta\left(v+(v_1-v_2)\right)\right].$$

Note that the product of two δ functions is not defined but their convolution product is. The amplitude spectrum of f consists of two lines (side bands), the signal $f(t)$ is a superposition of two sinusoidal oscillations.

References

Chapter 1

[1.1] I. Newton: *Opticks* (Dover, New York 1952)
[1.2] A. Einstein: "Zur Elektrodynamik bewegter Körper", Ann. Phys. **17**, 891–921 (1905)
[1.3] Lucretius: *De rerum natura* (Clarendon, Oxford 1984)
[1.4] M. Planck: "Zur Theorie des Gesetzes der Energieverteilung im Normalspektrum", Verh. Dtsch. phys. Ges. Berlin **2**, 202 (1900)
[1.5] A. Einstein: "Über einen die Erzeugung und Verwandlung des Lichtes betreffenden heuristischen Gesichtspunkt", Ann. Physik **17**, 132 (1905)
[1.6] D. Gabor: "A new microscopic principle", Nature **161**, 777 (1948)
[1.7] R. Menzel: *Photonics* (Springer, Berlin 2001)
B. E. A. Saleh, M. C. Teich: *Fundamentals of Photonics* (Wiley, New York 1991)
[1.8] M. A. Nielsen, I. L. Chuang: *Quantum Computation and Quantum Information* (Cambridge University Press, Cambridge 2000)
C. P. Williams, S. H. Clearwater: *Ultimate Zero and One. Computing at the Quantum Frontier* (Copernicus, New York 2000)
[1.9] W. K. Wootters, W. H. Zurek: "A single quantum cannot be cloned", Nature **299**, 802 (1982)
D. Dieks: "Communication by EPR devices", Phys. Lett. A **92**, 271 (1982)
A. Lamas-Linares, Ch. Simon, J. C. Howell, D. Bouwmeester: "Experimental quantum cloning of single photons", Nature **296**, 712 (2002)

Chapter 2

[2.1] T. Sauter, W. Neuhauser, R. Blatt, P. E. Toschek: "Observation of quantum jumps", Phys. Rev. Lett. **57**, 1696–1698 (1986)
J. C. Bergquist, R. G. Hulet, W. M. Itano, D. J. Wineland: "Observation of quantum jumps in a single atom", Phys. Rev. Lett. **57**, 1699–1702 (1986)
[2.2] R. Hanbury Brown, R. Q. Twiss: "Correlation between photons in two coherent beams of light", Nature **177**, 27–29 (1956)

Chapter 3

[3.1] R. P. Feynman, R. B. Leighton, M. Sands: *The Feynman Lectures on Physics, Vol. II* (Addison–Wesley, Reading, Mass. 1964)
M. Alonso, E. J. Finn: *Physics* (Addison–Wesley, Reading, Mass. 1992)
[3.2] R. Hasegawa: *Optical Solitons in Fibers* (Springer, Berlin, Heidelberg 1990)

[3.3]   J. Durnin, J. J. Miceli Jr., J. H. Eberly: "Diffraction-free beams", Phys.
        Rev. Lett. 58, 1499–1501 (1987)
        J. Durnin: "Exact solutions for nondiffracting beams. I. The scalar the-
        ory", J. Opt. Soc. Am. A4, 651–654 (1987)
        F. Gori, G. Guattari. C. Padovani: "Bessel–Gauss beams", Opt. Commun.
        64, 491–495 (1987)
[3.4]   A. Vasara, J. Turunen, A. T. Friberg: "Realization of general nondiffract-
        ing beams with computer-generated holograms", J. Opt. Soc. Am. A6,
        1748–1754 (1989)
        G. Scott, N. McArdle: "Efficient generation of nearly diffraction-free
        beams using an axicon", Opt. Eng. 31, 2640–2643 (1992)
[3.5]   M. Abramowitz, I. A. Stegun: Handbook of Mathematical Functions, Ap-
        plied Mathematics Series, vol. 55 (National Bureau of Standards, Wash-
        ington, reprinted 1968 by Dover Publications, New York)
[3.6]   R. Oron, S. Blit, N. Davidson, A. A. Friesem, Z. Bomzon, E. Hasman: "The
        formation of laser beams with pure azimuthal or radial polarization",
        Appl. Phys. Lett. 77, 3322–3324 (2000)
[3.7]   E. M. Purcell: Berkeley Physics Course, Vol. II: Electricity and Magnetism
        (McGraw–Hill, New York 1963)
        F. S. Crawford: Berkeley Physics Course, Vol. III: Waves (McGraw–Hill,
        New York 1967)

Chapter 4

[4.1]   H. Paul: "Interference between independent photons", Rev. Mod. Phys.
        58, 209–231 (1986)
[4.2]   J. W. Goodman: Introduction to Fourier Optics (McGraw Hill, New York
        1988)
[4.3]   M. J. Beran, G. B. Parrent Jr.: Theory of Partial Coherence (Prentice-Hall,
        Englewood Cliffs 1964)
[4.4]   R. J. Glauber: "The quantum theory of optical coherence", Phys. Rev. 130,
        2529–2539 (1963)
        R. J. Glauber: "Coherent and incoherent states of the radiation field",
        Phys. Rev. 131, 2766–2788 (1963)
[4.5]   H. Haken: Light (Elsevier North–Holland, New York 1981)
[4.6]   K. Rohlfs, T. L. Wilson: Tools of Radio Astronomy, 3rd edn. (Springer,
        Berlin 2000)

Chapter 6

[6.1]   J. C. Dainty (Ed.): Laser Speckle and Related Phenomena (Springer,
        Berlin, Heidelberg 1975)
[6.2]   A. Vogel, W. Lauterborn: "Time-resolved particle image velocimetry used
        in the investigation of cavitation bubble dynamics", Appl. Opt. 27, 1869–
        1876 (1988)
[6.3]   F. Durst, A. Melling, J. H. Whitelaw: Principle and Practice of Laser-
        Doppler-Anemometry (Academic Press, London 1981)
        T. S. Durrani, C. A. Greated: Laser Systems in Flow Measurements
        (Plenum, New York 1977)
[6.4]   A. Labeyrie: "Attainment of diffraction limited resolution in large tele-
        scopes by Fourier analysing speckle patterns in star images", Astron.
        Astrophys. 6, 85–87 (1970)
[6.5]   G. Weigelt, B. Wirnitzer: "Image reconstruction by the speckle-masking
        method", Opt. Lett. 8, 389–391 (1983)

[6.6]   G. Weigelt: "Triple-correlation imaging in optical astronomy", in E. Wolf
(Ed.): *Progress in Optics*, Vol. XXIX, S. 293–319 (North–Holland, Amsterdam 1991)

Chapter 7

[7.1]   D. Gabor: "A new microscopic principle", Nature **161**, 777 (1948)
[7.2]   G. Haussmann, W. Lauterborn: "Determination of size and position of fast
moving gas bubbles in liquid by digital 3-D image processing of hologram
reconstruction", Appl. Opt. **19**, 3529–3535 (1980)
[7.3]   W. Lauterborn, A. Judt, E. Schmitz: "High-speed off-axis holographic cinematography with a copper-vapor-pumped dye laser", Optics Letters **18**,
4–6 (1993)
[7.4]   W. Hentschel, W. Lauterborn: "New speed record in long series holographic cinematography", Appl. Opt. **23**, 3263–3265 (1984)
[7.5]   W. Lauterborn, A. Koch: "Holographic observation of period-doubled and
chaotic bubble oscillations in acoustic cavitation", Phys. Rev. A **35**, 1974–
1976 (1987)
[7.6]   W. Lauterborn, T. Kurz, R. Mettin, C. D. Ohl: "Experimental and theoretical bubble dynamics", in I. Prigogine, S. A. Rice (Ed.): *Advances in Chemical Physics*, Vol. 110, pp. 295–380 (Wiley, New York 1999)
[7.7]   W. H. Lee: "Computer-generated holograms: Techniques and applications", in E. Wolf (Ed.): *Progress in Optics*, Vol. XVI, pp. 119–232 (North–
Holland, Amsterdam 1978)
O. Bryngdahl, F. Wyrowski: "Digital holography – computer-generated
holograms", in E. Wolf (Ed.): *Progress in Optics*, Vol. XXVIII, pp. 1–86
(North–Holland, Amsterdam 1989)
[7.8]   J. W. Goodman: *Introduction to Fourier Optics* (McGraw–Hill, New York
1988)
[7.9]   C. David, J. Thieme, P. Guttmann, G. Schneider, D. Rudolph, G. Schmahl:
"Electron-beam generated x-ray optics for high resolution microscopy
studies", Optik **91**, 95–99 (1992)
G. Schmahl,   D. Rudolph,   B. Niemann,   P. Guttmann,   J. Thieme,
G. Schneider, C. David, M. Diehl, T. Wilhein: "X-ray microscopy studies",
Optik **93**, 95–102 (1993)
G. Schmahl, D. Rudolph (Ed.): *X-ray Microscopy*, Springer Ser. Opt. Sci.,
Vol. 43 (Springer, Berlin, Heidelberg 1984)
D. Sayre, M. Howells, J. Kirz, H. Rarback (Ed.): *X-ray Microscopy II*,
Springer Ser. Opt. Sci., Vol. 56 (Springer, Berlin, Heidelberg 1988)
A. G. Michette, G. R. Morrison, C. J. Buckley (Ed.): *X-ray Microscopy III*,
Springer Ser. Opt. Sci., Vol. 67 (Springer, Berlin, Heidelberg 1992)
[7.10]   A. Ritter, J. Böttger, O. Deussen, M. König, T. Strothotte: "Hardware-
based rendering of full-parallax synthetic holograms", Appl. Opt. **38**,
1364–1369 (1999)
[7.11]   U. Schnars, T. Kreis, W. Jueptner: "Digital recording and numerical reconstruction of holograms: reduction of the spatial frequency spectrum",
Opt. Eng. **35**, 977–982 (1996)
[7.12]   J. H. Milgram, Weichang Li: "Computational reconstruction of images
from holograms", Appl. Opt. **41**, 853–864 (2002)

Chapter 8

[8.1]   P. I. Hariharan: *Basics of Interferometry* (Academic Press, Boston 1992)
[8.2]   E. J. Post: "Sagnac effect", Rev. Mod. Phys. **39** 475–493 (1967)

[8.3]   G. E. Stedman, H. R. Bilger, Z. Li, M. P. Poulton, C. H. Rowe, L. Vetha-
         raniam, P. V. Wells: "Canterbury ring laser and tests for nonreciprocal
         phenomena", Australian J. Phys. **46** 87–101 (1993)
         H. R. Bilger, G. E. Stedman, M. P. Poulton, C. H. Rowe, Z. Li, P. V. Wells:
         "Ring laser for precision measurements of nonreciprocal phenomena",
         IEEE Trans. Instrumen. Meas. **43** 407–411 (1993)
[8.4]   C. M. Vest: *Holographic Interferometry* (Wiley, New York 1979)

Chapter 9

[9.1]   M. Born, E. Wolf: *Principles of Optics* (Pergamon, Oxford 1980)
[9.2]   J. W. Goodman: *Introduction to Fourier Optics* (McGraw–Hill, New York
         1968)

Chapter 10

[10.1]   A. E. Siegmann: *Lasers* (University Science Books, Mill Valley 1986)
         K. Shimoda: *Introduction to Laser Physics*, Springer Ser. Opt. Sci., Vol. 44
         (Springer, Berlin, Heidelberg 1990)
         O. Svelto: *Principles of Lasers* (Plenum Press, New York 1976)
         F. P. Schäfer (Ed.): *Dye Lasers* (Springer, Berlin, Heidelberg 1973)
         W. Koechner: *Solid-State Laser Engineering*, Springer Ser. Opt. Sci., Vol. 1
         (Springer, Berlin, Heidelberg 1992)
[10.2]   C. O. Weiss, R. Vilaseca: *Dynamics of Lasers* (VCH, Weinheim 1991)
[10.3]   D. A. B. Miller: "Quantum-well optoelectronic switching devices", Int. J.
         High Speed Electron. **1**, 19–46 (1990)
[10.4]   W. Hentschel, W. Lauterborn: "High speed holographic movie camera",
         Opt. Engineering **24**, 687–691 (1985)
[10.5]   J. Guckenheimer, P. Holmes: *Nonlinear Oscillations, Dynamical Systems,
         and Bifurcations of Vector Fields* (Springer, New York 1973)
         H. G. Schuster: *Deterministic Chaos – An Introduction* (VCH, Weinheim
         1989)
[10.6]   W. Lauterborn, R. Steinhoff: "Bifurcation structure of a laser with pump
         modulation", J. Opt. Soc. Am. **B5**, 1097–1104 (1988)
[10.7]   C. Scheffczyk, U. Parlitz, T. Kurz, W. Knop, W. Lauterborn: "Comparison
         of bifurcation structures of driven dissipative nonlinear oscillators", Phys.
         Rev. **43A**, 6495–6502 (1991)
[10.8]   C. Huygens: *Christiaan Huygens' The pendulum clock or Geometrical
         demonstrations concerning the motion of pendula as applied to clocks*,
         transl. with notes by R. J. Blackwell, Introduction by H. J. M. Bos (Ames:
         Iowa State Univ. Pr., 1986)
[10.9]   U. Parlitz: *Synchronization of Uni-Directionally Coupled Chaotic Dynam-
         ical Systems* (Cuvillier, Göttingen 1999)
[10.10] G. P. Agrawal, N. K. Dutta: *Semiconductor Lasers* (Van Nostrand Rein-
         hold, New York 1993)
[10.11] R. Lang, K. Kobayashi: "External optical feedback effects on semiconduc-
         tor injection laser properties", IEEE J. Quantum Electron. **16**, 347–355
         (1980)
[10.12] V. Ahlers, U. Parlitz, W. Lauterborn: "Hyperchaotic dynamics and syn-
         chronization of external-cavity semiconductor lasers", Phys. Rev. E **58**,
         7208–7213 (1998)
[10.13] L. M. Pecora, T. L. Carroll: "Synchronization in chaotic systems",
         Phys. Rev. Lett. **64**, 821–824 (1990)

[10.14] W. H. Press, S. A. Teukolsky, W. T. Vetterling, B. P. Flannery: *Numerical Recipes in C* (Cambridge University Press, Cambridge 1992)

Chapter 11

[11.1]   P. M. Paul, E. S. Toma, P. Breger, G. Mullot, F. Augé, Ph. Balcon, H. G. Muller, P. Agostini: "Observation of a train of attosecond pulses from high harmonic generation", Science **292**, 1689–1692 (2001)

[11.2]   S. Backus, Ch. G. Durfee III, M. M. Murnane, H. C. Kapteyn: "High power ultrafast lasers", Rev. Sci. Instr. **69**, 1207–1223 (1998)
         T. Brabec, F. Krausz: "Intense few-cycle laser fields: Frontiers of nonlinear optics", Rev. Mod. Phys. **72**, 545–591 (2000)
         M. Protopapas, C. H. Keitel, P. L. Knight: "Atomic physics with superhigh intensity lasers", Rep. Prog. Phys. **60**, 389–486 (1997)
         Ch. J. Joshi, P. B. Corkum: "Interactions of ultra-intense laser light with matter", Phys. Today **48**, 36–43 (1995)
         M. D. Perry, G. Mourou: "Terawatt to petawatt subpicosecond lasers", Science **264**, 917–925 (1994)

[11.3]   P. M. W. French: "The generation of ultrashort laser pulses", Rep. Prog. Phys. **58**, 169–267 (1995)

[11.4]   K. Morgner, F. X. Kärtner, S. H. Cho, Y. Chen, H. A. Haus, J. G. Fujimoto, E. P. Ippen, V. Scheurer, G. Angelow, T. Tschudi: "Sub-two-cycle pulses from a Kerr-lens mode locked Ti:sapphire laser", Opt. Lett. **24**, 411–413 (1999)

[11.5]   D. T. Reid: "Toward Attosecond Pulses", Science **291**, 1911–1912 (2001)

[11.6]   I. Walmsley, L. Waxer, Ch. Dorrer: "The role of dispersion in ultrafast optics", Rev. Sci. Instr. **72**, 1–29 (2001)

[11.7]   L. E. Hargrove, R. L. Fork, M. A. Pollack, "Locking of He-Ne laser modes induced by synchronous intracavity modulation", Appl. Phys. Lett. **5**, 4–5 (1964)

[11.8]   H. W. Mocker, R. J. Collins: "Mode competition and self-locking effects in a Q-switched ruby laser", Appl. Phys. Lett. **7**, 270–273 (1965)

[11.9]   D. E. Spence, P. N. Kane, W. Sibbett: "60-fsec pulse generation from a self-mode-locked Ti:sapphire laser", Opt. Lett. **16**, 42–44 (1991)

[11.10] B. Gompf, R. Günther, G. Nick, R. Pecha, W. Eisenmenger: "Resolving sonoluminescence pulse width with time-correlated single photon counting", Phys. Rev. Lett. **79**, 1405–1408 (1997)

[11.11] M. P. Brenner, S. Hilgenfeldt, D. Lohse: "Single-bubble sonoluminescence", Rev. Mod. Phys. **74**, 425–484 (2002)

[11.12] D. Strickland, G. Morou, "Compression of amplified chirped optical pulses", Opt. Commun. **56**, 219–221 (1985)

[11.13] E. P. Ippen, C. V. Shank: "Techniques for measurement", in: S. L. Shapiro (Ed.): *Ultrashort Light Pulses*, Chap. 3 (Springer, Berlin 1977)

[11.14] D. J. Bradley, G. H. New: "Ultrashort pulse measurements", Proc. IEEE bf 62, 313–345 (1974)

[11.15] J. A. Giordmaine, P. M. Rentzepis, S. L. Shapiro, K. W. Wecht: "Two-photon excitation of fluorescence by picosecond light pulses", Appl. Phys. Lett. **11**, 216 (1967)

[11.16] D. J. Kane, R. Trebino: "Characterization of arbitrary femtosecond pulses using frequency-resolved optical gating", J. Quantum Electr. **29**, 571–579 (1993)

[11.17] R. Trebino, K. W. DeLong, D. N. Fittinghoff, J. N. Sweetser, M. A. Krum-bügel, B. A. Richman: "Measuring ultrashort laser pulses in the time-frequency domain using frequency-resolved optical gating", Rev. Sci. Instrum. **68** (9), 3277–3295 (1997)

[11.18] M. A. Duguay, J. W. Hansen: "An ultrafast light gate", Appl. Phys. Lett. **15**, 192–194 (1969)

[11.19] M. A. Duguay, A. T. Mattick: "Ultrahigh speed photography of picosecond light pulses", Appl. Optics **10**, 2162–2170 (1971)

[11.20] M. A. Duguay: "The ultrafast optical Kerr shutter", in: E. Wolf (Ed.): *Progress in Optics*, Vol. XIV, pp. 161–193 (North Holland, Amsterdam 1976)

[11.21] L. Wang, P. P. Ho, C. Liu, G. Zhang, R. R. Alfano: "Ballistic 2-D imaging through scattering walls using an ultrafast optical Kerr gate", Science **253**, 769–771 (1991)

[11.22] D. Huang, E. A. Swanson, C. P. Lin, J. S. Schuman, W. G. Stinson, W. Chang, M. R. Hee, T. Flotte, K. Gregory, C. A. Puliafito, J. G. Fujimoto: "Optical coherence tomography", Science **254**, 1178–1181 (1991)

[11.23] E. Beaurepaire, A. C. Boccara, M. Lebec, L. Blanchot, H. Saint-Jalmes: "Full-field optical coherence microscopy", Opt. Lett. **23**, 244–246 (1998)

[11.24] C. K. Hitzenberger, A. F. Fercher: "Differential phase contrast in optical coherence tomography", Opt. Lett. **24**, 622–624 (1999)

[11.25] B. E. Bouma, G. J. Tearney: "Power-efficient nonreciprocal interferometer and linear-scanning fiber-optic catheter for optical coherence tomography", Opt. Lett. **24**, 531–533 (1999)

[11.26] G. J. Tearney, M. E. Brezinski, B. E. Bouma, S. A. Boppart, C. Pitris, J. F. Southern, J. G. Fujimoto: "In vivo endoscopic optical biopsy with optical coherence tomography", Science **276**, 2037–2039 (1997)

Chapter 12

[12.1] M. Goeppert–Mayer: "Über Elementarakte mit zwei Quantensprüngen", Ann. Phys. **9**, 273–294 (1931)

[12.2] W. Kaiser, C. G. B. Garrett: "Two-photon excitation in $CaF_2:Eu^{2+}$", Phys. Rev. Lett. **7**, 229–231 (1961)

[12.3] J. J. Hopfield, J. M. Worlock, K. Park: "Two-quantum absorption spectrum of KI", Phys. Rev. Lett. **11**, 414–417 (1963)

[12.4] H. Mahr: "Two-Photon absorption spectroscopy", in: H. Rabin, C. L. Tang (Ed.): *Quantum Electronics*, 285–361 (Academic Press, New York 1975)

[12.5] W. Zernik: "Two-photon ionization of atomic hydrogen", Phys. Rev. **135**, A51–A57 (1964)

[12.6] P. A. Franken, A. E. Hill, C. W. Peters, G. Weinreich: "Generation of optical harmonics", Phys. Rev. Lett. **7**, 118–119 (1961)

[12.7] M. Bass, P. A. Franken, A. E. Hill, C. W. Peters, G. Weinreich: "Optical mixing", Phys. Rev. Lett. **8**, 18–18 (1962)

[12.8] J. Warner: "Difference frequency generation and up-conversion", in: H. Rabin, C. L. Tang (Ed.): *Quantum Electronics*, 703–737 (Academic Press, New York 1975)

[12.9] A. W. Smith, N. Braslau: "Optical mixing of coherent and incoherent light", IBM J. Res. Develop. **6**, 361–362 (1962)
F. Zernike, P. R. Berman: "Generation of far infrared as a difference frequency", Phys. Rev. Lett. **15**, 999–1001 (1965)

[12.10] M. Bass, P. A. Franken, J. F. Ward, G. Weinreich: "Optical rectification", Phys. Rev. Lett. **9**, 446–448 (1962)

[12.11] M. A. Duguay, J. W. Hansen: "An ultrafast light gate", Appl. Phys. Lett.
         15, 192–194 (1969)
         M. A. Duguay, A. T. Mattick: "Ultrahigh speed photography of picosecond
         light pulses and echoes", Appl. Opt. 10, 2162–2170 (1971)
[12.12] J. A. Giordmaine, R. C. Miller: "Tunable coherent parametric oscillation
         in LiNbO at optical frequencies", Phys. Rev. Lett. 14, 973–976 (1965)
[12.13] R. L. Byer: "Optical parametric oscillators", in: H. Rabin, C. L. Tang (Ed.):
         Quantum Electronics, 587–702 (Academic Press, New York 1975)
[12.14] R. A. Fischer (Ed.): Optical Phase Conjugation (Academic Press, Orlando
         1983)
[12.15] R. W. Terhune, P. D. Maker, C. M. Savage: "Optical harmonic generation
         in calcite", Phys. Rev. Lett. 8, 404–406 (1962)
[12.16] Z. Chang, A. Rundquist, H. Wang, M. M. Murnane, H. C. Kapteyn: "Gen-
         eration of coherent soft X rays at 2.7 nm using high harmonics", Phys.
         Rev. Lett. 79, 2967–2970 (1997)
[12.17] M. Lewenstein, P. Balcou, M. Yu. Ivanov, A. L'Huillier, P. B. Corkum:
         "Theory of high-harmonic generation by low-frequency laser fields",
         Phys. Rev. A 49, 2117–2132 (1994)
[12.18] S. Singh, L. T. Bradley: "Three-photon absorption in naphtalene crystals
         by laser excitation", Phys. Rev. Lett. 12, 612–614 (1964)
[12.19] P. B. Corkum: "Plasma perspective on strong-field multiphoton ioniza-
         tion", Phys. Rev. Lett. 71, 1994–1997 (1993)
[12.20] S. Tzortzakis, L. Bergé, A. Couairon, M. Franco, B. Prade, A. Mysyrowicz:
         "Breakup and fusion of self-guided femtosecond light pulses in air", Phys.
         Rev. Lett. 86, 5470–5473 (2001)
[12.21] A. V. Sokolov, D. R. Walker, D. D. Yavuz, G. Y. Yin, S. E. Harris, „Raman
         generation by phased and antiphased molecular states", Phys. Rev. Lett.
         85, 562–565 (2000)
[12.22] T. Ditmire, J. Zweiback, V. P. Yanovsky, T. E. Cowan, G. Hays, K. B. Whar-
         ton: "Nuclear fusion from explosion of femtosecond laser-heated deu-
         terium clusters", Nature 398, 489–492 (1999)
[12.23] C. Scheffczyk, U. Parlitz, T. Kurz, W. Knop, W. Lauterborn: "Compari-
         son of bifurcation structures of driven dissipative nonlinear oscillators",
         Phys. Rev. 43A, 6495–6502 (1991)
[12.24] P. D. Maker, R. W. Terhune, M. Nisenhoff, C. M. Savage: "Effects of disper-
         sion and focusing on the production of optical harmonics", Phys. Rev. Lett.
         8, 21–22 (1962)
[12.25] A. Yariv: Quantum Electronics (Wiley, New York 1975)
[12.26] J. M. Manley, H. E. Rowe: "General energy relations in nonlinear react-
         ances", Proc. IRE 47, 2115–2116 (1959)

Chapter 13

[13.1]   J. C. Palais: Fiber Optic Communications (Prentice Hall, Englewood Cliffs
         1988)
[13.2]   H. Kogelnik: "High-capacity optical communication", IEEE J. Sel. Topics
         Quantum Electr. 6, 1279–1286 (2000)
[13.3]   J. Staudenraus, W. Eisenmenger: "Fibre-optic probe hydrophone for ul-
         trasonic and shock-wave measurements in water", Ultrasonics 31, 267–
         273 (1993)
[13.4]   H. E. Engan: "Acoustic field sensing with optical fibers", in A. Alippi,
         (Ed.), Acoustic Sensing and Probing, p. 57–75 (World Scientific, Singa-
         pore 1992)

328   References

[13.5]   B. Culshaw: "Fiber optics in sensing and measurement", IEEE J. Sel.
         Topics Quantum Electr. 6, 1014–1021 (2000)
[13.6]   E. Udd: "An overview of fiber-optic sensors", Rev. Sci. Instrum. 66, 4015–
         4030 (1995)
[13.7]   N. J. Zabusky, M. D. Kruskal: "Interaction of "solitons" in a collisionless
         plasma and the recurrence of initial states", Phys. Rev. Lett. 15, 240–243
         (1965)
[13.8]   A. Hasegawa: Optical Solitons in Fibers, Springer Trends Mod. Phys.,
         Vol. 116 (Springer, Berlin, Heidelberg 1990)
[13.9]   G. P. Agrawal: Nonlinear Fiber Optics, 3rd ed. (Academic Press,
         San Diego, London 2001)
[13.10] H. J. Doran, D. Wood: "Nonlinear-optical loop mirror", Opt. Lett. 13, 56–
         58 (1988)
[13.11] P. M. Butcher, D. Cotter: The Elements of Nonlinear Optics (Cambridge
         University Press, Cambridge 1991)
[13.12] E. Desurvire: Erbium-doped fiber amplifiers (John Wiley, New York 1994)
[13.12] M. N. Islam: Ultrafast Fiber Switching Devices and Systems (Cambridge
         University Press, Cambridge 1992)

Appendix

[A.1]   R. Bracewell: The Fourier Transform and Its Applications (McGraw–Hill,
        New York 1965)
[A.2]   J. W. Goodman: Introduction to Fourier Optics (McGraw–Hill, San Fran-
        cisco 1965)
[A.3]   A. Papoulis: The Fourier Integral and its Applications (McGraw–Hill, New
        York 1962)
[A.4]   M. J. Lighthill: Introduction to Fourier Analysis and Generalized Func-
        tions (Cambridge University Press, Cambridge 1959)
[A.5]   E. C. Titchmarsh: Introduction to the Theory of Fourier Integrals (Claren-
        don, Oxford 1962)
[A.6]   D. C. Champeney: A Handbook of Fourier Theorems (Cambridge Univer-
        sity Press, Cambridge 1987)

Additional Reading

Chapter 1

Asakura, T. (Ed.): International Trends in Optics and Photonics ICO IV (Springer,
    Heidelberg 1999)
Born, M., Wolf, E.: Principles of Optics (Pergamon, Oxford 1984)
Goodman, J. W. (Ed.): International Trends in Optics (Academic, Boston 1991)
Hecht, E.: Optics, 4th edn. (Addison-Wesley, Reading, Mass. 2001)
Horgan, J.: "Quantum philosophy", Scientific American, July 1992, 72–80.
Thompson, L. A.: "Adaptive Optics in Astronomy", Physics Today, December
    1994, 24–31.

Chapter 2

Goodman, J. W.: Statistical Optics (Wiley, New York 1985)
Hecht, E.: Optics, 4th edn. (Addison–Wesley, Reading, Mass. 2001)
Jenkins, F. A., H. E. White: Fundamentals of Optics (McGraw–Hill, New York
    1976)

Lipson, S. G., H. Lipson, D. S. Tannhauser: *Optical Physics*, 3rd edn. (Cambridge University Press, Cambridge 1995)

Loudon, R.: *The Quantum Theory of Light* (Clarendon, Oxford 1983)

Louisell, W. H.: *Quantum Statistical Properties of Radiation* (Wiley, New York 1973)

Mandel, L., E. Wolf: *Optical Coherence and Quantum Optics* (Cambridge University Press, Cambridge 1995)

Meystre, P., M. Sargent III: *Elements of Quantum Optics* (Springer, Berlin, Heidelberg 1991)

Saleh, B.: *Photoelectron Statistics* (Springer, Berlin, Heidelberg, New York 1978)

Wolf, E. (Ed.): *Progress in Optics*, Vol. I to Vol. 43 (North–Holland, Amsterdam 1961 to 2002)

Young, M.: *Optics and Lasers*, 5th edn. (Springer, Berlin, Heidelberg, New York 2000)

Chapter 3

Born, M., Wolf, E.: *Principles of Optics* (Pergamon, Oxford 1984)

Hecht, E.: *Optics*, 4th edn. (Addison–Wesley, Reading, Mass. 2001)

Young, M.: *Optics and Lasers*, 5th edn. (Springer, Berlin, Heidelberg, New York 2000)

Chapter 4

Born, M., E. Wolf: *Principles of Optics* (Pergamon, Oxford 1980)

Mandel, L.: "Fluctuations of light beams", in E. Wolf (Ed.): *Progress in Optics*, Vol. II, S. 181–248 (North–Holland, Amsterdam 1963)

Mandel, L.: "Coherence and indistinguishability", Opt. Lett. **16**, 1882–1883 (1991)

Mandel, L., E. Wolf: *Optical Coherence and Quantum Optics* (Cambridge University Press, Cambridge 1995)

Marathay A. S.: *Elements of Optical Coherence Theory* (Wiley, New York 1982)

Miyamoto, Y., T. Kuga, M. Baba, M. Matsuoka: "Measurement of ultrafast optical pulses with two-photon interference", Opt. Lett. **18**, 900–902 (1993)

Ou, Z. Y., E. C. Gage, B. E. Magill, L. Mandel: "Fourth-order interference technique for determining the coherence time of a light beam", J. Opt. Soc. Am. **B6**, 100–103 (1989)

Chapter 5

Born, M., Wolf, E.: *Principles of Optics* (Pergamon, Oxford 1984)

Hecht, E.: *Optics*, 2nd edn. (Addison–Wesley, Reading, Mass. 1987)

Chapter 6

Goodman, J. W.: *Statistical Optics* (Wiley, New York 1985)

Ostrowski, Y. I.: "Correlation holographic and speckle interferometry", in E. Wolf (Ed.): *Progress in Optics*, Vol. XXX, S. 87–135 (North–Holland, Amsterdam 1992)

Chapter 7

Collier, R. J., C. B. Burckhardt, L. H. Lin: *Optical Holography* (Academic Press, New York 1971)

Hariharan, P.: *Optical Holography* (Cambridge University Press, Cambridge 1984)

Hariharan, P.: *Basics of Holography* (Cambridge University Press, Cambridge 2002)

Hariharan, P.: "Colour holography", in E. Wolf (Ed.): *Progress in Optics*, Vol. XX, S. 263–324 (North–Holland, Amsterdam 1983)

W. Lauterborn (Ed.): *Cavitation and Inhomogeneities in Underwater Acoustics*, Springer Ser. Electrophys., Vol. 4 (Springer, Berlin, Heidelberg, New York 1980)

Marwitz, H. (Ed.): *Praxis der Holografie* (expert-Verlag, Ehningen 1990)

Ostrowski, Y. I., V. P. Shchepinov: "Correlation holographic and speckle interferometry", in E. Wolf (Ed.): *Progress in Optics*, Vol. XXX, S. 87–135 (North–Holland, Amsterdam 1992)

Chapter 8

Collier, R. J., C. B. Burckhardt, L. H. Lin: *Optical Holography* (Academic Press, New York 1971)

Dändliker, R.: "Heterodyne holographic interferometry", in E. Wolf (Ed.): *Progress in Optics*, Vol. XVII, S. 1–84 (North–Holland, Amsterdam 1980)

Ezekiel, S., H. J. Arditty: *Fiber-Optic Rotation Sensors and Related Technologies* Springer Ser. Opt. Sci., Vol. 32 (Springer, Berlin, Heidelberg, New York 1982)

Ostrovsky Y.I., V. P. Shchepinov, V. V. Yakovlev: *Holographic Interferometry in Experimental Mechanics* Springer Ser. Opt. Sci., Vol. 60 (Springer, Berlin, Heidelberg, New York 1991)

Ostrowski, Y. I., V. P. Shchepinov: "Correlation holographic and speckle interferometry", in E. Wolf (Ed.): *Progress in Optics*, Vol. XXX, S. 87–135 (North–Holland, Amsterdam 1992)

Rastogi, P. K. (Ed.): *Holographic Interferometry* Springer Ser. Opt. Sci., Vol. 68 (Springer, Berlin, Heidelberg, New York 1994)

Chapter 9

Champeney, D. C.: *Fourier Transforms and their Physical Applications* (Academic, Boston 1973)

Papoulis, A.: *Systems and Transforms with Applications in Optics* (Wiley, New York 1968)

VanderLugt, A.: *Optical Signal Processing* (Wiley, New York 1992)

Chapter 10

Arecchi, F. T., R. G. Harrison: *Instabilities and Chaos in Quantum Optics* (Springer, Berlin, Heidelberg 1987)

Vohra, S., M. Spano, M. Schlesinger, L. Pecora, W. Ditto (Ed.): *1st Experimental Chaos Conference* (World Scientific, Singapore 1991)

Chapter 11

Diels, J.-C., Rudolph, W.: *Ultrashort Laser Pulse Phenomena: Fundamentals, Techniques, and Applications on a Femtosecond Time Scale* (Academic Press, San Diego 1997)

Haus, H. A. "Mode-locking of lasers", IEEE J. Select. Topics Quantum Electron. **6**, 1173–1185 (2000)

Kaiser, W. (Ed.): *Ultrashort Laser Pulses. Generation and Applications*, Topics in Appl. Phys. Vol. 60, 2nd ed. (Springer, Berlin 1993)

Rullière, C. (Ed.): *Femtosecond Laser Pulses: Principles and Experiment* (Springer, Berlin 1998)

Chapter 12

Bloembergen, N.: *Nonlinear Optics* (Benjamin, New York 1965)

Boyd, R. W.: *Nonlinear Optics* (Academic Press, Boston 1992)

Butcher, P. N., D. Cotter: *The Elements of Nonlinear Optics* (Cambridge University Press, Cambridge 1991)

Gibbs, H. M.: *Optical Bistability: Controlling Light with Light* (Academic Press, London 1985)

Mills, P. L.: *Nonlinear Optics* (Springer, Berlin, Heidelberg 1991)

Shen, Y. R.: *Principles of Nonlinear Optics* (Wiley, New York 1984)

Yariv, A.: *Quantum Electronics* (Wiley, New York 1975)

Chapter 13

Agrawal, G. P.: *Applications of Nolinear Fiber Optics* (Academic Press, San Diego 2001)

Consortini, A. (Ed.): *Trends in Optics* (Academic Press, San Diego 1996)

Culshaw, B.: *Optical Fibre Sensing and Signal Processing* (Peregrinus, Stevenage 1984)

Dainty, J. C. (Ed.): *Current Trends in Optics* (Academic Press, London 1994)

Goodman, J. W. (Ed.): *International Trends in Optics* (Academic Press, Boston 1991)

Kumar, A.: "Soliton dynamics in a monomode optical fibre", Phys. Reports **187**, 63–108 (1990)

Midwinter, J. E.: *Optoelectronics and Lightwave Technology* (Wiley, New York 1992)

Mills, D. L.: *Nonlinear Optics* (Springer, Berlin, Heidelberg 1991)

Snyder, A. W., Love, J. D.: *Optical Waveguide Theory*, (Chapman & Hall, London 1996)

Taylor, J. R. (Ed.): *Optical Solitons – Theory and Experiment* (Cambridge University Press, Cambridge 1992)

Udd, E.: *Fiber Optic Sensors* (Wiley, New York 1991)

Index